Nonbenzenoid Aromatics

Volume II

This is Volume 16-II of
ORGANIC CHEMISTRY
A series of monographs
Editors: ALFRED T. BLOMQUIST and HARRY WASSERMAN

A complete list of the books in this series appears at the end of the volume.

Nonbenzenoid Aromatics

Edited by

James P. Snyder

Belfer Graduate School of Science
Yeshiva University
New York, New York

Volume II

ACADEMIC PRESS   New York and London 1971

ACADEMIC PRESS, INC.
111 Fifth Avenue, New York, New York 10003

United Kingdom Edition published by
ACADEMIC PRESS, INC. (LONDON) LTD.
Berkeley Square House, London W1X 6BA

LIBRARY OF CONGRESS CATALOG CARD NUMBER: 77-162937

PRINTED IN THE UNITED STATES OF AMERICA

To Jennifer

Contents

1. Nonalternant Hydrocarbons, Radical Ions, and Their Heteroanalogs; Characteristics of Ground and Excited States

R. Zahradník

2. ESR Spectra of Radical Ions of Nonbenzenoid Aromatics

F. Gerson and J. H. Hammons

3. Diamagnetic Susceptibility Exaltation as a Criterion of Aromaticity

Hyp J. Dauben, Jr., James D. Wilson, and John L. Laity

4. Monocyclic and Polycyclic Aromatic Ions Containing Six or More π-Electrons

P. J. Garratt and M. V. Sargent

5. Chemical Binding and Delocalization in Phosphonitrilic Derivatives

D. P. Craig and N. L. Paddock

6. Cyclobutadiene-Metal Complexes

P. M. Maitlis and K. W. Eberius

List of Contributors

Numbers in parentheses indicate the pages on which the authors' contributions begin.

D. P. CRAIG (273), *Research School of Chemistry, Institute of Advanced Studies, The Australian National University, Canberra, A.C.T., Australia*

HYP J. DAUBEN, Jr.* (167), *University of Washington, Seattle, Washington*

K. W. EBERIUS (359), *Department of Chemistry, McMaster University, Hamilton, Ontario, Canada*

P. J. GARRATT (207), *Department of Chemistry, University College, London, England*

F. GERSON (81), *Physikalisch-Chemisches Institut der Universität, Basel, Switzerland*

J. H. HAMMONS† (81), *Physikalisch-Chemisches Institut der Universität, Basel, Switzerland*

JOHN L. LAITY (167), *Shell Development Co., Emeryville, California*

P. M. MAITLIS (359), *Department of Chemistry, McMaster University, Hamilton, Ontario, Canada*

N. L. PADDOCK (273), *Department of Chemistry, The University of British Columbia, Vancouver, B.C., Canada*

M. V. SARGENT † (207), *University Chemical Laboratories, Canterbury, Kent, England*

JAMES D. WILSON (167), *Central Research Department, Monsanto Company, St. Louis, Missouri*

R. ZAHRADNÍK (1), *Institute of Physical Chemistry, Czechoslovak Academy of Sciences, Prague, Czechoslovakia*

* Deceased.

† Present address, Department of Chemistry, Swarthmore College, Swarthmore, Pennsylvania.

† Present address, Department of Organic Chemistry, University of Western Australia, Nedlands, Western Australia.

Preface

For one hundred and fifty years chemical species possessing a cyclic array of parallel π orbitals occupied by six electrons have been catalogued as "aromatic" or "benzenoid." Remarkably, in one-tenth that time, the field of nonbenzenoid aromatics has undergone a minor revolution. Conjugated π-systems from two to thirty electrons have recently been generated as radicals, cations, carbanions, and a wide range of unusual neutral compounds. In addition, the ambiguities associated with the historical designation "aromatic" appear to be experiencing a clarification.

The last major attempt to survey this field was made in 1959 in the now classic compilation "Non-Benzenoid Aromatic Compounds."[1] No monocyclic nonbenzenoid aromatic species with other than 6π electrons was available for discussion in that volume. Although a number of excellent reviews have appeared in the interim,[2] this treatise is intended to provide an in-depth multiauthored evaluation of activity as it has developed in the last fifteen years. In view of the spectrum of interests represented by the individual contributions, the reader might anticipate a particular organization from volume to volume. Practical considerations on the contrary have resulted in a generous heterogeneity within each book. Nevertheless the theme that threads its way through the chapters is that of "aromaticity"; each author making an effort to evaluate this concept in light of his own work. It is with this in mind that this treatise was initiated with an historical account tracing the development of the idea up to the discovery of the electron.

An expression of gratitude is owed the authors who made this volume possible and Marie Kouirinis whose patience and cooperation lightened

[1] David Ginsburg (ed.), "Non-Benzenoid Aromatic Compounds." Wiley (Interscience), New York, 1959.

[2] G. M. Badger, "Aromatic Character and Aromaticity." Cambridge Univ. Press, London and New York, 1969.

W. Baker, The widening outlook in aromatic chemistry, Part 1, *Chemistry in Britain*, p. 191, May, 1965.

W. Baker, The widening outlook in aromatic chemistry, Part 2, *Chemistry in Britain*, p. 250, June, 1965.

K. Hafner, Structure and aromatic character of non-benzenoid cyclically conjugated systems, *Angew. Chem. Internat. Ed.* 3 [No. 3] (1964).

M. E. Vol'pin, Non-benzenoid aromatic compounds and the concept of aromaticity, *Russian Chem. Rev.*, March, 1960.

A. J. Jones, Criteria for aromatic character, *Rev. Pure Appl. Chem.* 18, 253 (1968).

the editorial task considerably. Thanks are due the staff of Academic Press for efficiency, patience, and accessibility. A final note of appreciation goes to C.G.D. for her gentle charm, enthusiasm, and lightheartedness, an indispensable influence in the consolidation of the work.

JAMES P. SNYDER

Contents of Volume I

Nonalternant Hydrocarbons, Radical Ions, and Their Heteroanalogs; Characteristics of Ground and Excited States

R. Zahradník

I. Introduction

A. Subject and Scope

In this article special attention will be paid to two problems: (1) What must be known in order to estimate the thermochemical and kinetic stability of a system which has not yet been synthesized. Accordingly, heats of formation, theoretical indices of reactivity and other molecular characteristics will be discussed. The aim is to develop a sufficiently deep knowledge of the relationships between these quantities and of the experimental behavior of representative groups of compounds in order to allow interpolation. (2) Attention will also be paid to the calculation of electronic spectra, although the latter are certainly not as specifically characteristic as nuclear magnetic resonance (NMR) and electron spin resonance (ESR) spectra. Nevertheless, electronic spectra provide essential information, important for both theoretical and experimental purposes. The rapidly growing interest in photochemical processes is relevant in this connection.

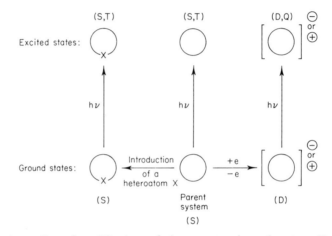

FIG. 1. An outline of modifications of the parent-conjugated system. Designation: (S) singlet, (D) doublet, (T) triplet, (Q) quartet; e represents an electron.

We shall also be interested in how properties such as stability and electron distribution are changed by modifications which preserve the basic atomic skeleton: (i) Electronic excitation: We shall be concerned mostly with the first excited singlet (S_1) and triplet (T_1) states. These are of primary interest in photochemistry. (ii) Addition or removal of a certain number of π-electrons: A one-electron change is most important. For neutral systems it corresponds to the formation of radical ions. (iii) Introduction of a heteroatom or attachment of a substituent.

Processes (i)–(iii) are schematically presented in Fig. 1. The investigation of these changes is interesting not only from a strictly theoretical point of view, but also for purely practical reasons. Modifications (i) and (ii) are frequently associated with a redistribution of atomic positions possessing highest and lowest π-electron densities. This is significant for preparative reasons. Alteration (iii) is of importance because sometimes even a relatively small

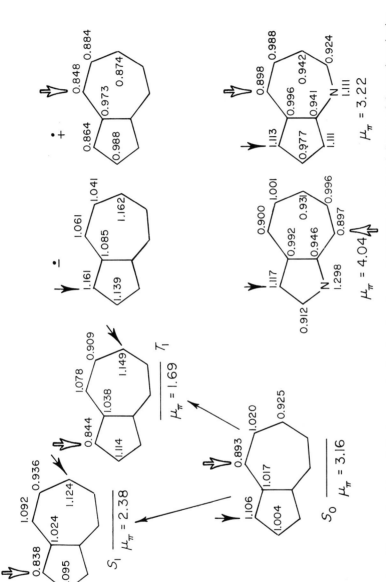

FIG. 2. π-Electron densities and dipole moments of azulene based on an LCI–SCF calculation; influence of electronic excitation, of addition and removal of an electron, and of introduction of a heteroatom. Positions of highest and lowest π-electron densities are visualized by arrows (↓,⇓).

modification (e.g., the introduction of alkyl groups or the substitution of =CH— by =N—, or —CH=CH— by —S—) can provide a considerable increment in stability. This is particularly topical for parent systems which are rather unstable. MO theory, in its simple version, can indicate for example which atomic position should be substituted in order to increase stability. In order to provide more specific insight, calculations for azulene as a consequence of processes (i)–(iii) are shown in Fig. 2.

B. REMARKS ON THE CLASSIFICATION OF THE CHEMICAL SYSTEMS AND ON THEORETICAL METHODS

The classification of nonalternant systems can be based either on a structural formalism such as the number or rings,[1,2] or on some physical characteristic

TABLE I

CLASSIFICATION OF SYSTEMS

Group	$\Delta p_{\mu\nu}(max)^a$	Presence of a NBMO[b]	Examples	Methods[c]
1	Relatively small (0.20–0.35)	No		PPP or HMO
2[d]	Relatively large (0.30–0.45)	No	CH$_2$	PPP including β^c-variation or HMO with β-variation
3	Relatively small (0.13–30)	Yes		e

[a] The parentheses indicate the approximate interval of the maximum values of the bond order differences.
[b] Or the presence of an occupied antibonding or unoccupied bonding molecular orbital.
[c] More detailed methodical data are given in Tables II–V.
[d] A single Kekulé structure is appropriate to the compounds of this group.
[e] See text.

such as strong bond-length alternation and/or the presence of a nonbonding MO; *vide infra*. Both these notions will be applied.

[1] R. Zahradník and J. Michl, *Collection Czech. Chem. Commun.* **30**, 515 (1965).
[2] R. Zahradník, *Angew. Chem. Intern. Ed. Engl.* **4**, 1039 (1965).

An alternative classification utilizes HMO characteristics. For example, a maximum value of the difference between two bond orders in a given system ($\Delta p_{\mu\nu}$) serves as a rough estimate of the measure of bond alternation. A small value of $\Delta p_{\mu\nu}$ in conjunction with the presence of a nonbonding molecular orbital (NBMO), or the presence of an unoccupied bonding or an occupied antibonding MO of the electro-neutral form is considered to be an empirical symptom for a nonaromatic system. The HMO classification thus provides three subsets of compounds (Table I).

The most widely known subset is the first. Its theoretical treatment is least exciting but satisfactory. Reasonable theoretical data may be obtained for the second group of substances only if values of the resonance integrals reflect the true molecular geometry. Meaningful applications of theoretical methods to the third set of substances is a difficult task. It seems that utilization of a semiempirical method in which the true molecular geometry is respected can yield reasonable results for one of several possible resonance structures. Systems of this group do not seem to have maximum possible symmetry; *vide infra*.

Numerous sources of information are available on quantum chemical procedures for calculating electron distributions, electronic spectra, etc.[3-11] For this reason, no specific details on this subject will be presented here. However, a short outline of the various versions of MO methods used (or which could be used) for investigating nonalternant systems is given. In Table II the treatments are classified according to the number of electrons which are explicitly taken into consideration, and according to the level of sophistication. In Tables III and IV, expressions for the Hamiltonian, the wave function, the matrix elements of the Hamiltonian, and the scope of the application and the limitations are presented for the semiempirical and empirical methods of groups I and II.1 (see Table II).

[3] B. Pullman and A. Pullman, "Les théories électroniques de la chimie organique." Masson, Paris, 1952.

[4] R. Daudel, R. Lefebvre, and C. Moser, "Quantum Chemistry." Wiley (Interscience), New York, 1959.

[5] A. Streitwieser, Jr., "Molecular Orbital Theory for Organic Chemists." Wiley, New York, 1961.

[6] J. N. Murrell, "The Theory of the Electronic Spectra of Organic Molecules." Methuen, London, 1963.

[7] L. Salem, "The Molecular Orbital Theory of Conjugated Systems." Benjamin, New York, 1966.

[8] J. N. Murrell, S. F. A. Kettle, and J. M. Tedder, "Valence Theory." Wiley, New York, 1965.

[9] H. Suzuki, "Electronic Absorption Spectra and Geometry of Organic Molecules." Academic Press, New York, 1967.

[10] R. G. Parr, "Quantum Theory of Molecular Electronic Structure." Benjamin, New York, 1964.

[11] H. H. Jaffé and M. Orchin, "Theory and Applications of Ultraviolet Spectroscopy." Wiley, New York, 1962.

TABLE II

OUTLINE OF VARIOUS VERSIONS OF MOLECULAR ORBITAL METHODS

I. All (or at least all valence-shell) electrons included; simple methods useful for σ-electron systems	II. π-Electron approximation	
	1. Closed-shell in the ground state	2. Open-shell in the ground state

Level of sophistication: *ab initio* methods		
$\longleftarrow$ Clementi[a] $\longrightarrow$ Preuss[b]		

	Level of sophistication: semiempirical methods Craig[m]	
CI: Del Bene, Jaffé[c] SCF: Dewar, Klopman[d,e]; Pople, Santry, Segal[f]; Katagiri, Sandorfy[g]; Jungen, Labhart[h]	CI: Pariser, Parr[n] SCF: Roothaan[o] SCF: Pople[p]	CI: Ishitani, Nagakura[t] SCF: Roothaan[u] SCF: Longuet-Higgins, Pople[v]

	Level of sophistication: empirical methods	
Hoffmann,[i] Sandorfy,[j] Brown,[k] Del Re[l]	Improved HMO[q] (e.g., Coulson–Golebiewski) HMO[r] Perturbation[s] treatment	McLachlan[w] HMO[r]

[a] E. Clementi, *Chem. Rev.* **68**, 341 (1968).
[b] H. Preuss, *Fortschr. Chem. Forsch.* **9**, 325 (1968).
[c] J. Del Bene and H. H. Jaffé, *J. Chem. Phys.* **48**, 1807 and 4050 (1968).
[d] M. J. S. Dewar and G. Klopman, *J. Am. Chem. Soc.* **89**, 3089 (1967).
[e] G. Klopman, *J. Am. Chem. Soc.* **87**, 3300 (1965).
[f] J. A. Pople, D. P. Santry, and G. A. Segal, *J. Chem. Phys.* **43**, S129 (1965).
[g] S. Katagiri and C. Sandorfy, *Theoret. Chim. Acta* **4**, 203 (1966).
[h] M. Jungen and H. Labhart, *Theoret. Chim. Acta* **9**, 345 (1968).
[i] R. Hoffmann, *J. Chem. Phys.* **39**, 1397 (1963); **40**, 2480 (1964).
[j] C. Sandorfy, *Can. J. Chem.* **33**, 1337 (1955).
[k] R. D. Brown, *J. Chem. Soc.* p. 2615 (1953).
[l] G. Del Re, *J. Chem. Soc.* p. 4031 (1958).
[m] D. P. Craig, *Proc. Roy. Soc.* **A200**, 474 (1950).
[n] R. Pariser and R. G. Parr, *J. Chem. Phys.* **21**, 466 and 767 (1953).
[o] C. C. J. Roothaan, *Rev. Mod. Phys.* **23**, 69 (1951).
[p] J. A. Pople, *Trans. Faraday Soc.* **49**, 1375 (1953); A. Brickstock and J. A. Pople, *ibid.* **50**, 901 (1954).
[q] G. W. Wheland and D. E. Mann, *J. Chem. Phys.* **17**, 264 (1949); C. A. Coulson and A. Golebiewski, *Proc. Phys. Soc.* **78**, 1310 (1961); M. J. Janssen and J. Sandström, *Tetrahedron* **20**, 2339 (1964).
[r] E. Hückel, *Z. Physik* **70**, 204 (1931); **76**, 628 (1932).

The most important feature of the semiempirical methods is that electronic repulsion is explicitly taken into consideration. On the other hand, the Hamiltonian used for empirical and SCF methods is expressible as a mere sum of the one electron *effective* Hamiltonians.

The matrix elements $F_{\mu\nu}$ or $H_{\mu\nu}$, presented in Tables III and IV, are necessary for solving the secular equations for the unknown expansion coefficients c_ν:

$$\sum_{\nu=1}^{n} c_\nu(M_{\mu\nu} - ES_{\mu\nu}) \qquad \mu = 1, 2, \ldots n \tag{1}$$

where $M_{\mu\nu}$ stands either for $F_{\mu\nu}$ or $H_{\mu\nu}$. Methods for solving these equations by matrix diagonalization and expressions for calculating electron densities, bond orders, etc., from the expansion coefficients, are well known and available in standard texts.[3-11] Obviously, for our purposes the semiempirical methods of group II.1 are of the greatest importance. Therefore, in Table V approximations which have been used in the Pariser–Parr–Pople-type calculations are outlined. The approximations *A*, *B*, and *C* have to do with the Hamiltonian, the core integrals, and the electronic-repulsion integrals, respectively. This material is designed for those who are familiar with methods like the HMO and with the ω-technique, and at least roughly acquainted with the LCI–SCF procedure. Expressions for open-shell systems are available in the literature.[12, 13]

A somewhat complicated expression for calculating bond orders within the framework of the LCI method should be mentioned.[14] Only mono-excited configurations are included.

$$
\begin{aligned}
p_{\mu\nu}^{\text{(LCI)}} = {} & 2 \sum_{i,occ} c_{i\mu} c_{i\nu} + \sum_{i}^{occ} \sum_{j}^{em} A_{a,ij}[(c_{j\mu} c_{j\nu} - c_{i\mu} c_{i\nu}) + \\
& \sqrt{2}\, A_{a,\theta}(c_{i\mu} c_{j\nu} + c_{j\mu} c_{i\nu})] + \\
& \sum_{k>j}^{em} A_{a,ik}(c_{j\mu} c_{k\nu} + c_{k\mu} c_{j\nu}) - \\
& \sum_{h>i}^{occ} A_{a,hj}(c_{i\mu} c_{h\nu} + c_{h\mu} c_{i\nu})
\end{aligned}
\tag{2}
$$

[12] R. Zahradník and P. Čársky, *J. Phys. Chem.* **74**, 1235 (1970).
[13] H. C. Longuet-Higgins and J. A. Pople, *Proc. Phys. Soc.* **A68**, 591 (1955).
[14] E. F. McCoy and I. G. Ross, *Australian J. Chem.* **15**, 573 (1962).

---

[s] C. A. Coulson, *Proc. Phys. Soc.* **A65**, 933 (1952); H. C. Longuet-Higgins and R. G. Sowden, *J. Chem. Soc.* p. 1404 (1952).
[t] A. Ishitani and S. Nagakura, *Theoret. Chim. Acta* **4**, 236 (1966); see also R. Zahradník and P. Čársky, *J. Phys. Chem.* **74**, 1235 (1970).
[u] C. C. J. Roothaan, *Rev. Mod. Phys.* **32**, 179 (1960).
[v] H. C. Longuet-Higgins and J. A. Pople, *Proc. Phys. Soc.* **A68**, 591 (1955).
[w] A. D. McLachlan, *Mol. Phys.* **3**, 233 (1960).

TABLE III

HAMILTONIAN, WAVE FUNCTIONS, AND MATRIX ELEMENTS OF VARIOUS VERSIONS OF THE MO–LCAO METHODS[a]

Method	$\hat{H}$	ψ	Hamiltonian matrix elements	Scope of application and limitation
1. CNDO/1 (Pople, Santry, Segal)[b]	$\sum_\mu H^c_\mu + \sum_{\mu<\nu} \sum e^2/r_{\mu\nu}$	Δ_0	$F_{\mu\mu} = U_{\mu\mu} + (P_{AA} - 1/2 P_{\mu\mu})\gamma_{AA} + \sum_{A\neq B} (P_{BB}\gamma_{AB} - V_{AB})$ $F_{\mu\nu} = \beta^0_{AB} S_{\mu\nu} - 1/2 P_{\mu\nu}\gamma_{AB}$ $(\mu\neq\nu)$	Ground-state properties of systems of various types (inorganic and organic containing up to 60–80 AO's); valence angles, deformation vibrations, dipole moments, barriers of inner rotation, chemical shifts; does not work for bond-length and dissociation energy calculations
2. CNDO/2 (Pople, Segal)[c]	$\sum_\mu H^c_\mu + \sum_{\mu<\nu} \sum e^2/r_{\mu\nu}$	Δ_0	$F_{\mu\mu} = -1/2(I_\mu + A_\mu) + [(P_{AA} - Z_A) - 1/2(P_{\mu\mu} - 1)]\gamma_{AA} + \sum_{B\neq A} (P_{BB} - Z_B)\gamma_{AB}$ $F_{\mu\nu} = \beta^0_{AB} S_{\mu\nu} - P_{\mu\nu}\gamma_{AB}$	See 1; this version uses a more convenient parametrization
3. Extended HMO (Hoffmann)[d]	$\sum_\mu H^{eff}_\mu$	Product function	$H_{\mu\mu} = $ I.P. (valence state) $H_{\mu\nu} = 0.5K(H_{\mu\mu} + H_{\nu\nu})S_{\mu\nu}$ (mostly $K = 1.75$)	Ground-state properties of aliphatic, aliphatic-aromatic, and aromatic systems: conformation of cyclic compounds, geometrical isomers, barriers of inner rotation, deformation vibrations; does not work for calculations of bond-lengths and stretching vibrations; qualitative and semi-quantitative estimates

4. MO, a very simple version (Del Re)[e]	$\sum_{\mu} H_{\mu}^{\text{eff}}$	Product function	$H_{\mu\mu} = \alpha + \delta_\mu \beta$ $H_{\mu\nu} = \epsilon_{\mu\nu}\beta$ (α, β: parameters of the method)	An extremely simple method (which does not require a computer) useful for rough estimation of electron-distribution dipole moments and equilibrium constants of aliphatic compounds

[a] Methods in which all valence-shell electrons are considered and simple methods which do not represent π-electron approximation; closed-shell systems in the ground state. Symbols used have their usual meaning. Index μ (ν) denotes an atomic orbital situated on the atom A (B); $P_{\mu\nu}$ defines the usual bond order; P_{AA} is the total charge of the atom A. γ_{AB} and V_{AB} represent interaction energy; β_{AB}° are quantities determined empirically. Δ_0 denotes a Slater determinant for the ground state.

[b] J. A. Pople, D. P. Santry, and G. A. Segal, *J. Chem. Phys.* **43**, S129 (1965).

[c] J. A. Pople and G. A. Segal, *J. Chem. Phys.* **43**, S136 (1965); **44**, 3289 (1966).

[d] R. Hoffmann, *J. Chem. Phys.* **39**, 1397 (1963); **40**, 2480 (1964).

[e] G. Del Re, *J. Chem. Soc.* p. 4031 (1958).

TABLE IV

HAMILTONIAN, WAVE FUNCTIONS, AND MATRIX ELEMENTS OF VARIOUS VERSIONS OF THE MO–LCAO METHOD (π-ELECTRONIC APPROXIMATION; CLOSED-SHELL SYSTEMS IN THE GROUND STATE)[a]

Method	$\hat{H}$	ψ	Hamiltonian matrix elements	Scope of application and limitations
1. HMO (Hückel)[b]	$\sum_\mu H_\mu^{eff}$	Product function	$H_{\mu\mu} = \alpha + \delta_\mu \beta$ $H_{\mu\nu} = \rho_{\mu\nu} \beta$	Qualitative discussions of properties of ground and excited states; it is impossible to distinguish between S and T states; relative values of various characteristics can be obtained which are useful for correlating experimental data
2. HMO–SC (Wheland, Mann[c]; ω-technique)	$\sum_\mu H_\mu^{eff}$	Product function	$H_{\mu\mu} = \alpha + (1 - q_\mu) \omega\beta$ (for hydrocarbons $\omega = 1.4$) $H_{\mu\nu} = \rho_{\mu\nu} \beta$	See 1; especially useful for calculating dipole moments, μ, and IP
3. HMO–SC (Coulson, Golebiewski)[d]	$\sum_\mu H_\mu^{eff}$	Product function	$H_{\mu\mu} = \alpha + \xi_{\mu\mu} \beta$ $H_{\mu\nu} = \rho_{\mu\nu} \exp - [2.683(1.517 - 0.180 \, p_{\mu\nu})]\beta$	See 1; valuable for bond-lengths–bond-order correlations
4. HMO–SC (Janssen, Sandström)[e]	$\sum_\mu H_\mu^{eff}$	Product function	$H_{\mu\mu} = \alpha + [\xi_{\mu\mu} + \omega(n_\mu - q_\mu)]\beta$ $H_{\mu\nu} = \rho_{\mu\nu}(1 + 0.5\, p_{\mu\nu})\beta$	See 1, 2, and 3
5. SCF (Pople)[f]	$\sum_\mu H_\mu^c + \sum_{\mu<\nu} \sum e^2/r_{\mu\nu}$	Δ_0	$F_{\mu\nu} = \beta_{\mu\nu}^c - 1/2 p_{\mu\nu} \gamma_{\mu\nu} +$ $\left[\sum_{\sigma \neq \mu} (p_{\sigma\sigma} - Z_\sigma)\gamma_{\mu\sigma} + \gamma_{\mu\mu} \right] \delta_{\mu\nu}$	Ground state properties: π-electron energy, IP, EA, $q(\mu,$ ESR$)$, p (geometry); does not work with systems having a strong bond alternation (e.g., polyenes, fulvenes)

6. SCF–SC (Brown, Heffernan[g]; VESCF)	$\sum_\mu H_\mu^c + \sum_{\mu<\nu} e^2/r_{\mu\nu}$	Δ_0	$F_{\mu\nu}$, see method 5 and $\gamma_{\mu\mu} = f(Z_\mu)$ $I_\mu = f'(Z_\mu)$ $(Z_\mu = N_\mu + C_1 - 0.35 q_\mu)$	Ground state properties: see 5; especially useful for systems with a strong charge separation
7. SCF–SC (β^c-variation)[h]	$\sum_\mu H_\mu^c + \sum_{\mu<\nu} e^2/r_{\mu\nu}$	Δ_0	$F_{\mu\nu}$, see method 5 and $\beta_{\mu\nu}^c = f(l_{\mu\nu})$	Ground state properties: see 5; works also for systems with strong bond alternation; useful for bond-length estimation
8. SCF–SC (α^c- and β^c-variation)[i]	$\sum_\mu H_\mu^c + \sum_{\mu<\nu} e^2/r_{\mu\nu}$	Δ_0	$F_{\mu\nu}$, see method 5 and $\gamma_{\mu\mu} = f(Z_\mu)$ $I_\mu = f'(Z_\mu)$ $\beta_{\mu\nu} = f(l_{\mu\nu})$	Ground state properties: see 5; this method should be applicable rather generally
9. LCI (Pariser, Parr)[j]	$\sum_\mu H_\mu^c + \sum_{\mu<\nu} e^2/r_{\mu\nu}$	$C_0\Delta_0 + \sum_n C_{n,i\to j}\Delta_{i\to j}$	$F_{J1}\delta_{1k} = F_{ik}\delta_{J1} + 2(kj\|G\|li) - (kj\|G\|il)$ $F_{J1} = \sum_{\mu\neq\nu} c_{J\mu} F_{\mu\nu} c_{1\nu};$ $(kj\|G\|li) = \sum_\mu\sum_\nu c_{1\mu} c_{J\nu} c_{k\mu} c_{1\nu}\gamma_{\mu\nu}$	Properties of the electronically excited states, π-electronic energies of S and T states (electronic spectra); $q(\mu)$, p (geometry) in the excited states; the LCI calculations should be performed with SCF–MO in the case of nonalternant systems and heterocyclic compounds

[a] Symbols have their usual meaning. α and β are the Coulomb and resonance integrals of the HMO theory, $p_{\mu\nu}$ is the bond order, $\gamma_{\mu\nu}$ the electronic repulsion integral. Δ_0 (Δ_{l_j}) represents the Slater determinants of the ground ($i \to j$ excited) states.

[b] E. Hückel, *Z. Physik* **70**, 204 (1931); **76**, 628 (1932).

[c] G. W. Wheland and D. E. Mann, *J. Chem. Phys.* **17**, 264 (1949).

[d] C. A. Coulson and A. Golebiewski, *Proc. Phys. Soc.* **78**, 1310 (1961).

[e] M. J. Janssen and J. Sandström, *Tetrahedron* **20**, 2339 (1964).

[f] J. A. Pople, *Trans. Faraday Soc.* **49**, 1375 (1953).

[g] R. D. Brown and M. L. Heffernan, *Trans Faraday Soc.* **54**, 757 (1958).

[h] P. Hochmann, Ph.D. Thesis, Inst. Phys. Chem. ČSAV, Prague, 1967.

[i] R. Zahradník, A. Kröhn, J. Pancíř, and D. Šnobl, *Collection Czech. Chem. Commun.* **34**, 2553 (1969).

[j] R. Pariser and R. G. Parr, *J. Chem. Phys.* **21**, 466 and 767 (1953).

A.1. Relativistic corrections are not included[a]

A.2. Adiabatic approximation[a]

A.3. Electron correlation is considered only through empirical parameters[a]

A.4. Assumption of σ-π separability[a,b]

B.1. $\mu = \nu$
  (i) Approximation by I'_μ (neglect of penetration integral)[c]
  (ii) Calculation of penetration integral[d]

B.2. $\mu \neq \nu$
  (i) Adjacent atoms (μ–ν)
    (α) Constant empirical values (2–3 eV)
    (β) Values depending on the bond lengths[e,f] ($\beta_{\mu\nu} = f(r_{\mu\nu}$ or $p_{\mu\nu})$)
  (ii) Nonadjacent atoms ($\mu - \ldots - \nu$)
    (α) $\beta_{\mu\nu} = 0 \ldots$ tight-binding approximation
    (β) $\beta_{\mu\nu} \sim S_{\mu\nu}{}^g$

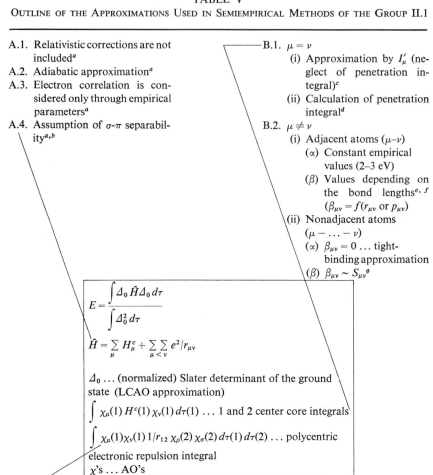

$$E = \frac{\int \Delta_0 \hat{H} \Delta_0 \, d\tau}{\int \Delta_0^2 \, d\tau}$$

$$\hat{H} = \sum_\mu H_\mu^c + \sum_{\mu < \nu} e^2/r_{\mu\nu}$$

$\Delta_0 \ldots$ (normalized) Slater determinant of the ground state (LCAO approximation)

$\int \chi_\mu(1) H^c(1) \chi_\nu(1) \, d\tau(1) \ldots$ 1 and 2 center core integrals

$\int \chi_\mu(1) \chi_\nu(1) \, 1/r_{12} \, \chi_\rho(2) \chi_\sigma(2) \, d\tau(1) \, d\tau(2) \ldots$ polycentric electronic repulsion integral

χ's $\ldots$ AO's

C.1. "N^4 difficulty"[a] (ZDO assumption: $\chi_\mu(1) \chi_\nu(1) \, d\tau(1) = 0 \ldots \delta_{\mu\nu} \delta_{\rho\sigma}(\mu\nu|\rho\sigma)$)

C.2. Monocentric electronic repulsion integral: $(\mu\mu|\mu\mu) \equiv \gamma_{\mu\mu} = I_\mu - A_\mu{}^h$

C.3. Bicentric electronic repulsion integral: $(\mu\mu|\nu\nu) \equiv \gamma_{\mu\nu}$, approximated by various formulas[i,j]

[a] R. G. Parr, "Quantum Theory of Molecular Electronic Structure." Benjamin, New York, 1964.

[b] P. G. Lykos and R. G. Parr, *J. Chem. Phys.* **24**, 1166 (1956).

[c] M. Goeppert-Mayer and A. L. Sklar, *J. Chem. Phys.* **6**, 646 (1938).

[d] I. Fischer-Hjalmars, *J. Chem. Phys.* **42**, 1962 (1965).

[e] H. Hosoya and S. Nagakura, *Theoret. Chim. Acta* **8**, 319 (1967).

[f] R. L. Flurry, Jr., E. W. Stout, and J. J. Bell, *Theoret. Chim. Acta* **8**, 203 (1967).

[g] R. L. Flurry, Jr. and J. J. Bell, *J. Am. Chem. Soc.* **89**, 523 (1967).

[h] R. Pariser, *J. Chem. Phys.* **21**, 568 (1953).

[i] N. Mataga and K. Nishimoto, *Z. Physik. Chem.* (*Frankfurt*) [N.S.] **13**, 140 (1957).

[j] R. G. Parr, *J. Chem. Phys.* **20**, 1499 (1952).

In this expression i,h (j,k) denote molecular orbitals which are occupied (empty) in the ground state configuration; A represents the expansion coefficient of the LCI functions (subscripts a, ij, and θ denote the ath energy level, the $i \rightarrow j$ transition and the ground state, respectively), $c_{k\sigma}$ are the expansion coefficients of the one-electron function (HMO– and SCF–MO's), and $p_{\mu\nu}$ is the bond order between μ and ν (if $\mu = \nu$, then $p_{\mu\mu} \equiv q_\mu$).

II. Ground State Properties

A. AROMATICITY

"Aromaticity" is one of the terms used in chemistry on which opinions differ considerably. Some individuals are inclined to it, others have cast it off. There is no doubt that it is rather vague in concept. However, if the problem is defined specifically, part of the trouble is removed. Following a period when aromatic substances were catalogued as a result of their aroma, Kekulé attributed "aromaticity" to compounds according to their structural type. Erlenmeyer connected it with their chemical properties.[15, 15a] In this connection we shall consider the latter aspect.

Chemists, frequently, intuitively associate "aromaticity" with stability in the simplest sense of the word. A substance is presumed stable,[16] if its composition remains unchanged over long periods when stored at room temperature in the presence of air oxygen, humidity, and carbon dioxide. This, of course, is not a proper definition, but there is no doubt regarding its utility. Such an emphasis clearly indicates a kinetic problem and suggests that stability depends on the values of the free energies of activation of the individual processes to be considered. Energy gain due to delocalization (delocalization energy) is not, strictly speaking, important in this connection. The significant role of the delocalization energy in considering aromaticity follows from the concept of the delocalization energy itself. However, in recent years, some authors have interpreted the various properties of conjugated systems in terms of hybridization rather than delocalization energies.[17-19] Whatever development these suggestions might undergo, it seems that the endeavor to exclude delocalization energy represents an extreme (probably one not useful).

The attempt to specify terms in this area, however, is undoubtedly valuable. First of all, it is necessary to stress that there exist two independent and different views as regards stability, and thus also aromaticity. For quite some

[15] E. Heilbronner, in "Non-Benzenoid Aromatic Compounds" (D. Ginsburg, ed.), p. 171. Wiley (Interscience), New York, 1959.

[15a] J. P. Snyder, in "Nonbenzenoid Aromatics," Vol. I, p. 1. Academic Press, New York, 1969.

[16] In this connection we are interested entirely in fully conjugated systems.

[17] M. J. S. Dewar and H. N. Schmeising, Tetrahedron 5, 166 (1959).

[18] M. J. S. Dewar and H. N. Schmeising, Tetrahedron 11, 96 (1960).

[19] C. J. McGinn, Tetrahedron 18, 311 (1962).

time no distinction has been made between these two aspects. Thus MO-values[20] of the delocalization energy have been unjustifiably used as the basis for aromaticity estimates. The aromaticity of a particular system has not been regarded simply as the difference between the enthalpies of formation of a given Kekulé structure and the corresponding delocalized system, but rather a measure of its reactivity. This procedure is unacceptable for several reasons. In the first place, there is no theoretical or experimental basis for drawing conclusions as to the magnitude of activation barriers by means of ground state thermodynamic parameters. It seems however, *and it must be stressed that this is only an empirical fact*, that systems with high specific values of the delocalization energy (values obtained by respecting the bond lengths being considered) nearly always exhibit very favorable theoretical MO-values of the reactivity indices. Thus these compounds are "aromatic" from both points of view. Rasch[21] mentioned the fact that the highest delocalization energies correspond to the chemically most stable species in groups of structurally related substances. Without attempting to interpret this empirical finding, we should at least like to state that a high DE_{sp}-value is conditioned by the most thorough removal of the difference between the single and double bonds. Due to this uniformity it is not easy to localize electrons on a certain bond of the system. Thus the electrophilic addition of a benzene derivative (predominance of substitution over addition is one of the experimental criteria for aromaticity) is rarely observed.[21a]

The definition of aromaticity was placed on a firmer foundation as a result of Hückel's studies.[22] Although the Hückel $(4n + 2)$ π-electron rule is limited in scope and based on questionable assumptions, it has played a significant role in chemistry and is useful even today, when correctly used. It is necessary to mention that the rule was formulated for monocyclic systems (for systems with large cycles, see Polansky[23, 24]), the atoms of which contribute p_z-orbitals to the conjugation (different rules are valid for p_π–d_π-systems).[25, 26] Application to cata-condensed systems (which may be considered as monocycles with weak perturbations, represented by crossing bonds) seems acceptable. A naive extension of Hückel's rule according to the size of the perimeter appears less

---

[20] The effect of the bond lengths on the resonance integral values was frequently disregarded in the computations. Consequently unreal values especially in cases with a single Kekulé structure have resulted.

[21] G. Rasch, *Wiss. Z. Friedrich-Schiller-Univ. Jena, Math. Naturwiss. Reihe* **13**, 275 (1964).

[21a] Under these conditions even localization of 0, 1, or 2 electrons in the individual position is rather difficult.

[22] E. Hückel, *Z. Elektrochem.* **43**, 752 (1937).

[23] O. E. Polansky, *Monatsh.* **91**, 203 (1960).

[24] O. E. Polansky, *Monatsh.* **91**, 916 (1960).

[25] D. P. Craig and N. L. Paddock, *Nature* **181**, 1052 (1958).

[26] D. P. Craig and N. L. Paddock, *J. Chem. Soc.* p. 4118 (1962).

justified. For example, one might be tempted on structural grounds to consider pyrene a cyclotetradecaheptaen, "internally" perturbed by ethylene. It also seems dubious to consider fluoranthene as a perturbed system consisting of benzene and naphthalene, because the appropriate bonds cannot be considered as weak perturbations even in the ground state.[19, 27] It is necessary to bear in mind that the reactivity of the systems under investigation and the effect of the solvent medium are neglected in the Hückel treatment. Moreover, no difference is made between the stability of hydrocarbons and ions. This leads, for example, to a considerable overestimation of the similarity of systems 1–3. Neglect of

chemical reactivity (in a general sense, viz. kinetic and equilibrium aspects) prevents one from comprehending, within the framework of the rule, why the anti-Hückel fluorenylium cation 4 is much more stable under current laboratory conditions than the Hückel anion 5. The relative stability of compounds

such as furan and tropone and the instability of azepine and of cyclopentadienone, however, are understandable from the point of view of the $(4n + 2)$ rule, if the heteroatom or the substituent is considered to be a perturbation of the monocyclic system. An SCF interpretation of Hückel's rule which includes internuclear repulsion[28] has been applied to the stability of related compounds. In order to characterize derivatives of monocyclic and bicyclic systems, the sums of the π-electron densities at positions corresponding to the individual cycles have also been used. It remains to be added that from the time of the introduction of the Hückel rule, aromaticity has tended to be characterized in a fashion as condensed as possible. Often a single number, for example a magic sum of π-electrons, has served this purpose. If one

[27] J. Michl and R. Zahradník, *Collection Czech. Chem. Commun.* **31**, 3478 (1966), and preceding papers.

[28] K. Fukui, A. Imamura, T. Yonezawa, and C. Nagata, *Bull. Chem. Soc. Japan* **33**, 1591 (1960).

considers, however, the numerous aspects involved in assessing aromaticity, it becomes clear that this simplistic trend is not exactly favorable. Nevertheless, such indices do exist and they provide useful information. However one should not rely too much on any single one of them.

Craig[29, 30] has attempted to formulate precise definitions of the terms aromaticity and pseudoaromaticity. He found that there are certain conjugated hydrocarbons, the ground states of which are not totally symmetric in terms of a valence-bond description. These systems were labeled pseudoaromatic. Most of the conjugated systems characterized by their symmetry properties, however, are catagorized as aromatic. As is known, for symmetric systems it is very easy to decide to which group the investigated species belongs by means of two indices, p and q. Their values are determined by employing a symmetry operation (C_{2v}) which transforms one (VB) structure into another belonging to the same canonical set. The number of interchanged positions (p_z-orbitals) is denoted by p, while q designates the number of spin symbol interchanges necessary for restoring the original labeling with respect to that after the transformation. If the character χ is positive, the system under investigation is aromatic, otherwise pseudoaromatic:

$$\chi = (-1)^{p+q} \tag{3}$$

The first shortcoming is that this definition encompasses symmetric compounds only. Aromaticity, however, is not a property solely of symmetric systems. Second and more seriously, the "aromaticity", or "pseudoaromaticity", of a variety of chemical types depends on the labeling of the positions (see, for example, Abramowitch and McEwen[31]). This is a serious deficiency. Let it be added that cyclobutadiene, pentalene, heptalene, and s-indacene are representatives of the pseudoaromatic systems in the sense of the Craig rule.

It is the author's opinion that significant progress was achieved by Peters,[32] who pointed out that substances evidencing aromaticity are conditioned by low chemical reactivity with neutral systems, and by a favorable equilibrium with ionic ones. Therefore, the stability of an aromatic species is not a direct result of a high delocalization energy. Part of the numerical data discussed in Peters' paper[32] must be considered as insufficient, however, because HMO data on fulvenoid and pentalenoid systems fit no reasonably interpreted pattern.[33]

[29] D. P. Craig, *Proc. Roy. Soc.* **A202**, 498 (1950).

[30] D. P. Craig, *J. Chem. Soc.* p. 3175 (1951).

[31] R. A. Abramowitch and K. L. McEwen, *Can. J. Chem.* **43**, 261 (1965).

[32] D. Peters, *J. Chem. Soc.* p. 1274 (1960).

[33] Moreover, in comparing delocalization energies it is not satisfactory to work either with absolute or relative values (E_D per electron or per σ C—C bond). Even the latter quantities depend on the extent of the system, *vide infra*.

Lately, it has been found[34,35] that one cannot draw conclusions about stability or stable configuration on the basis of π-electron energy information alone. The effect of σ-bond compression has to be considered. In accordance with the latter, an HMO modification was introduced.[36-38] First of all, the bond alternation parameter k was defined as:

$$k = \beta_s/\beta_d \qquad (4)$$

where β_s and β_d represent the resonance integrals for a single and double bond, respectively. The HMO calculation is initiated with a certain value of k. Calculated values of the bond orders ($p_{\mu\nu}$) are then transformed into bond lengths ($r_{\mu\nu}$):

$$r_{\mu\nu}(\text{Å}) = 1.520 - ap \qquad (5)$$

where a is a constant. The values of $\beta(r)$ for the next step of the iterative calculation are obtained from the following expression:

$$\beta(r) = \beta^0 \exp[b(1.397 - r_{\mu\nu})] \qquad (6)$$

where b is a constant. The total bond energy of the molecule is expressed as the sum of the π- and σ-bond energies[34]:

$$V = E_\pi + E_\sigma = (2/ab) \sum_{\mu<\nu} \beta_{\mu\nu} + \sum_\mu q_\mu \alpha_\mu + \text{const.} \qquad (7)$$

where $\alpha_\mu = \alpha^0 + (1 - q_\mu)\beta^0$. Values of V as a function of k (0–1) were calculated for the individually studied systems. The equilibrium bond alternation ($k_{\min}$) is defined by the minimum in the V vs k graph. The $k_{\min}$ values are in the range of 0.54–0.60. For a number of physical quantities (dipole moments, delocalization energy, magnetic properties) this method has yielded substantially more correct results than the HMO technique. However, except in one regard, this improvement can be achieved by the usual π-electron approximation in which α- and β-variations are considered. A specific advantage of the above procedure is that with some systems (pentalene, heptalene) it indicates that a structure with a lower degree of symmetry has a lower total electron energy than the structure with the highest degree of symmetry. Thus for the pentalene structures **6a** and **6b** $V_1 < V_2$. This interesting finding is overshadowed by the following unfortunate result. The same method predicts a similar conclusion for other molecules having C_{2v} symmetry, e.g., azulene in contradiction to

[34] H. C. Longuet-Higgins and L. Salem, *Proc. Roy. Soc.* **A251**, 172 (1959).
[35] C. A. Coulson and W. T. Dixon, *Tetrahedron* **17**, 215 (1962).
[36] P. C. Boer-Veenendaal and D. H. W. Boer, *Mol. Phys.* **4**, 33 (1961).
[37] T. Nakajima and S. Katagiri, *Bull. Chem. Soc. Japan* **35**, 910 (1962).
[38] T. Nakajima, *in* "Molecular Orbitals in Chemistry, Physics, and Biology" (P.-O. Löwdin and B. Pullman, eds.), p. 451. Academic Press, New York, 1964.

$V_1(C_{2h})$ $V_2(D_{2h})$

6a   **6b**

experiment. The value of the method is consequently diminished (see, however, Francois and Julg[39]).

Progress in this area is represented by the recent introduction of a new method for studying bond fixation.[40, 41] Bond fixing of the first order from which relations between $r_{\mu\nu}$ and $p_{\mu\nu}$, which do not *decrease* the molecular symmetry, may be derived are differentiated from those of the second order, which can decrease the symmetry. A theoretical analysis of the problem has shown that it is possible to decide whether or not distortion will occur with the help of the highest eigenvalue of the bond–bond polarizability matrix:

$$\|\pi_{\rho\sigma,\kappa\lambda} - \delta_{\rho\sigma,\kappa\lambda}\Lambda\| = 0 \qquad (8)$$

If Λ_{max} is larger than the critical (estimated) value Λ_{crit} $(1.8\,\beta^{-1})$ then the second-order effects in the π-energy overcome the σ-compression energy and the molecule will lose its original symmetry. The method is applicable also to excited states and to radical ions.

An attempt had been made to ascribe to all first row atoms a numerical constant k in such a way that the sum of these values over the whole aromatic system gives a total aromaticity constant K of the system.[42] The following holds for k:

$$k = (0.478Z^*/r - 1.01 - n_\pi)\,100 \qquad (9)$$

where Z^* depends, among other things, on the atomic number, r stands for the co-valent radius (Å), and n_π denotes the number of π-electrons (which equals 0, 1, or 2).

It should be mentioned that the term antiaromatic has been introduced for compounds, the $4n$ π-electron cyclic form of which has a higher energy than the corresponding acyclic form (e.g., cyclopropenyl anion–allyl anion).[43, 44]

[39] P. Francois and A. Julg, *Theoret. Chim. Acta* **11**, 128 (1968).
[40] G. Binsch, E. Heilbronner, and J. N. Murrell, *Mol. Phys.* **11**, 305 (1966).
[41] G. Binsch and E. Heilbronner, *Tetrahedron* **24**, 1215 (1968).
[42] A. T. Balaban and Z. Simon, *Tetrahedron* **18**, 315 (1962).
[43] R. Breslow, *Chem. Brit.* **4**, 100 (1968).
[44] R. Breslow and M. Douek, *J. Am. Chem. Soc.* **90**, 2698 (1968).

Our work can be summarized as follows. In order to be able to estimate the degree of "aromaticity" of a system, we have postulated[45] that the necessary conditions are: (i) a high value of the specific delocalization energy ($E_{D,sp}$); (ii) favorable values of the frontier orbital energies (theoretical measure of ionization potentials and electron affinities); (iii) favorable extreme values of chemical reactivity indices. A high $E_{D,sp}$ value favors the formation of the conjugated system with regard to less conjugated precursors. As not only E_D,

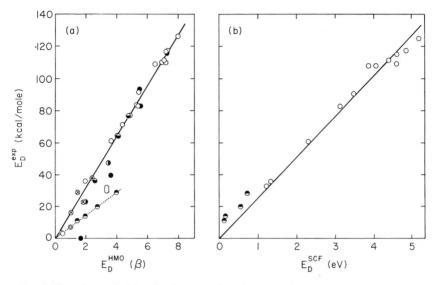

FIG. 3. Experimental delocalization energies plotted against HMO (a) and SCF–SC (b) delocalization energies [R. Zahradník, unpublished results (1965)]. Symbols: ◎ polyenes, ○ benzenoid hydrocarbons, ◑ nonalternant hydrocarbons with a single Kekulé structure, ◑ pyridine-like heterocyclic compounds, ⊗ pyrrole-like heterocyclic compounds, ◒ monosubstituted hydrocarbons, ● quinones.

but even $E_{D,sp}$ depends on the extent of conjugation, a comparison of the calculated $E_{D,sp}$ value for the systems under investigation with $E_{D,sp}$ values for some standard compounds of a similar size is desirable.[2] Figure 3 shows that there exists a fair agreement between experimental and calculated delocalization energies. For the HMO data points corresponding to systems for which a single Kekulé structure is appropriate lie on a separate straight line. SCF–SC data lead to a single straight line[46, 47] (for SCF data, see Dewar and Gleicher[48]).

[45] R. Zahradník and J. Michl, *Collection Czech. Chem. Commun.* **30**, 520 (1965).
[46] R. Zahradník, unpublished results (1967).
[47] More specific information on data used in Fig. 3 is available on request from the author.
[48] M. J. S. Dewar and G. J. Gleicher, *J. Am. Chem. Soc.* **87**, 685 (1965).

An attempt has been made[49] to use a proper combination of the indices in the form of a complex "stability" index (P), instead of employing each of them separately. Obviously it is difficult to assess the relative importance of the indices selected. In addition to $E_{D,sp}$, we have chosen the difference between the maximum and minimum π-electron densities, Δq, related to the reactivity toward nucleophilic and electrophilic agents, the maximum free valence F, related to the radical reactivity, and the maximum radical superdelocalizability S_r, related to radical reactivity, oxidizability, and reducibility. The following expression was formulated (for details, see Zahradník et al.[49]):

$$P = \frac{273(E_{D,sp} - 0.2796)(0.552 - \Delta q)(0.586 - F)}{S_r} \tag{10}$$

We applied this index initially to peri-condensed tetracyclic hydrocarbons. All hydrocarbons known at the time had relatively high P-values. Additional compounds considered, 7 and 8, have since been synthesized,[50, 51] and even

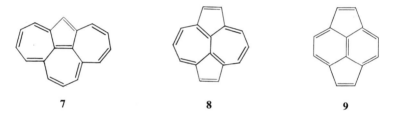

7   8   9

solutions of pyracylene (9),[52] have been prepared. We have also made successful use of the P index with other types of nonalternant hydrocarbons.

It is even more difficult to use the term aromatic with ions than with neutral systems.[53, 54] It is usually associated with those cations which are weakly acidic and electrophilic (formed from carbinols by the action of comparatively mild acids, or from hydrocarbons relatively easily by the hydride ion transfer reaction, etc.), and with weakly basic and nucleophilic anions (formed from hydrocarbons by the action of relatively weak bases). Another requirement is the stability of the ion once it has been formed and the absence of rearrangements or polymerization. In view of the omnipresence of oxygen, the high sensitivity to oxidation is always serious; easy reducibility (high electron affinity) much less so.

[49] R. Zahradník, J. Michl, and J. Pancíř, *Tetrahedron* **22**, 1355 (1966).
[50] K. Hafner, R. Fleischer, and K. Fritz, *Angew. Chem.* **77**, 42 (1965).
[51] A. G. Anderson, Jr., A. A. MacDonald, and A. F. Montana, *J. Am. Chem. Soc.* **90**, 2993 (1968).
[52] B. M. Trost and G. M. Bright, *J. Am. Chem. Soc.* **89**, 4244 (1967).
[53] R. Zahradník and J. Michl, *Collection Czech. Chem. Commun.* **30**, 1060 (1965).
[54] This paragraph as a whole has been adopted from Zahradník and Michl. [53]

To conclude it is necessary to cite several comprehensive papers on aromaticity.[55-65]

As far as the experimental characterization of aromaticity is concerned, it seems that there are now three means available. There is the classic determination of the experimental delocalization energy[66] by measuring heats of combustion, hydrogenation, or halogenation. The techniques of calorimetry have achieved remarkable improvement. Most important is the availability of microcalorimetric methods. Small amounts of material can yield accurate information. Requirements as to the extreme purity of the substances studied, of course, remain unchanged.

In view of its ready accessibility, NMR spectroscopy would seem an attractive method for estimating aromaticity. Elvidge and Jackman[67] have defined an aromatic compound as one which sustains an induced ring current. The extent of π-electron delocalization manifests itself by the magnitude of the ring current and is, therefore, a measure of aromaticity. Shielding contributions due to ring currents have been computed. The ability to estimate line positions in the absence of a ring current is necessitated. This may be accomplished with the help of model structures in which the π-electrons are regarded as rigidly localized. From the differences of computed and measured values of proton chemical shifts, aromaticity is expressed directly as a percentage in terms of the π-bond localized models. Sometimes the presence of certain lines in the NMR spectra is used to make a qualitative decision as to whether the substance is aromatic or nonaromatic.[68] In this way it was concluded that [18]annulene is an aromatic substance, whereas the [14] and [24]annulenes are not. However, the reactivity of [18]annulene is well known. This is easily understood theoreti-

[55] R. Robinson, *Tetrahedron* **3**, 323 (1958), and papers following (Paper Symposium on Aromatic Character).

[56] Aromaticity (an international symposium held at Sheffield on 6th–8th July 1966). *Chem. Soc.* (London), *Spec. Publ.* **21** (1967).

[57] M. J. S. Dewar, *in* "Non-Benzenoid Aromatic Compounds" (D. Ginsburg, ed.), p. 177. Wiley (Interscience), New York, 1959.

[58] M. E. Vol'pin, *Usp. Khim.* **29**, 298 (1960).

[59] A. W. Johnson, *J. Roy. Inst. Chem.* **84**, 90 (1960).

[60] T. G. Harrington, *School Sci. Rev.* **43**, 361 (1962).

[61] R. Breslow, *Chem. Eng. News* **43**, 90 (1965).

[62] A. T. Balaban and Z. Simon, *Rev. Roumaine Chim.* **10**, 1059 (1965).

[63] A. Courtin and H. Siegel, *Experientia* **19**, 407 (1965).

[64] R. Zahradník, *Advan. Heterocyclic Chem.* **5**, 1 (1965).

[65] G. M. Badger "Aromatic Character and Aromaticity." Cambridge Univ. Press. London and New York, 1969.

[66] G. W. Wheland, "Resonance in Organic Chemistry." Wiley, New York, 1955.

[67] J. A. Elvidge and L. M. Jackman, *J. Chem. Soc.* p. 859 (1961).

[68] L. M. Jackman, F. Sondheimer, Y. Amiel, D. A. Ben-Efraim, Y. Gaoni, R. Wolovsky, and A. A. Bothner-By, *J. Am. Chem. Soc.* **84**, 4307 (1962).

cally by a consideration of the rapidity with which atom and bond localization energies in the $(4n + 2)$ annulene series decrease with increasing n (see Fig. 4), and the magnitude of the highest occupied MO energy. Clearly, the NMR characterization of aromaticity does not include kinetic aspects. It is similar in concept to aromaticity characterization by delocalization energy. This is also the case if exaltation of diamagnetic susceptibility[69] is applied to the problem. The significant anisotropy of magnetic susceptibility $[\Delta\chi = \chi_z - 1/2(\chi_x + \chi_y)$, where χ_i is the component of the molar susceptibility, χ_M, in the direction of the i-axis in the (xy)-plane, where the planar conjugated system is located$]$ is

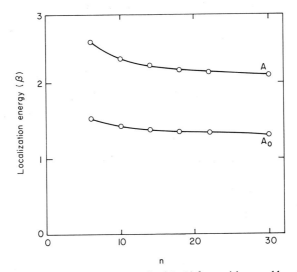

FIG. 4. Atom and ortho localization energies (A, A_0) for positions and bonds in annulenes with $4n + 2$ carbon atoms; n stands for the total number of carbon atoms.

due to the cyclic delocalization of π-electrons and is thus a significant characteristic of such systems. The experimental determination of anisotropy is difficult, but luckily enough it appears that the effective cyclic delocalization is also displayed by the increase of the average susceptibility χ_μ with respect to the value calculated from a group of contributions for a localized structure. The magnitude of this exaltation is considered by Dauben et al.[70] to be a criterion of aromatic character (see also Lonsdale and Toor[71] and Nakajima and Kohda[72]). A parallelism between magnetic susceptibility exaltation and

[69] P. W. Selwood, "Magnetochemistry," 2nd ed. Wiley (Interscience), New York, 1956.

[70] H. J. Dauben, Jr., J. D. Wilson, and J. L. Laity, J. Am. Chem. Soc. **90**, 811 (1968); ibid. Chapter X, this volume.

[71] K. Lonsdale and E. W. Toor, Acta Cryst. **12**, 1048 (1959).

[72] T. Nakajima and S. Kohda, Bull. Chem. Soc. Japan **39**, 804 (1966).

delocalization energy suggests itself. In fact a fair linear relationship correlating these quantities obtains.

B. CHEMICAL REACTIVITY

Although there is no doubt that the theory of chemical reactivity is the most important theory in chemistry, comparatively little attention will be devoted to it here. This is not due to the lack of interest in the subject on the part of the author. There are two reasons for this: (i) the amount of experimental data collected on the representative series of nonalternant compounds are very small; (ii) there is sufficient information in the literature concerning the quantum-chemical theory of reactivity.[73-81] It is the author's opinion that the lack of experimental observations for both nonalternant and alternant systems is the most significant obstacle to developing the theory of chemical reactivity. Until it is removed, meaningful calculations can hardly be made.

At present, since the possibility for computing activation entropy in solutions in the near future is small, the quantum-chemical treatment is limited essentially to the computation of π-electron changes of activation enthalpy. The values derived are compared with velocity constant logarithms. A similar procedure is utilized in discussing equilibrium data: computed π-electron enthalpy differences are compared with logarithms of the equilibrium constants. These comparisons are meaningful only if certain conditions are fulfilled:

(1) The changes of activation entropy, $\Delta S^{\ddagger}$ (entropy, ΔS), are correlated with the changes of the activation enthalpy, $\Delta H^{\ddagger}$; (enthalpy, ΔH).

(2) Steric interactions are similar for all members of a given reaction series.

(3) A constancy must exist in the changes of σ-electron energies accompanying the formation of the transition complex and the products and in the variations of solvation energies.

(4) The computations should be carried out at a semiempirical level taking into account electron repulsion. In the alternant series it has been demonstrated that the data splits into several subsets[81-83] even with a series of

[73] R. D. Brown, *Quart. Rev. (London)* **6**, 63 (1952).

[74] R. Daudel, *Advan. Chem. Phys.* **1**, 165 (1958).

[75] S. F. Mason, *Progr. Org. Chem.* **6**, 214 (1964).

[76] R. McWeeny and H. H. Greenwood, *Advan. Phys. Org. Chem.* **4**, 73 (1966).

[77] M. J. S. Dewar, *Advan. Chem. Phys.* **8**, 65 (1965).

[78] M. J. S. Dewar and C. C. Thompson, Jr., *J. Am. Chem. Soc.* **87**, 4414 (1965).

[79] G. Klopman, *J. Am. Chem. Soc.* **90**, 223 (1968).

[80] O. Chalvet, R. Daudel, and F. Peradejordi, *in* "Electronic Aspects of Biochemistry" (B. Pullman, ed.), p. 283. Academic Press, New York, 1964.

[81] R. Zahradník and O. Chalvet, *Collection Czech. Chem. Commun.* **34**, 3402 (1969).

[82] R. Zahradník and J. Koutecký, *Tr. Konf. po probl. primeneniya korrelyatsion. Uravneni v Organ. Khim, Tartusk. Gos. Univ., Tartu,* **1**, 89 (1962).

[83] R. Zahradník, *Chem. Listy* **60**, 289 (1966).

structurally homogeneous substances. Electrophilic substitution of benzenoid hydrocarbons illustrated by bromination[84] exemplifies the trend (see Fig. 5). It was clearly shown that the threefold subdivision of the HMO-correlated experimental data disappears upon application of the SCF method. Data dissection is, therefore, not a response to steric effects.

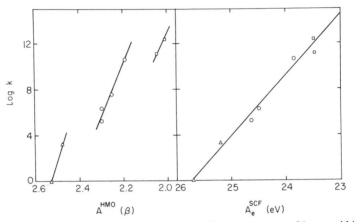

FIG. 5. Logarithms of rate constants for electrophilic bromination of benzenoid hydrocarbons [L. Altschuler and E. Berliner, *J. Am. Chem. Soc.* **88**, 5837 (1966)] plotted against HMO and SCF localization energies, *A*. Positions: class 0 (benzene-like) △, class 1 (α-naphthalene-like) ○, class 2 (*meso*-anthracene-like) □ [R. Zahradník and O. Chalvet, *Collection Czech. Chem. Commun.* **34**, 3402 (1969)].

Let us now return to the troublesome lack of quantitative experimental data in the nonalternant series. This situation, and not the state of the theory, is the cause of unsatisfactory progress in the area. One should not be misled by the apparently more extensive data, for example, the partial rate factors of fluoranthene[85] or the p*K*-values of five isomeric aminofluoranthenes.[86] From

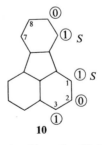

10

[84] L. Altschuler and E. Berliner, *J. Am. Chem. Soc.* **88**, 5837 (1966).

[85] A. Streitwieser, Jr. and R. C. Fahey, *J. Org. Chem.* **27**, 2352 (1962).

[86] J. Michl, R. Zahradník, and W. Simon, *Collection Czech. Chem. Commun.* **31**, 3464 (1966).

formula **10** it may be seen that of the five nonequivalent positions of fluoranthene, two are class zero positions[87] (benzene-like positions) and three are class 1 positions (α-naphthalene-like). Two of the latter are, however, sterically less accessible (marked S) than the third. They resemble, somewhat, position 4 in phenanthrene. The data for the five nonequivalent positions in **10** consequently break down to three sets.

Regarding the correlation of equilibrium data, it is necessary to mention the acidity of hydrocarbons (e.g., cyclopentadienyl, tropilidene)[88-90] and the pseudobasicity of carbinols (e.g., troponol, florenol).[91] Great attention has been devoted to the protonation of azulene (**11**) in the ground as well as in the excited state[92]:

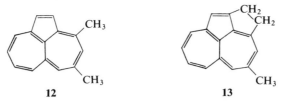

$$(11)$$

11   H H

The situation for simple derivatives of cycloheptazulene, hydrocarbons **12** and **13**, is interesting.[93] Absorption curves of the protonated hydrocarbons were

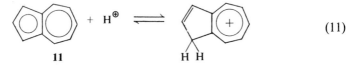

12   **13**

recorded in 70% $HClO_4$ and in 50% H_2SO_4. A comparison of LCI–SCF spectral data with experimental absorption curves indicates that the protonation products have structures **14** and **15**. The influence of CH_2 and CH_3

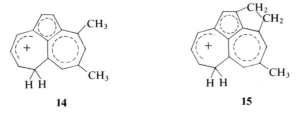

14   **15**

[87] J. Koutecký, R. Zahradník, and J. Čížek, *Trans. Faraday Soc.* **57**, 169 (1961).

[88] J. B. Conant and G. W. Wheland, *J. Am. Chem. Soc.* **54**, 1212 (1932).

[89] G. W. Wheland, *J. Chem. Phys.* **2**, 474 (1934).

[90] A. Streitwieser, Jr., *Tetrahedron Letters* p. 23 (1960).

[91] D. Meuche, H. Strauss, and E. Heilbronner, *Helv. Chim. Acta* **41**, 57 and 414 (1958); **43**, 1221 (1960).

[92] R. Hagen, E. Heilbronner, W. Meier, and P. Seiler, *Helv. Chim. Acta* **50**, 1523 (1967).

[93] K. Hafner and R. Zahradník, unpublished results (1966).

groups was neglected[94] (Table VI). On the other hand, if protonation had taken place in the five-membered ring, the LCI–SCF theory would have predicted the first band at the very edge of the visible region (approximately 690 mμ). Recently it has been found[95] that protonation of dimethylcyclo-heptazulene is a more complicated process. The structure of the first kinetically

TABLE VI

Experimental and Theoretical Positions
of Absorption Bands of Ions **14** and **15**
(in mμ)

λ_{exp}		λ_{th}[a]	
14	**15**	LCI–HMO	LCI–SCF
385–389	398	373	374
—	—	455	437
478–492	496	487	471

[a] These values apply to the parent system.

controlled protonation product is **16**. The latter equilibrates to the thermo-dynamically more stable isomer **14**.

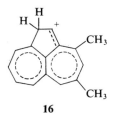

16

As mentioned above, the literature contains pK values for the five possible aminofluoranthene isomers.[86] Experimental and theoretical data (inclusive of electron densities on the nitrogen atom for excited states) are given in Table VII. Not even qualitative agreement between empirical and calculated values is found. Analysis of the data in connection with pK values of other aromatic amines leads to a reasonable result however. A correlation of pK and ΔW^{HMO}

[94] Protonation is possible in the substituted 7-membered ring as well. A qualitative consideration of alkyl inductive effects and of the more favorable steric situation in the unsubstituted ring indicates protonation occurs as pictured.

[95] E. Haselbach, personal communication (1968).

TABLE VII

BASICITIES OF X-AMINOFLUORANTHENES[a,b]

X (see formula 10)	π-Electron energy		Experimental values[c]		q_N^{SCF}[d]		
	SCF (eV)	HMO (β)	± 0.1 pK	± 0.05 pK$^+_{\text{AcOH}}$	S_0	S_1	T_1
1	-725.0807	2.51	1.6	7.69	1.771 [1.765]	1.593 (93.0)	1.643 (90.9)
2	-721.1800	2.44	3.9	9.25	1.807 [1.818]	1.553 (72.8)	1.629 (90.0)
3	-722.4336	2.52	2.8_5	8.76	1.770 [1.754]	1.641 (89.0)	1.605 (94.0)
7	-730.1242	2.46	2.9_5	—	1.795 [1.802]	1.536 (86.6)	1.679 (85.3)
8	-730.9751	2.44	4.2	9.30	1.804 [1.810]	1.604 (87.2)	1.711 (76.4)

[a] J. Michl, R. Zahradník, and W. Simon, *Collection Czech. Chem. Commun.* **31** 3464 (1966).

[b] Parameters used, HMO: $\alpha_N = \alpha + \beta$; SCF: $I_N = 25.40$ eV, $A_N = 9.10$ eV, $\beta_{CN}^c = 1.0138\beta$.

[c] pK, 50% methanol; pK$^+$, 99.94% acetic acid.

[d] SCF π-electron densities on the nitrogen atom for the ground state (S_0) and for the first excited singlet (S_1) and triplet (T_1) states. Values in brackets stand for HMO data. Values in parentheses mean the weight of the 1,1′ ($N \to V_1$) configuration in the excited state.

values leads to the result that points for the sterically hindered amines are on a separate straight line.[96]

A wide variety of scattered data concerning the course of substitution reactions for different types of nonalternant hydrocarbons and their hetero-analogs (see Hafner and Moritz[97]) can be found in the literature. The author does not intend to analyze these data. In summary it may be said that there is fair agreement between the values of the static and dynamic indices and the data on the course of substitution and addition reactions. Therefore, theoretical data may be used to estimate the structure of substitution products in cases when experimental evidence is lacking. For sake of illustration, the electro-philic nitration of pleiadiene (**17**) and acepleiadylene (**18**)[98] is instructive.

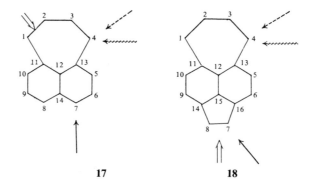

The constitution of the products is not known. From the data in Table VIII (static indices have been previously[99] published) it may be judged that the main product of nitration or other electrophilic substitution of hydrocarbon **17** will be the 7-derivative; the 5-derivative ranks second. Hydrocarbon **18** would seem to prefer substitution at the 7-position also. Probable centers of sub-stitution reactions (electrophilic, nucleophilic, radical; ↓, ↓, ↓ respectively) and of addition reactions   (⇓)   are shown in formulas **17** and **18**. An attempt to interpret the partial rate factors for nitration of fluoranthene in terms of HMO[85] localization energies was not really successful. The causes of these difficulties have already been mentioned.

In connection with a study[100] on the importance of steric effects at aryl-methyl carbon atoms, attention was devoted to the role of change in angle strain upon reactivity. This is a particularly important feature for some nonalternant systems.

[96] S. Guha and R. Zahradník, *Collection Czech. Chem. Commun.* **32**, 2448 (1967).
[97] K. Hafner and K. L. Moritz, *in* "Friedel-Crafts and Related Reactions" (G. A. Olah, ed.), p. 127. Wiley (Interscience), New York, 1965.
[98] V. Boekelheide and G. V. Vick, *J. Am. Chem. Soc.* **78**, 653 (1956).
[99] B. Pullman, A. Pullman, G. Berthier, and J. Pontis, *J. Chim. Phys.* **49**, 20 (1952).
[100] G. J. Gleicher, *J. Am. Chem. Soc.* **90**, 3397 (1968).

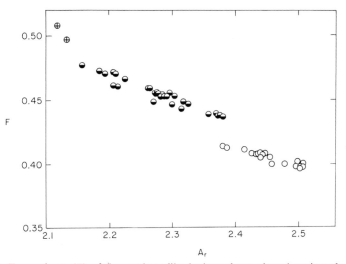

FIG. 6. Free valence (F) of fluoranthene-like hydrocarbons plotted against the radical atom localization energy (A_r). Positions: class 0 ○, class 1 ◓, class 2 ⊕ [J. Michl and R. Zahradník, *Collection Czech. Chem. Commun.* **31**, 3453 (1966); reproduced by permission].

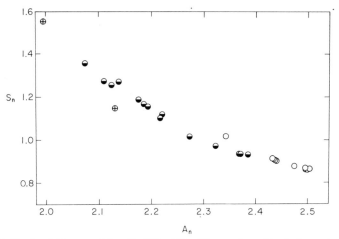

FIG. 7. Nucleophilic superdelocalizability (S_n) of fluoranthene-like hydrocarbons plotted against nucleophilic atom localization energy (A_n). Positions: class 0 ○, class 1 ◓, class 2 ⊕ [J. Michl and R. Zahradník, *Collection Czech. Chem. Commun.* **31**, 3453 (1966); reproduced by permission].

Considerable attention has been devoted to the connection between indices of various types (e.g., A_e, q_{max}; A_n, q_{min}; A_r, F_{max}, and similar pairs including superdelocalizability instead of atom localization energy).[101] Two of these are illustrated in Figs. 6 and 7 (for fluoranthene and its derivatives). Moreover, in

[101] J. Michl and R. Zahradník, *Collection Czech. Chem. Commun.* **31**, 3453 (1966).

TABLE VIII

An Output from a Computer[a,b]

Pleiadiene (17)

N	Q	F	S(E)	S(R)	S(N)	A(E)	A(R)	A(N)	I	J	P	I	J	P
1	0.9573	0.4991	1.3737	1.3737	1.3737	2.0139	2.0139	2.0139	1	2	0.7634	1	11	0.4695
2	0.9662	0.4411	1.1377	1.1377	1.1377	2.2119	2.2119	2.2119	2	3	0.5275			
3	0.9662	0.4411	1.1377	1.1377	1.1377	2.2119	2.2119	2.2119	3	4	0.7634			
4	0.9573	0.4991	1.3737	1.3737	1.3737	2.0139	2.0139	2.0139	4	13	0.4695			
5	1.0733	0.4715	1.7180	1.2735	.82908	1.9918	2.2010	2.4103	5	6	0.6346	5	13	0.6259
6	0.9953	0.3952	.85245	.85245	.85245	2.5155	2.5155	2.5155	6	7	0.7022			
7	1.0633	0.4804	1.7856	1.3412	.89671	1.9614	2.1439	2.3264	7	14	0.5494			
8	1.0633	0.4804	1.7856	1.3412	.89671	1.9614	2.1439	2.3264	8	9	0.7022	8	14	0.5494
9	0.9953	0.3952	.85245	.85245	.85245	2.5155	2.5155	2.5155	9	10	0.6346			
10	1.0733	0.4715	1.7180	1.2735	.82909	1.9918	2.2010	2.4103	10	11	0.6259			
11	0.9772	—	.84375	.84375	.84375	2.8305	2.8305	2.8305	11	12	0.5109			
12	0.9338	—	.66156	.88378	1.1060	3.0345	2.7934	2.5523	12	13	0.5109	12	14	0.5330
13	0.9772	—	.84375	.84375	.84375	2.8305	2.8305	2.8305						
14	1.0011	—	.69959	.69959	.69959	3.0671	3.0671	3.0671						

Acepleiadylene (18)

N	Q	F	S(E)	S(R)	S(N)	A(E)	A(R)	A(N)	bond	P	bond	P
1	0.9337	0.4876	.98057	1.1806	1.3806	2.2334	2.1296	2.0257	1 2	0.7145	1 11	0.5299
2	0.9379	0.4376	.84592	1.0459	1.2459	2.4310	2.2961	2.1612	2 3	0.5799		
3	0.9379	0.4376	.84592	1.0459	1.2459	2.4309	2.2961	2.1612	3 4	0.7145		
4	0.9337	0.4876	.98056	1.1806	1.3806	2.2334	2.1296	2.0257	4 13	0.5299		
5	1.0549	0.4494	1.2137	1.0137	.81366	2.1642	2.3233	2.4824	5 6	0.7127	5 13	0.5700
6	0.9429	0.4482	.80406	1.0041	1.2041	2.5004	2.3353	2.1702	6 16	0.5712		
7	1.0953	0.4703	1.2851	1.0851	.88506	2.0799	2.2038	2.3277	7 8	0.7373	7 16	0.5244
8	1.0953	0.4704	1.2851	1.0851	.88506	2.0799	2.2038	2.3277	8 14	0.5244		
9	0.9429	0.4482	.80406	1.0041	1.2041	2.5004	2.3353	2.1702	9 10	0.7127	9 14	0.5712
10	1.0549	0.4494	1.2137	1.0137	.81366	2.1642	2.3233	2.4824	10 11	0.5700		
11	0.9438	—	.66130	.86130	1.0613	3.0559	2.8551	2.6544	11 12	0.4979		
12	0.9339	—	.65243	.85243	1.0524	3.0616	2.8270	2.5925	12 13	0.4979	12 15	0.5669
13	0.9438	—	.66130	.86130	1.0613	3.0559	2.8551	2.6544				
14	1.0801	—	1.0721	.87210	.67210	2.6326	2.8079	2.9833	14 15	0.5003		
15	1.0889	—	1.0140	.81405	.61405	2.6059	2.8738	3.1416	15 16	0.5003		
16	1.0801	—	1.0721	.87210	.67210	2.6326	2.8079	2.9832				

[a] These calculations were performed on an Elliott 503 computer with a program also permitting calculation of localization energies in one step.

[b] HMO Indices of chemical reactivity (symbols: position (N), π-electronic density (Q), free valence (F), superdelocalizability (S), atom localization energy (A), designation of a bond (II), bond order (P); E, R, and N in parentheses stand for electrophilic, radical, and nucleophilic attack.

Fig. 8 the relationship between the localization energy, computed in the usual way, and the energy difference between complex **19** and the parent system **20** are shown. The final example is concerned with the correlation between ortho-localization energy and bond order (Fig. 9). These last two examples likewise deal with fluoranthene-like derivatives. In addition, correlations between the above mentioned quantities have been constructed at various levels of sophistication. For example A_e^{SCF} and A_e^{HMO} in fluoranthene are very

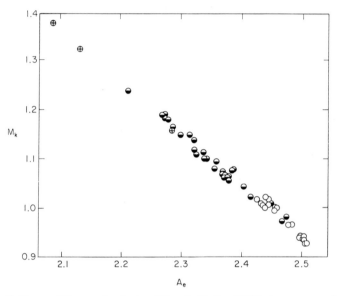

FIG. 8. Energy difference for electrophilic substitution (M_k) between a complex of type **19** and the parent hydrocarbon **20** plotted against A_e of the parent hydrocarbon at the respective position. Positions: class 0 ○, class 1 ◒, class 2 ⊕ [J. Michl and R. Zahradník, *Collection Czech. Chem. Commun.* **31**, 3453 (1966); reproduced by permission].

closely correlated, while A_n^{SCF} and A_n^{HMO} are not. The values q^{SCF} and q^{HMO} are parallel; only the order at positions 2 and 8 differ. The parameters A_e^{ω}(ω refers to the ω-technique) and A_e^{HMO} are also closely related. A similar result was obtained in the case of azulene and its benzo derivatives.[102] Encouraging is the fact that the position of maximum reactivity is usually the same regardless of the level of refinement of the quantum chemical method. However, electrophilic localization energies for anion **21**, for example, indicate caution must be exercised. According to HMO data protonation should lead to the derivative of sesquilfulvalene **22**. By the iteration method (the effect of bond-lengths on β-values is taken into account), the substituted vinylazulene **23** is predicted to be the product. Experimentally it has been shown beyond doubt that azulene

[102] J. Sandström, personal communication (1967).

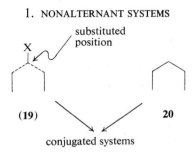

substituted
position

X

(19)   20

conjugated systems

23 is produced. Thus the HMO method leads to an incorrect prediction. It is necessary to bear in mind that this failure of the HMO technique will always occur when the parent system and the π-framework generated by extracting

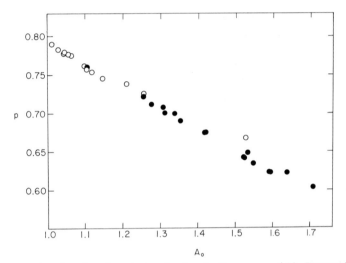

FIG. 9. Bond order plotted against ortho localization energy (A_0). Benzenoid hydro-carbons, ○; fluoranthene-like hydrocarbons, ● [J. Michl and R. Zahradník, *Collection Czech. Chem. Commun.* **31**, 3453 (1966); reproduced by permission].

a p_z orbital from it represent systems of different types. We have an example of a substrate with "good delocalization" (group 1) and a system to which there is only a single appropriate Kekulé structure (group 2) (see Table I).

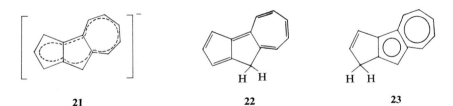

21   22   23

In connection with the study of ion reactivity in compounds with an odd number of carbon atoms, it has been shown that the sequence of electrophilic and nucleophilic reactivities for similar positions in the anion and the cation, respectively (derived from the same parent system) is theoretically identical.[103]

III. Excited State Properties

A. REDISTRIBUTION OF ELECTRONS ACCOMPANYING EXCITATION

Most organic compounds are characterized by a singlet ground state. This nearly always corresponds to an electronic configuration in which all bonding molecular orbitals are occupied by two electrons with antiparallel spin. By absorbing a suitable amount of energy, the ground state is promoted to an excited state with the same or altered state multiplicity. Spin multiplicity is related to the total spin angular momentum S and is defined by $2S + 1$. The first case represents the singlet and the second, the triplet excited state. For a number of reasons (see, e.g., Turro[104]) the first excited singlet (S_1) and triplet states (T_1) are the most interesting. Both states differ considerably as to lifetime. The former ranges from 10^{-9} to 10^{-7} sec; the latter, from 10^{-3} to 10 sec. Molecules which have a single odd electron in the ground state (radicals) exist as a doublet state. The corresponding excited states are either doublets or quartets. Triplet and quintet states are appropriate to biradicals with triplet ground states.

The redistribution of electrons accompanying excitation is usually extensive. Thus considerable changes in physical and chemical properties frequently results. The dipole moment change corresponds to alterations of the electron densities. A reversal in the order of nucleophilic and electrophilic centers usually takes place. Bond-length changes manifest themselves in bond-order changes. The latter has, of course, a consequence of free valence variation.

Even within the HMO framework one can understand why changes accompanying the $N \rightarrow V_1$ excitation $(S_0 \rightarrow S_1$ and $S_0 \rightarrow T_1)$ are connected with such dramatic effects. Frontier orbitals frequently determine the character of the molecule. This result is the foundation of a simple theory of chemical reactivity.[105] It should be realized that within the limits of the simple HMO method, changes due to the $N \rightarrow V_1$ excitation correspond to the simultaneous formation of the respective radical anion and radical cation. Obviously, even a cursory examination of electron distributions derived from the HOMO and LFMO treatments makes it possible to estimate the changes accompanying

[103] R. Fleischer, K. Hafner, J. Wildgruber, P. Hochmann, and R. Zahradník, *Tetrahedron* **24**, 5943 (1968).

[104] J. N. Turro, "Molecular Photochemistry." Benjamin, New York, 1965.

[105] K. Fukui, T. Yonezawa, C. Nagata, and H. Shingu, *J. Chem. Phys.* **22**, 1433 (1954).

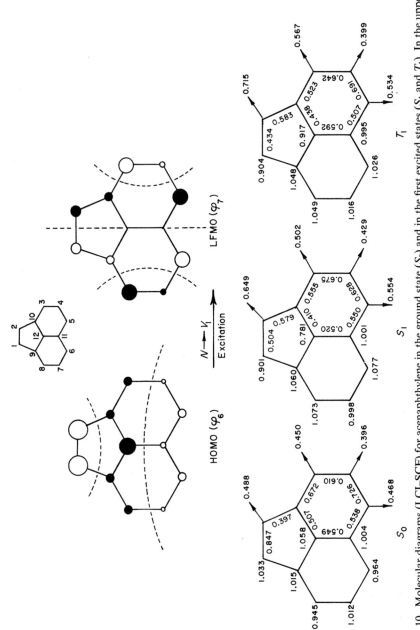

FIG. 10. Molecular diagrams (LCI–SCF) for acenaphthylene in the ground state (S_0) and in the first excited states (S_1 and T_1). In the upper part information is given concerning the distribution of an electron in the highest occupied (HOMO) and lowest free or unoccupied (LFMO) π-molecular orbitals. The radii of circles are proportional to the absolute values of expansion coefficients; white and black circles correspond to their positive or negative values. Nodal planes are given by dotted lines.

electronic excitation. For sake of illustration let us examine acenaphthylene[106] (see Fig. 10). From the magnitude of the absolute values of the expansion coefficients, a pronounced shift of electrons may be observed from the region of the five-membered ring (the density of π-electrons is greatest at positions 1, 2, and 12) to the region of the naphthalene part of the molecule (positions 3, 5, 6, and 8). This is accompanied not only with a change in the magnitude but also in the direction of the dipole moment.[106a] In the same way, from the expansion coefficients and their signs, one may immediately infer a substantial decrease in the order of the 1–2 bond and an increase in the order of the 2–10 bond. The order of a bond, through which a nodal plane passes (HOMO), increases during the $N \rightarrow V_1$ excitation and vice versa.

Let us now attempt to characterize the overall changes accompanying the transition from the ground state to the first excited states (S_1 and T_1) by means of LCI molecular diagrams.[107] (i) In contrast to benzenoid hydrocarbons, the determinant of the 1,1'-configuration for nonalternant hydrocarbons (which corresponds to the $N \rightarrow V_1$ excitation) frequently dominates the CI wave function of the S_1 and T_1 states. (ii) The lowest empty molecular orbital always has one more nodal plane than highest occupied molecular orbital. Thus the sum of the bond orders in the excited state is always lower than in the ground state. This may be seen from the expression for the total π-electron energy expressed in terms of bond orders. (iii) The average changes accompanying excitation decrease with increasing size of the system. (iv) Differences in electron distribution in the S_1 and T_1 states are, of course, much smaller than the differences between S_0 and S_1 (T_1). The qualitative features of the distribution in S_1 and T_1 are for the most part identical. (v) In small systems positions with maximum π-electron density in the S_0 state are usually the positions with minimum π-electron density in the first excited states.

B. ELECTRONIC SPECTRA

Recently, considerable progress has been achieved in explaining the electronic spectra of nonalternant hydrocarbons. The number of systems studied has increased to such an extent that it is possible to give a more detailed explanation of only a few compounds and to indicate other applications. Moreover, with a few exceptions, the information cited in previous articles[2, 108] is not repeated here. Data from earlier studies can be found in the relevant references. For older papers concerning the application of simple methods see

[106] R. Zahradník, calculations performed for this article.
[106a] K. Seibold, R. Zahradník, and H. Labhart, *Helv. Chim. Acta* **53**, 805 (1970).
[107] R. Zahradník and J. Pancíř, unpublished results (1968).
[108] R. Zahradník, *Fortschr. Chem. Forsch.* **10**, 1 (1968).

Streitwieser *et al.* [3, 5, 109] The nonalternant structures will be discussed according to the system mentioned above.

1. *Monocyclic Systems*

The structure of HMO energy levels of odd monocycles (degeneration of HOMO and LFMO) clearly indicates that the Hückel arrangement[22] (the "natural ion") corresponds either to an anion or a cation.

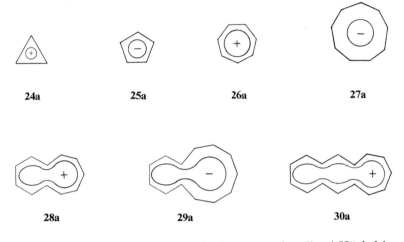

If two electrons are added or removed, the appropriate "anti-Hückel ions" are created:

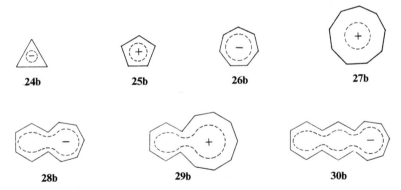

All these ions have a degenerate frontier orbital pair occupied by two electrons. A triplet ground state may thus be anticipated, provided Jahn–Teller distortion is not operating.

[109] C. A. Coulson and A. Streitwieser, Jr., "Dictionary of π-Electron Calculations." Pergamon Press, Oxford, 1965.

Compounds **24a–28a** and **24b–26b** or their simple derivatives are known. The highest members of the series have not yet been prepared. It is the author's opinion that expectations as to their preparation are good. Therefore, it is reasonable to forecast their spectra.[109a] The higher members of the even cycles may be named as follows: e.g., [9]annulenide anion (**27a**) and [9]annulenium cation (**27b**). Figure 11 shows schematically the absorption data for hydrocarbons **24a–30a**. In accord with the calculation, the tri-*tert*-butylcyclopropenium cation displays only end absorption in the 200 mμ region.[110]

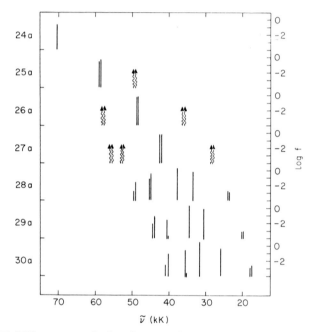

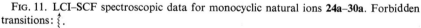

FIG. 11. LCI–SCF spectroscopic data for monocyclic natural ions **24a–30a**. Forbidden transitions: {.

Fair agreement, on the whole, has been found between LCI–HMO[111] values and the experimental data[112] for the series of benzo derivatives of the cyclopentadienide anion. The absorption curves are unfortunately available only for a small range of wavelengths due to experimental difficulties.

A few benzo and naphtho derivatives of the cyclopentadienyl cation **25b** are known as stable solutions in CH_2Cl_2 containing $AlCl_3$ and in sulfuric

[109a] Synthesis of "Vogel-type" systems, compounds with a CH_2 or a heteroatom bridge ought to prove useful for these large rings.

[110] J. Ciabattoni and E. C. Nathan, III, *J. Am. Chem. Soc.* **90**, 4495 (1968).

[111] J. Koutecký, P. Hochmann, and J. Michl, *J. Chem. Phys.* **40**, 2439 (1964).

[112] A. Streitwieser, Jr. and S. I. Brauman, *J. Am. Chem. Soc.* **85**, 2633 (1963).

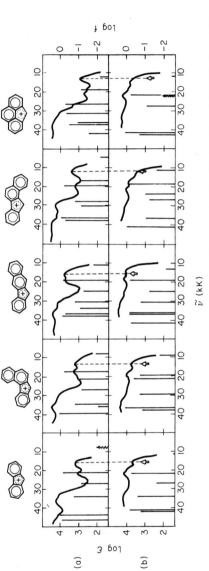
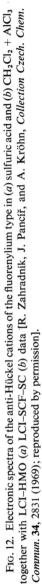

Fig. 12. Electronic spectra of the anti-Hückel cations of the fluorenylium type in (a) sulfuric acid and (b) CH₂Cl₂ + AlCl₃ together with LCI-HMO (a) LCI-SCF-SC (b) data [R. Zahradník, J. Pancíř, and A. Kröhn, *Collection Czech. Chem. Commun.* **34**, 2831 (1969); reproduced by permission].

acid.[113, 114] Standard LCI–HMO and LCI–SCF procedures provide for the fluorenylium cation a first band which is too "long-wave." The calculated values are favorably affected by the introduction of the β^c-variation (Fig. 12). This is clearly in response to the comparatively low order of the π-bond, directly joining the six-membered rings. On the other hand, the introduction of the "α^c-variation" (a VESCF procedure[115]) has no significant influence on these particular compounds.

It is also worth displaying the molecular diagrams of the indenyl cation (Fig. 13). Transition to the electronically excited state is associated with a

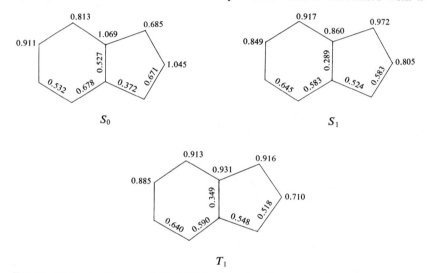

FIG. 13. Molecular diagrams (LCI–SCF) for the indenylium cation in the ground (S_0) and excited states (S_1 and T_1).

certain equalization of electron densities, with a very pronounced decrease of the central bond order (8–9) and with an equalization of bond orders along the perimeter. As regards electron distribution this cation resembles calculated results for the cyclononatetraenyl cation. The situation is quite the opposite with the Hückel evaluated indenyl anion. Excitation is accompanied by an increase in the nonuniformity of the electron distribution and by an increase in the central bond order. Values of the latter amount to 0.46, 0.58, and 0.53 for the S_0, S_1, and T_1 states, respectively.

The tropylium cation and its derivatives have been studied very

[113] J. Michl, R. Zahradník, and P. Hochmann, *J. Phys. Chem.* **70**, 1732 (1960).

[114] R. Zahradník, J. Pancíř, and A. Kröhn, *Collection Czech. Chem. Commun.* **34**, 2831 (1969).

[115] R. D. Brown and M. L. Heffernan, *Trans. Faraday Soc.* **54**, 757 (1958).

thoroughly[116-119] (see also Koutecký *et al.*[111]). Comparison of computed LCI–HMO spectra with experimental absorption curves lead to satisfactory agreement.

The first LCI–SCF band[106] for the cyclononatetraenide anion ($\tilde{\nu} \doteq 28.5$ kK) roughly mimics the experimental[120, 121] value ($\tilde{\nu} = 31.0 - 31.5$ kK). The methylene derivative of cation **28a** has been prepared, and its properties have been successfully interpreted theoretically.[122] It may be safely assumed that higher members of this series will ultimately be prepared.[109a] Accordingly results of the LCI–SCF computation of the spectral characteristics for systems with 13 and 15 carbon atoms are shown in Fig. 11.

Agreement between theoretical and experimental positions of the first absorption bands of trication **31** is not especially good. It would be interesting

31

exp: 28.7 kK (log $\epsilon = 4.32$)[123]
LCI–SCF: 31.4 kK (log $f = 0.14$)[106]

32

exp: 31.2 kK (log $\epsilon = 4.62$)[124]
LCI–SCF: 30.3 kK (log $f = -0.23$)[106]

to have more detailed data concerning the spectra of the triphenylcyclopropenium cation (**32**).

Fulvenes are formally simple derivatives of odd monocyclic systems. In contrast to the parent cycles (displaying ideal delocalization) these compounds exhibit very strong bond alternation.[3, 125, 126] The parent member of the series, fulvene (**33**), has been subject to considerable attention on the part of

[116] G. Naville, H. Strauss, and E. Heilbronner, *Helv. Chim. Acta* **43**, 1221 and 1243 (1960).

[117] E. Heilbronner and J. N. Murrell, *Mol. Phys.* **6**, 1 (1963).

[118] G. V. Boyd and N. Singer, *Tetrahedron* **22**, 547 (1966).

[119] G. Hohlneicher, R. Kiessling, C. Jutz, and P. A. Straub, *Ber. Bunsenges. Phys. Chem.* **70**, 60 (1966).

[120] T. J. Katz and P. J. Garratt, *J. Am. Chem. Soc.* **85**, 2852 (1963).

[121] E. A. LaLancette and R. E. Benson, *J. Am. Chem. Soc.* **87**, 1941 (1965).

[122] W. Grimme, E. Heilbronner, G. Hohlneicher, E. Vogel, and J. P. Weber, *Helv. Chim. Acta* **51**, 225 (1968).

[123] H. Volz, private communication (1966).

[124] R. Breslow and C. Yuan, *J. Am. Chem. Soc.* **80**, 5991 (1958).

[125] E. D. Bergmann *Progr. Org. Chem.* **3**, 81 (1955).

[126] E. D. Bergmann, *Chem. Rev.* **68**, 41 (1968).

theoreticians (e.g., references[3, 111, 127–129]). Only recently has a good preparatory method made **33** comparatively accessible in larger amounts.[130] Figure 14 compares the results of various versions of the LCI–SCF method with the experimental data on fulvene. The LCI–SCF–SC method applied to the latter, and heptafulvene (**34**) (Fig. 15) leads to very good agreement with experiment.[131] Theoretical (MO) characteristics of the higher members of this series are not discouraging. Provided synthetic difficulties are overcome, it is reasonable to expect the homologous fulvenes to be long-lived enough to allow physical measurements to be determined. Figure 16 shows the LCI–SCF spectroscopic characteristics for nona- and undecafulvene **35**, **36**, and **37**.[109a] Differences in the two nonafulvene goemetries (**35**, **36** in Fig. 16) do not

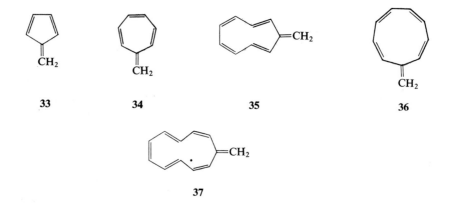

significantly alter the results. The 1 → 1' transition dominates the LCI function of the first excited singlet state in all the fulvenes.

Recently, bisdimethylaminononafulvene was successfully prepared.[132] For steric reasons the dimethylamino groups are twisted from the ring plane,[132a] Therefore, application of the standard values for changes of the Coulomb integral at the perturbed (substituted) sites is of uncertain value. In view of the LCI–SCF computation for nonafulvene (**36**) it is to be noted that the first band for the bisdimethylamino derivative is expected at 22.7 kK. A first-order

[127] P. Hochmann, Ph.D. Thesis, Inst. Phys. Chem. ČSAV, Prague, 1967.
[128] P. A. Straub, D. Meuche, and E. Heilbronner, *Helv. Chim. Acta* **49**, 517 (1966).
[129] A. Julg and P. Francois, *Compt. Rend.* **258**, 2067 (1964).
[130] H. Schaltegger, M. Neuenschwander, and D. Meuche, *Helv. Chim. Acta* **48**, 955 (1965).
[131] W. E. von Doering, *Theoret. Org. Chem., Papers Kekule Symp. London*, 1958 p. 35. Academic Press, New York, 1959.
[132] K. Hafner and H. Tappe, *Angew. Chem.* **81**, 564 (1969).
[132a] According to K. Hafner, the 9-membered ring is not planar (private communication, 1969).

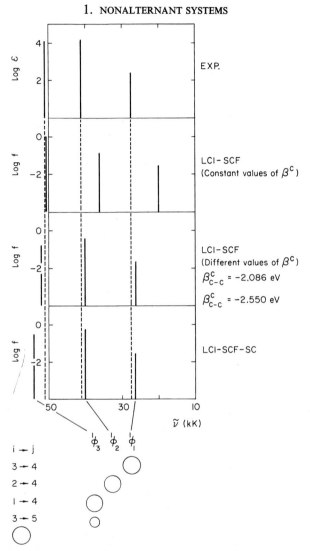

FIG. 14. Absorption maxima of fulvene [H. Schaltegger, M. Neuenschwander, and D. Meuche, *Helv. Chim. Acta* **48**, 955 (1965)] and results of various versions of LCI–SCF calculations. Information on the composition of the LCI wave functions ($^1\phi_i$) is presented in the lower part; $i \rightarrow j$ denotes a transition between the ith and jth MO's. The radii of circles are proportional to values of the expansion coefficients in the LCI functions. Beneath the data on $i \rightarrow j$, a circle with unit radius is presented for comparison.

perturbation treatment ($\delta\alpha = 13.75$ kK) for one $N(CH_3)_2$ group leads to a value of 28.7 kK; for two $N(CH_3)_2$ groups (planar), to 34.7 kK. The experimental measurement affords 30.3 kK.

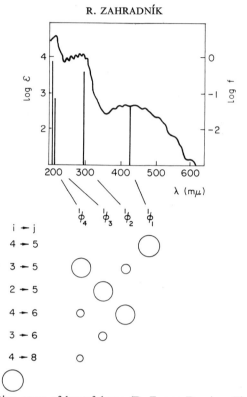

FIG. 15. Absorption curve of heptafulvene (D. E. von Doering, *Theoret. Org. Chem.*, *Papers Kekule Symp. London*, 1958 p. 35. Academic Press, New York, 1959) and results of an LCI–SCF–SC calculation. Information on the composition of the LCI wave function is presented beneath the spectrum (see legend to Fig. 14).

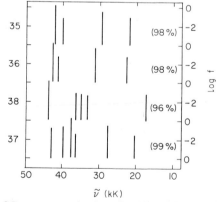

FIG. 16. LCI–SCF–SC spectroscopic data for higher fulvenes. The percentages (%) indicate the degree of participation of the $N \to V_1$ configuration in the LCI wave function of the first excited singlet state.

The effect of introducing a transannular bond into nonafulvene **35** is interesting. Whereas benzo annelation at the formal fulvene double bond produces a hypsochromic shift,[3] annelation at the formal single bond will probably display a significant bathochromic shift (Fig. 16). So far, it has not been possible to prepare isobenzofulvene (**38**), but the amino derivative **39** is yellow-brown while its diphenyl derivative **40** is blue.[133] This is easily understood, if one realizes that the first-order perturbation treatment predicts an

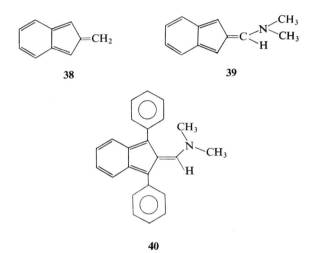

exceptionally high hypsochromic shift, nearly 1 eV (7400 cm^{-1}), when passing from isobenzofulvene (**38**) to the amine **39**. Obviously because of steric hindrance in **40** the dimethylamino group is removed from conjugation. From a π-electronic point of view compound **40** thus represents only a weakly perturbed parent isobenzofulvene.

The molecules in molecules approximation have proved accurate not only with fulvene but also with 6-vinylfulvene.[128, 129] A whole series of polyenefulvenes has been prepared. Most are phenyl and benzo derivatives of compound **41** (*n* commonly equals zero).[134] The LCI–SCF–SC procedure applied to structures **42–47** leads to reasonable values for the position of the first band (Fig. 17).[135]

Similarly, satisfactory results were also obtained for various fulvalene-like systems. Three of the latter, fulvalene, heptafulvalene, and sesquifulvalene, have been studied theoretically[3, 109, 127] many times by various methods.

[133] K. Hafner and W. Bauer, *Angew. Chem.* **80**, 312 (1968).
[134] C. Jutz, private communication (1965).
[135] C. Jutz and R. Zahradník, unpublished calculations (1966).

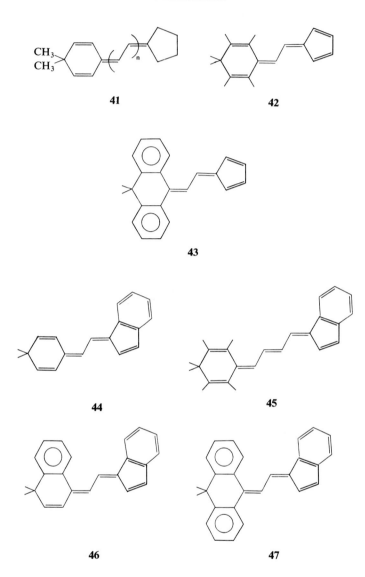

Figure 18 shows a comparison of the theoretical data (LCI–SCF–SC) with experimental absorption values[136, 137] for classical fulvalenes. Figure 19 indicates that calicene benzo annelations (on formal double bonds in the five-

[136] W. von E. Doering, *Theoret. Org. Chem., Papers Kekule Symp. London*, 1958 p. 35. Academic Press, New York, 1959.
[137] H. Prinzbach and W. Rosswog, *Tetrahedron Letters* p. 1217 (1963).

membered ring) provide a hypsochromic shift similar to fulvene. The same figure also registers fair agreement between theory and experiment[138, 139] for absorption in the UV region. Finally, Fig. 19 demonstrates how benzo annelation affects the π-electron distribution.[140]

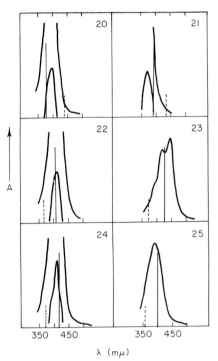

FIG. 17. Long-wave band of fulvene-polyenes [C. Jutz, personal communication (1966)] and theoretical positions of absorption bands calculated by the LCI–SCF–SC method (strong band, solid line; medium band, dotted line).

Bicyclic systems have long been attractive experimentally as well as theoretically. On the one hand, there are the pentalenoid and the heptalenoid systems, on the other, the azulenoid skeleton.[2] There is no purpose in repeating or summarizing older findings at this point. Only a few comments will be made.

Pentalene itself still remains to be conquered. However the number of man years devoted to the pentalene problem have not gone unrewarded. The dimethylamino derivatives **48** have been synthesized and their spectra interpreted.[141] In Fig. 20 the LCI–SCF data for heptalene are compared with the

[138] H. Prinzbach, D. Seip, and U. Fischer, *Angew. Chem.* **77**, 258 (1965).
[139] H. Prinzbach and U. Fischer, *Angew. Chem.* **77**, 621 (1965).
[140] M. J. S. Dewar and G. J. Gleicher, *Tetrahedron* **21**, 3423 (1965).
[141] K. Hafner, K. F. Bangert, and V. Orfanos, *Angew. Chem.* **79**, 414 (1967).

experimental absorption curve.[142] The prediction for pentalene is shown in the
same diagram. These computations were carried out for model systems in
which bond alternation is expected and which, therefore, only have a symmetry
axis.

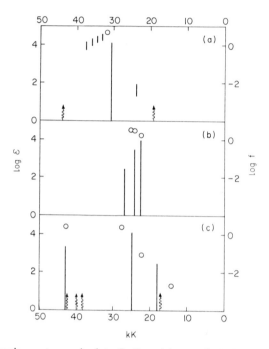

Fig. 18. Electronic spectroscopic data (indicated by small circles and short lines) for
fulvalene (*a*), heptafulvalene (*c*) (D. E. von Doering, *Theoret. Org. Chem., Papers Kekule
Symp. London,* 1958 p. 35. Academic Press, New York, 1959) and for a benzyl derivative of
sesquifulvalene (*b*) [H. Prinzbach and W. Rosswog, *Tetrahedron Letters* p. 1217 (1963)].
LCI–SCF–SC data are denoted by histogram; wavy lines refer to forbidden transitions.

As far as higher members of this series are concerned, it seems there is more
hope for preparing bis ions; dianions in the pentalenoid series and dications for
heptalenoids. The first band of the pentalene dianion (dilithium) was found[143]
at 296 mμ. An LCI–SCF computation[106] yielded the value 274 mμ. The
discrepancy might be due to ion-pair formation. Various heteroanalogs
(e.g., azapentalenes[144]) and derivatives (e.g.,1,2,3,4-tetraphenylbenzopental-

[142] H. J. Dauben, Jr. and D. J. Bertelli, *J. Am. Chem. Soc.* **83**, 4658 (1961).
[143] T. J. Katz and M. Rosenberger, *J. Am. Chem. Soc.* **84**, 865 (1962).
[144] V. Galasso and G. De Alti, *Theoret. Chim. Acta* **11**, 411 (1968).

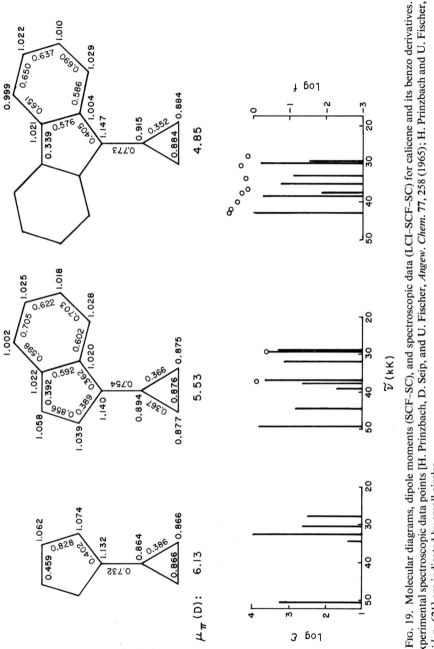

FIG. 19. Molecular diagrams, dipole moments (SCF–SC), and spectroscopic data (LCI-SCF–SC) for calicene and its benzo derivatives. Experimental spectroscopic data points [H. Prinzbach, D. Seip, and U. Fischer, *Angew. Chem.* 77, 258 (1965); H. Prinzbach and U. Fischer, *ibid.* p. 621] are indicated by small circles.

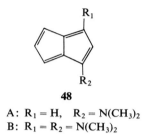

48

A: $R_1 = H$, $R_2 = N(CH_3)_2$
B: $R_1 = R_2 = N(CH_3)_2$

ene[145, 146]) of pentalene have been studied both experimentally and theoretically.

The hypothesis[2] that it should be possible to prepare "higher" azulenes has

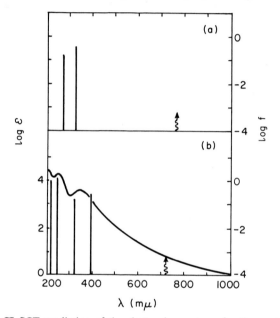

Fig. 20. (a) LCI–SCF prediction of the electronic spectrum for the presently unknown pentalene (b) comparison of calculated data (LCI–SCF) with the experimental absorption curve of heptalene [H. J. Dauben, Jr. and D. J. Bertelli, *J. Am. Chem. Soc.* **83**, 4658 (1961)]. Calculations were performed for models with alternant bonds (C_2 symmetry; β^c (single bond) = −2.086 eV; β^c (double bond) = −2.550 eV). Forbidden transitions are designated by wavy lines.

been verified for one species.[108, 147] Predicted spectra for the two unknown higher azulenes[109a] are given in Fig. 21 ([7, 9] and [9, 11]). In order to assess the

[145] W. Ried and D. Freitag, *Tetrahedron Letters* p. 3135, (1967).
[146] W. Ried and R. Zahradník, unpublished calculations (1968).
[147] F. Sondheimer, private communication (1966).

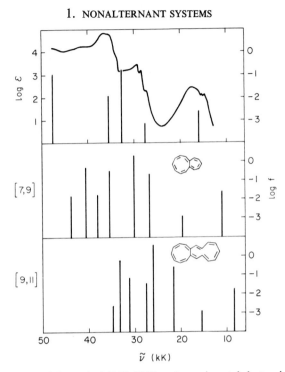

FIG. 21. Comparison of theoretical (LCI–SCF) and experimental electronic spectroscopic data for azulene [J. Koutecký, P. Hochmann, and J. Michl, *J. Chem. Phys.* **40**, 2439 (1964)]. Predictions for two higher azulene-like hydrocarbons are shown: ([7,9], [9,11]). The assumed geometry is indicated (R. Zahradník, calculations made for this article).

substance of this forecast, one may examine the theoretical and experimental data for azulene included in the same figure.

Recently the first cata-condensed tricyclic system **49** consisting of five- and seven-membered rings was prepared.[103] The unsatisfactory agreement of

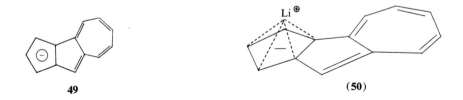

49   (**50**)

LCI–SCF data with experimental absorption bands was tentatively explained by assuming the occurrence of an ion pairing (**50**) between anion **49** and the lithium cation. Figure 22 shows that a decrease in the magnitude of the monocentric electronic repulsion integral (from 10.84 to 8.84 eV) yields

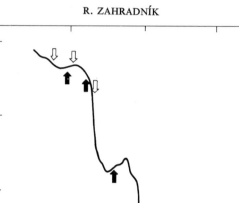

$\tilde{\nu}$ (kK)

FIG. 22. Electronic absorption curve for the lithium salt of anion **49**. The positions of calculated (LCI–SCF) excitation energies are given by arrows ($\Longrightarrow$ standard procedure; $\longrightarrow$ modified values of ionization potentials) [R. Fleischer, K. Hafner, J. Wildgruber, P. Hochmann, and R. Zahradník, *Tetrahedron* **24**, 5943 (1968); reproduced by permission].

satisfactory agreement with the experiment. The product of protonating pentaleno[2,1,6-*def*]heptalene (**51**) compound **52**,, also belongs to this group.[103] Carbonium ion **52** is a bridged derivative of the tropylium cation **53**. The first band is in the region of 13.8–15.7 kK. Calculation yields 13.0 kK.

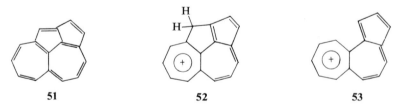

| **51** | **52** | **53** |

A systematic study of the properties of peri-condensed tricyclic systems[108] (see also Hochmann *et al.*[148]) has already been mentioned. At this point,

[148] P. Hochmann, R. Zahradník, and V. Kvasnička, *Collection Czech. Chem. Commun.* **33**, 3478 (1968).

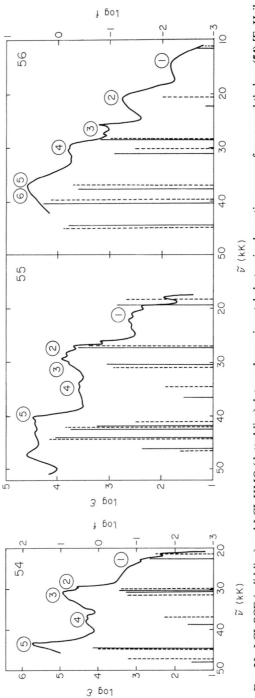
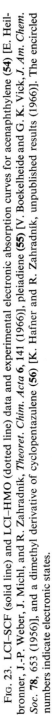

Fig. 23. LCI-SCF (solid line) and LCI-HMO (dotted line) data and experimental electronic absorption curves for acenaphthylene (**54**) [E. Heilbronner, J.-P. Weber, J. Michl, and R. Zahradník, *Theoret. Chim. Acta* **6**, 141 (1966)], pleiadiene (**55**) [V. Boekelheide and G. K. Vick, *J. Am. Chem. Soc.* **78**, 653 (1956)], and a dimethyl derivative of cyclopentazulene (**56**) [K. Hafner and R. Zahradník, unpublished results (1966)]. The encircled numbers indicate electronic states.

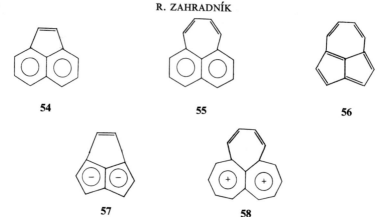

54   **55**   **56**

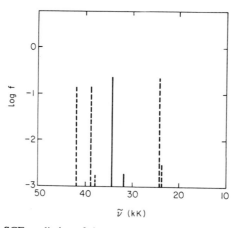

57   **58**

therefore, only a comparison of theoretical and experimental data for the three systems acenaphthylene (**54**), pleiadiene (**55**), and cyclopentazulene (**56**) will be displayed,[149] (Fig. 23). Spectroscopic predictions have been made for dianion **57** and dication **58** also (Fig. 24). If these ions can be prepared, they will probably display olefinic properties.

FIG. 24. LCI–SCF prediction of electronic spectra for dianion **57** (solid line) and dication **58** (dotted line).

The spectra of several tetracyclic peri-condensed systems have already been investigated.[108, 150] Therefore, only a few comments will be made. Computation of the electronic spectrum and the synthesis[50] of pentalenoheptalene **59** were carried out independently. Agreement of the data is reasonable (Fig. 25).

[149] K. Hafner and R. Zahradník, unpublished results (1966).

[150] P. Baumgarten, E. Weltin, G. Wagnière, and E. Heilbronner, *Helv. Chim. Acta* **48**, 751 (1965).

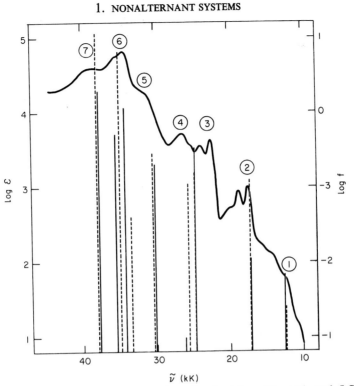

Fig. 25. Absorption curve of dimethylpentalenoheptalene (dioxane) and LCI–SCF (solid line) and LCI–HMO (dotted line) data for the parent hydrocarbon [K. Hafner and R. Zahradník, unpublished results (1966)]. The encircled numbers indicate electronic states.

For hydrocarbon **60** the theoretical data were successfully employed to verify the correctness of the indicated structure.[149] The author was pleased by the report of the synthesis[151] of dicyclopentheptalene **61**, since the relatively high stability observed was previously anticipated.[49] Moreover, fair agreement was found between the LCI–SCF data,[152] obtained earlier, and the experimental

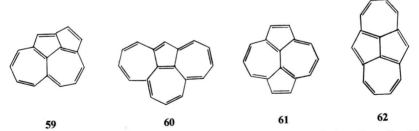

| **59** | **60** | **61** | **62** |

[151] A. G. Anderson, Jr., A. A. MacDonald, and A. F. Montana, *J. Am. Chem. Soc.* **90**, 2993 (1968).
[152] R. Zahradník and G. Hafner, unpublished results (1966).

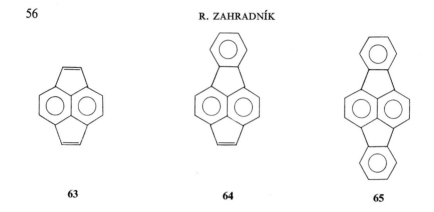

63   64   65

findings (Fig. 26). Figure 26 also contains the prediction for the isomeric dicycloheptapentalene **62**. The synthesis of pyracylene[52] (**63**) should be considered an important result. Because of the presence of a NBMO,[49] theory would not suggest this molecule to be an easy synthetic task. Data concerning

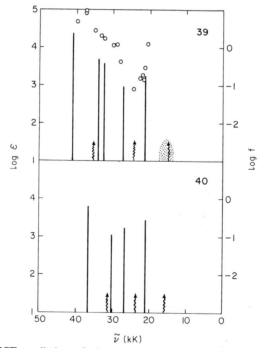

FIG. 26. LCI–SCF prediction of electronic spectra for hydrocarbons **61** and **62** [R. Zahradník and G. Hafner, unpublished results (1966)]. For dicyclopenta[*ef,kl*]heptalene recent experimental data are also given [A. G. Anderson, Jr., A. A. MacDonald, and A. F. Montana, *J. Am. Chem. Soc.* **90**, 2994 (1968)]: circles and the dotted area. Forbidden transitions are indicated by wavy lines.

this compound and its benzo derivatives, **63–65**, are shown in Fig. 27. Hydrocarbon **65** (dibenzopyracyclene) may be considered a kind of "doubled" fluoranthene (Fig. 28).

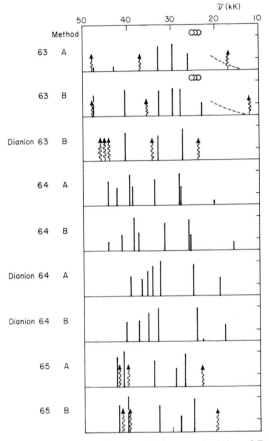

FIG. 27. LCI–SCF (with and without β^c variation, method A and B, respectively) electronic spectroscopic data for pyracylene and its benzo derivatives [R. Zahradník and J. Michl, unpublished results (1967)]. Experimental data (absorption maxima, circles; and tailing, dotted line, according to B. M. Trost and G. M. Bright, *J. Am. Chem. Soc.* **90**, 2732 (1968). Forbidden transitions are indicated by wavy lines.

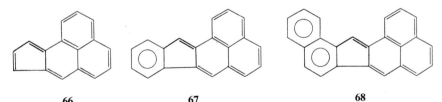

66   **67**   **68**

As far as tetracyclic hydrocarbons of other skeletal variations are concerned, new data on the benzo derivative of the Reid hydrocarbon **66**, i.e., **67**, may be mentioned. (Fig. 29).[153, 154] The naphtho derivative **68** gave good agreement between the calculated and measured position of the first band.[108]

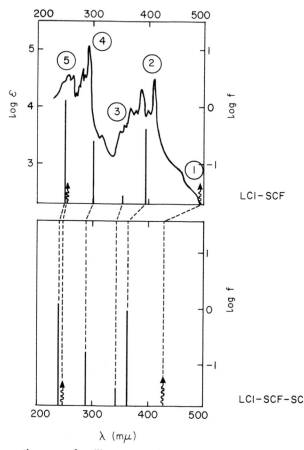

FIG. 28. Absorption curve for dibenzopyracylene (E. Clar, "Polycyclic Hydrocarbons," Vol. 2, p. 339. Academic Press, New York, 1964; reproduced by permission) and LCI–SCF and LCI–SCF–SC data [R. Zahradník and J. Michl, unpublished data (1968)]. The encircled numbers indicate electronic states.

Our final comments shall be directed to polycyclic systems. Data on the electronic spectrum of the interesting hydrocarbon[155] acenaphth[1,2-*a*]acenaphthylene (**69**) are perhaps incomplete, because the authors mention

[153] R. Zahradník and J. Michl, *Collection Czech. Chem. Commun.* **30**, 520 (1965).

[154] D. H. Reid, *Chem. Soc. (London), Spec. Publ.* **12**, 69 (1958).

[155] R. L. Letsinger and J. A. Gilpin, *J. Org. Chem.* **29**, 243 (1964).

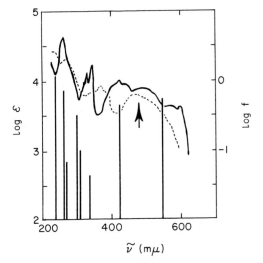

$\widetilde{\nu}$ (mμ)

Fig. 29. Absorption curve of indeno[2,1-a]perinaphthene, **67**, in cyclohexane (solid line) and of indolo[2,3-a]perinaphthene (dotted line in methanol [D. H. Reid, *Chem. Soc. (London), Spec. Publ.* **12**, 69 (1958); reproduced by permission]. LCI–SCF data for **67** and first-order perturbation treatment for its nitrogen heteroanalog. The estimated position of the first band of indolo[2,3-a]perinaphthene is visualized by a thick arrow.

purple-red crystals and give the first maximum at 24.5 kK. This is supported by LCI–SCF–SC and LCI–SCF computations[106] which lead to 21.9 and 18.5 kK. To conclude, we should like to say that encouraging agreement between

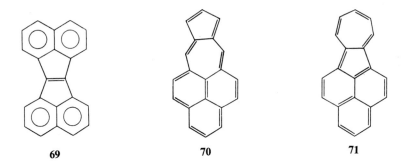

69	**70**	**71**

theory[106] and experiment[156] was achieved with azuleno[5,6,7-cd]phenalene (**70**). Figure 30 contains the corresponding data and also indicates the forecast for the isomeric system **71**.

[156] Ch. Jutz and R. Kirchlecher, *Angew. Chem.* **78**, 493 (1966).

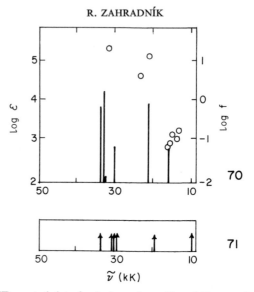

FIG. 30. LCI–SCF spectral data for hydrocarbons **70** and **71**, containing azulenyl and perinaphthenyl units, respectively [R. Zahradník, unpublished results (1966)]. Experimental data according to Jutz and Kirchlecher [C. Jutz and R. Kirchlecher, *Angew. Chem.* **78**, 493 (1966)].

IV. Effect of Addition or Removal of Electrons: Formation of Radicals and Radical Ions

A. General Remarks

Numerous nonalternant hydrocarbons, and, of course also their hetero-analogs, can be oxidized or reduced comparatively easily. Some of them are readily subject to both processes. The ionization potential and the electron affinity are quantities which allow for the estimation of ease of oxidation and reduction. If we proceed from neutral systems, the primary product is usually a radical cation or a radical anion, i.e., a relatively reactive particle. The reactivity of these compounds is connected with the presence of an odd electron, as well as with a positive or negative charge. Dipolar aprotic solvents (e.g., dimethylformamide, acetonitrile) are good solvents for hydrocarbon ions as well as suitable media for reducing reactivity.

Oxidation or reduction can be carried out either chemically or electro-chemically.[157, 158] The advantage of the electrochemical process is its usually good definition when successful. A serious disadvantage is that they are not generally applicable. Only exceptionally can one carry out oxidation or

[157] F. Gerson, "Hochanflösende ESR-Spektroskopie." Verlag Chemie, Weinheim, 1967.
[158] E. T. Kaiser and L. Kevan, eds., "Radical Ions." Wiley (Interscience), New York, 1968.

reduction quantitatively. From this point of view chemical reduction is more useful. Generally, it is best to carry out the preparation on a vacuum line. However, numerous radical ions can be generated in solutions deprived of air oxygen by means of purging with pure nitrogen. It has even been possible to prepare a hydrocarbon radical cation in the solid state.[159]

Because frontier orbital HMO energies are usually a valid measure for the relative values of the ionization potentials and the electron affinities (the highest

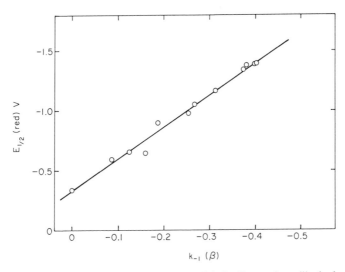

FIG. 31. Polarographic cathode half-wave potentials for fluoranthene-like hydrocarbons, $E_{1/2}$ [E. D. Bergmann, *Trans. Faraday Soc.* **52**, 690 (1956)] plotted against HMO energy of the lowest free molecular orbital (k_{-1}) [R. Zahradník and J. Michl, *Collection Czech. Chem. Commun.* **31**, 3442 (1966); reproduced by permission].

occupied and the lowest unoccupied MO), radical ion formation can be estimated easily. Qualitatively, systems with an HOMO energy higher than $\alpha + 0.4\beta$ are relatively easily oxidized, while systems with an LFMO energy lower than $\alpha - 0.4\beta$ are easily reduced. Experimentally determined ionization potentials and electron affinities for substances of this type are sporadic. However, the relationship of these quantities to the corresponding MO characteristics have been verified by means of charge-transfer absorption[160] and polarographic half-wave potentials.[5, 161] For the sake of illustration Fig. 31 shows the polarographic half-wave potentials of the cathodic waves of

[159] S. Hünig, D. Scheutzow, and H. J. Friedrich, *Angew. Chem.* **76**, 818 (1964).
[160] M. Nepraš and R. Zahradník, *Collection Czech. Chem. Commun.* **29**, 1545 (1964).
[161] R. Zahradník and C. Párkányi, *Talanta* **12**, 1389 (1965).

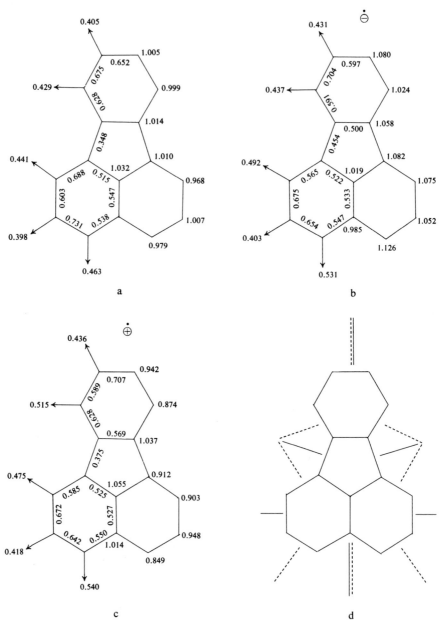

FIG. 32. SCF molecular diagrams for (*a*) fluoranthene, (*b*) the respective radical anion, and (*c*) the radical cation. (*d*) The nodal planes of the highest occupied and lowest unoccupied molecular orbitals are indicated by full and dotted lines [P. Čársky and R. Zahradník, unpublished calculations (1968)].

fluoranthene-like hydrocarbons[162] plotted against the energies of the lowest empty MO's.[163]

It seems that bonds of a relatively high order are significantly affected by the HOMO contribution, while the LFMO has nodal planes which pass through these bonds. Therefore, radical ions of both charge types clearly have a more uniform ("more aromatic") bond order set than the initial neutral form. The results for fluoranthene[164] are presented in Fig. 32.

Comparison of HMO and SCF molecular diagrams for members of several series of nonalternant hydrocarbons allows one to conclude that from a semiquantitative point of view the differences are not important.[164]

B. INHERENT STABILITY OF RADICALS (DISMUTATION EQUILIBRIA)[165]

The reactivity of radicals can be discussed in the same terms as the reactivity of closed-shell systems. There exists, however, besides dimerization, one exceptionally important process which determines whether a certain radical (displaying, perhaps, remarkably good reactivity indices) may be prepared at all: the dismutation of radicals to higher or lower oxidation forms. In general the oxidized form (Ox) of a certain system accepts an electron reversibly and is converted to a radical (semiquinone form, Sem). By accepting another electron the reduced form (Red) is generated. The following holds for the dismutation reaction:

$$Ox + Red \rightleftharpoons 2\ Sem \qquad (12)$$

$$K = \frac{a_{Sem}^2}{a_{Ox}\ a_{Red}} \qquad (13)$$

where a denotes the thermodynamic activity. The following equations likewise obtain.

$$-\Delta G = RT \ln K \qquad (14)$$

$$\Delta G = \Delta H - T\Delta S \qquad (15)$$

Provided ΔH and ΔS are mutually linearly dependent quantities, the π-electron contribution to the total change of enthalpy is decisive. The accompanying variation of solvation energy is usually constant over the series of compounds investigated. We may then write:

$$-\Delta E_\pi = E_\pi^{Ox} + E_\pi^{Red} - 2E_\pi^{Sem} \sim \ln K \qquad (16)$$

[162] I. Bergman, *Trans. Faraday Soc.* **52**, 690 (1956).

[163] R. Zahradník and J. Michl, *Collection Czech. Chem. Commun.* **31**, 3442 (1966).

[164] P. Čársky and R. Zahradník, unpublished results (1968); and (in part) *Collection Czech. Chem. Commun.*, in press (1971).

[165] P. Čársky, S. Hünig, D. Scheutzow, and R. Zahradník, *Tetrahedron* **25**, 4781 (1969).

Brdička[166] has shown that the value of the dismutation constant can be determined by analyzing the polarographic reduction curve of the oxidized form. Generally, the reduction takes place in two one-electron steps. If the differences in half-wave potentials is $\Delta E_{1/2}$ it holds that

$$E'_{1/2} - E''_{1/2} = \Delta E_{1/2}(V) = 0.06 \ln K \tag{17}$$

Figure 33 shows the two-step reduction curve of dimethylaceheptylene.[167] The first and second waves are clearly appropriate to the following processes:

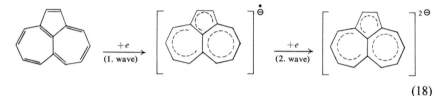

$$\tag{18}$$

$\Delta E_{1/2}$ is 0.92 V and thus $\ln K = 15.5$ at 23° [Eq. (17)]. The equilibrium concentration of the radical anion is, therefore, high and does not show a tendency toward dismutation. The situation is similar for numerous other hydrocarbons. Consequently the radical ion can be prepared by mixing the di-ion with the neutral hydrocarbon.

The question arises as to the possibility of calculating quantum-chemically the quantity ΔE_π [Eq. (16)].[165] Clearly the quantity ΔE_π^{HMO} has the same value for all hydrocarbons and for all heterosystems, provided the empirical parameters of the three oxidation forms have the same value; i.e.,

$$-\Delta E_\pi^{HMO} = 0 \tag{19}$$

Obviously, there is a serious discrepancy between this result and experience. If electron repulsion is introduced explicitly into π-electron energy calculations of the individual forms, it holds that

$$-\Delta E_\pi^{SCF} = J_{mm} \tag{20}$$

where J_{mm} is the electron repulsion integral, expressed either in terms of molecular orbitals,

$$J_{mm} = \int \varphi_m(1) \varphi_m(2) \frac{e^2}{r_{12}} \varphi_m(1) \varphi_m(2) \, d\tau(1) \, d\tau(2) \tag{21}$$

or of atomic orbitals,

$$J_{mm} = \sum_\mu \sum_\nu c_{m\mu}^2 c_{m\nu}^2 \gamma_{\mu\nu} \tag{22}$$

[166] R. Brdička, Z. Elektrochem. 47, 314 (1941).
[167] P. Čársky, Ph.D. Dissertation, Inst. Phys. Chem. ČSAV, Prague, 1968.

where m is the index of a singly occupied MO, and μ and ν are the indices of an AO. For $\gamma_{\mu\nu}$ it obtains that[168]

$$\gamma_{\mu\nu} = \frac{e^2}{a + r_{\mu\nu}} \qquad (23)$$

where $r_{\mu\nu}$ is the distance between the μ- and the νth AO in angstroms.

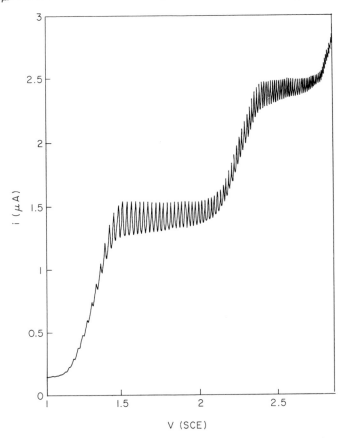

FIG. 33. Polarographic reduction of 3,5-dimethylacehepthylene (6×10^{-4} M solution in 0.1 M tetrabutylammonium perchlorate in dimethylformamide) (P. Čársky, Ph.D. Dissertation, Inst. Phys. Chem. ČSAV, Prague, 1968).

The validity of Eqs. (16) and (20) has been confirmed experimentally and theoretically for a series of molecules containing heteroatoms.[165] One may, therefore, write that

$$\ln K = aJ_{mm}^{\text{Sem}} + b \qquad (24)$$

[168] N. Mataga and K. Nishimoto, Z. Physik. Chem. (Frankfurt) [N.S.] 13, 140 (1957).

where a and b are empirical constants. The symbol Sem with J_{mm}^{Sem} in Eq. (24) means that SCF-MO expansion coefficients of the open-shell form were used to calculate it.

Table IX gives a few values of the integrals J_{mm}^{Sem}.

TABLE IX

VALUES OF THE INTEGRALS J_{mm}^{Sem} FOR
SELECTED RADICAL IONS[a]

System	J_{mm}^{Sem}
Azulene (A)	4.882
Azulene (C)	5.020
Aceheptylene (A)	4.370
Aceheptylene (C)	4.090
Fluoranthene (A)	3.967
Fluoranthene (C)	3.815

[a] A, anion; C, cation. Data from P. Čársky, S. Hünig, D. Scheutzow, and R. Zahradník, *Tetrahedron*, **25**, 4781 (1969).

C. ESR Spectra

In view of Gerson and Hammons' chapter[168a] in this volume, as well as other outstanding comprehensive reviews on ESR,[157,169] the author only intends to make a short comment.

The radical ions required for spectroscopic study are prepared directly in special vessels in a fashion similar to that mentioned in Section IV,A. Chemical agents are usually preferred. Alkaline metals are commonly used for reduction; H_2SO_4 and $SbCl_5$[170] are applied most frequently for oxidation.

Three relations are used for interpreting ESR spectra.[157] The original McConnell equation [Eq. (25)] has been used most widely.

$$a_i^H = Q\rho_i^\pi \qquad (25)$$

The hyperfine splitting constant, a_i^H in gauss, is determined experimentally. Q is a proportionality constant and ρ_i^π is the spin density at the conjugated center i. Two improvements have been introduced. The Colpa–Bolton

---

[168a] F. Gerson and J. H. Hammons, in "Nonbenzenoid Aromatics" (James P. Snyder, ed.), Vol. II, p. 81. Academic Press, New York, 1971.

[169] A. Carrington, *Quart. Rev. (London)* **17**, 67 (1963).

[170] I. C. Lewis and L. S. Sanger, *J. Chem. Phys.* **43**, 2712 (1965).

equation includes a term which depends on the excess charge density at a respective carbon atom, while the Giacometti–Nordia–Pavan equation accounts for contributions from spin densities in bonds between a respective carbon atom and its nearest neighbors.

It appears that both of the latter are somewhat more accurate than the original McConnell equation. However, simplicity plus accuracy are responsible for the more frequent use of Eq. (25). Moreover, there exists a rather close linear correlation between the hyperfine splitting constants and the squares of the Hückel expansion coefficients [which approximate the ρ_i^π values of Eq. (25)] provided that the system under investigation does not possess atomic positions with negative or small positive spin density. In the latter case the calculation is slightly more complicated, and it is carried out using the McLachlan formula[171] based on a configuration interaction treatment.

Measurement of an ESR spectrum provides the most clear-cut experimental information about the distribution of the odd electron. Interpretation of experimental data in terms of the molecular orbital treatment should be considered one of the greatest successes of this theory. ESR spectra have made it possible to verify experimentally the pairing properties of molecular orbitals in alternant hydrocarbons. Typically radical cations and anions have very similar spectra. In addition the ESR technique has demonstrated[172] the nonexistence of pairing properties for nonalternant hydrocarbons. Acepleiady-lene (72) and acenaphth[1,2-a]acenaphthylene (69) are good examples. The

72

width of the spectra of the anions and cations are completely different:

Width of ESR Spectra (gauss)

	⊖	⊕
72:	20.5	32.0
69:	29.4	12.0

[171] A. D. McLachlan, *Mol. Phys.* **3**, 233 (1960).
[172] F. Gerson and J. Heinzer, *Helv. Chim. Acta* **49**, 7 (1966).

Processing the radical ion data derived from compounds **72** and **69** yielded a Q value of 23.4 gauss for the anions [Eq. (23)] and 32.3 gauss for the cations. Similarly the radical cation of anthracene shows higher Q value than that found for the radical anion. The difference, however, is smaller.[173]

D. ELECTRONIC SPECTRA

Analogous to radical ions derived from alternant systems, nonalternant radical ions frequently display their first electronic absorption band in the vicinity of the infrared.[164] One of the reasons for this is the small gap between energy levels associated with the first transition. Such a circumstance arises because the lowest "unnatural" $N \rightarrow V$ transition energy is nearly always much lower than the "natural" $N \rightarrow V_1$ transition energy. By natural transition we mean excitation of an electron from the highest bonding to the lowest anti-bonding molecular orbital.

In contrast to electronic spectra of closed-shell systems, the lowest excited state of radicals has the same spin multiplicity as the ground state, i.e., it is a doublet.[174, 175] This may explain why phosphorescence (in contrast to fluorescence) has so far not been observed for radicals. In view of the unusual character of the term scheme in conjugated radicals (D_1 is situated lower than Q_1), the likelihood of observing $D_0 \rightarrow Q_1$ (and $Q_1 \rightarrow D_0$) transitions would seem to be poor. On the other hand, measurement of the hitherto uninvestigated $Q \rightarrow Q$ transitions appears as an attractive possibility. Another feature of radical electronic spectra is that differences in the electron distribution of the first excited singlet and triplet are usually only quantitative. In the wave functions of the S_1 and T_1 states determinants of the $N \rightarrow V_1$ configuration frequently predominate. By contrast differences in D_1 and Q_1 are usually very pronounced and qualitative, because in the D_1 state wave function the $m \rightarrow m + 1$ configuration dominates (m is the index of the singly occupied MO). For the Q_1 wave function, the $(m - 1) \rightarrow (m + 1)$ configuration is overriding.

The number of radicals derived from nonalternant systems for which electronic spectra are available is small. Upon electroreduction the violet-blue dimethylformamide solution of azulene fades to the yellow of the azulene radical anion.[167, 176] A hypsochromic shift occurs in the electronic spectrum. Dimethylaceheptylene[176] and fluoranthene[177] have been investigated in a similar manner. The diazulenylethylene radical cation **73** is remarkable because

[173] J. R. Bolton and G. K. Frankel, *J. Chem. Phys.* **40**, 3307 (1964).
[174] R. Zahradník and P. Čársky, *J. Phys. Chem.* **74**, 1240 (1970).
[175] R. Zahradník and P. Čársky, *J. Phys. Chem.* **74**, 1249 (1970).
[176] P. Čársky, K. Hafner, and R. Zahradník, unpublished results (1967).
[177] R. Pointean and J. Favede, *Compt. Rend.* **250**, 2556 (1960).

it is the first hydrocarbon radical cation to be isolated in the form of a per-chlorate salt.[159] Oxidation and reduction products, **74** and **75**, have likewise been isolated. Finally, anions obtained from highly acidic hydrocarbons by ferricyanide oxidation produce radicals whose spectra have been recorded.[178]

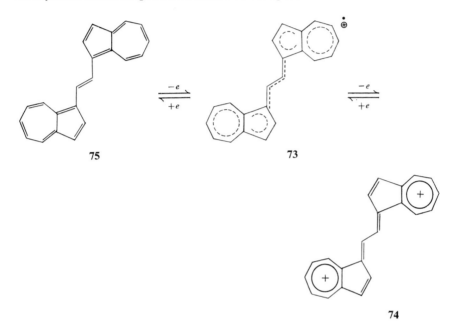

75   73

74

The interpretation of an absorption spectrum requires in general application of the configuration interaction method.[179, 180] Nevertheless, the first two or three bands of radical ions are usually conditioned by nearly pure transitions. A detailed description of the use of the CI method for open-shell systems is not within the scope of this paper.[174, 175] However, it should be recognized that the number of configurations which must be considered is higher for open than for closed-shell systems. Besides the ground state configuration there are four types of monoexcited doublet configurations which must be taken into account. These are the A, B, C_α and C_β type configurations. The ground state configuration, G, interacts with these monoexcited configurations even in the SCF approximation. In the theoretical treatment mentioned here, the LCI calculations usually considered four configurations of the A type, four of the B type, sixteen of the C_α type (the transitions from the four highest doubly occupied MO's to the four lowest unoccupied MO's), sixteen of the C_β type,

[178] R. Kuhn, H. Fischer, F. A. Neugebauer, and H. Fischer, *Ann.* **654**, 64 (1962).
[179] C. C. J. Roothaan, *Rev. Mod. Phys.* **32**, 179 (1960).
[180] H. C. Longuet-Higgins and J. A. Pople, *Proc. Phys. Soc.* **A68**, 591 (1955).

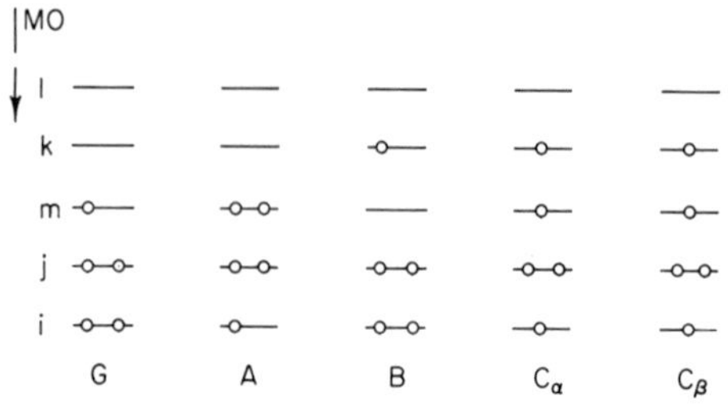

FIG. 34. Pictorial representation of the ground state (G) and excited state (A, B, C_α, C_β) configurations.

and the G configuration; a total of 41 configurations (Fig. 34). For example, the azulene radical anion study mentioned above concluded that the band near 23.5 kK, the cause of the yellow color of the solution, is apparently only the second band or a mixture of two bands, the second and the third. The first band

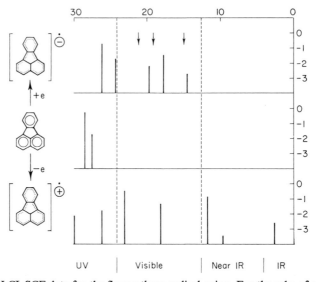

FIG. 35. LCI–SCF data for the fluoranthene radical anion. For the sake of comparison, data for the parent hydrocarbon and for the respective cation are also presented [P. Čársky and R. Zahradník, unpublished results (1968)]. Experimental points [R. Pointean and J. Favede, Compt. Rend. **250**, 2556 (1960)] for the radical anion are indicated by arrows.

is expected to fall near 7.5 kK. Similarly measurements and calculations on the radical anion of dimethylaceheptylene[176] indicate that the first band is still to be measured in the near infrared region (8–9 kK.) Rather poor agreement between the LCI–SCF theory and the experimental position of the first absorption band was achieved for radical cation **73** even when the interaction of 61 configurations was considered. The first calculated band was shifted hypsochromically by about 2 kK with respect to the observed band.

Figure 35 shows that the calculated excitation energies for the fluoranthene radical anion at least roughly parallel the experimental band positions.[164, 177] In this connection it should be mentioned that LCI–SCF theory applied to the fluoranthene radical cation indicates that the first electronic transition will take place in the infrared region at about 2.53 kK. Semiempirical calculations of the spectra were also carried out for heterocyclic radical cations **76** and **77**. Compound **76** is isoelectronic with the pentafulvalene radical cation. For cation **76** theoretical agreement with the first experimental band is encouraging ($\tilde{\nu}^{LCI} = 18$ kK, $\tilde{\nu}_{exp} = 18$ kK), but a large discrepancy was found for system **77** ($\tilde{\nu}^{LCI} = 20.7$ kK, $\tilde{\nu}_{exp} = 15.7$ kK).

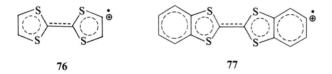

76   **77**

It is expected that the electronic spectral studies of nonalternant radical ions and their heteroanalogs will advance rapidly and yield interesting results. No less interesting is the study of the reactivity of radical ions. There is optimism that it may be possible to control product structure in substitution and other reactions by varying substituents on the radical ion substrate.

V. Introduction of a Heteroatom

A. Transition from a Parent Hydrocarbon to a Heteroanalog

In a recent study on azulene derivatives,[181] considerable attention was devoted to the effect of heteroatom introduction into the conjugated skeleton. Therefore, we shall limit ourselves to a brief summary. (i) In the first place it is necessary to point out that empirical values describing the orbitals of heteroatoms and the heteroatom carbon bonds have been found for integrals appearing in HMO[5] and PPP[108] theories. It is, therefore, possible to calculate theoretical characteristics of heterocycles as well as isocyclic compounds by standard MO methods. (ii) Considerable advantages are connected with the application of first-order perturbation[182–184] theory within the framework of

[181] K. Hafner and R. Zahradník, in preparation (1969).

[182] C. A. Coulson, *Proc. Phys. Soc.* **A65**, 933 (1952).

[183] H. C. Longuet-Higgins and R. G. Sowden, *J. Chem. Soc.* p. 1404 (1952).

[184] E. Heilbronner and H. Bock, "Das HMO-Modell und seine Anwendung." Verlag Chemie, Weinheim, 1968.

the HMO theory. It is the author's opinion that the advantage of this procedure is not that one spared the complete calculations of the HMO or PPP treatments. These manipulations are fairly straightforward even by both of the latter. However qualitative exploitation of the perturbation calculation allows a great many conclusions to be drawn regarding the consequence of introducing the perturbation (heteroatom, substituent) into the parent system. Further-

TABLE X

PERTURBATION TREATMENT[a]

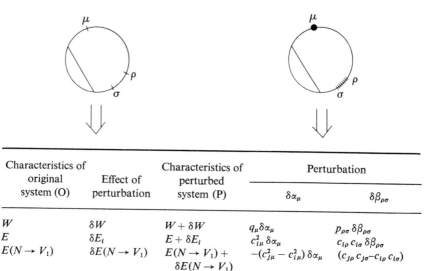

Characteristics of original system (O)	Effect of perturbation	Characteristics of perturbed system (P)	Perturbation	
			$\delta\alpha_\mu$	$\delta\beta_{\rho\sigma}$
W	δW	$W + \delta W$	$q_\mu \delta\alpha_\mu$	$P_{\rho\sigma}\,\delta\beta_{\rho\sigma}$
E	δE_i	$E + \delta E_i$	$c_{i\mu}^2\,\delta\alpha_\mu$	$c_{i\rho}c_{i\sigma}\,\delta\beta_{\rho\sigma}$
$E(N \to V_1)$	$\delta E(N \to V_1)$	$E(N \to V_1) +$ $\delta E(N \to V_1)$	$-(c_{j\mu}^2 - c_{i\mu}^2)\,\delta\alpha_\mu$	$(c_{j\rho}c_{j\sigma} - c_{i\rho}c_{i\sigma})$

[a] For explanation of symbols see text. $c_{i\mu}$, q_μ, $P_{\rho\sigma}$ are the expansion coefficients. π-electron densities, and bond orders of the parent system, respectively; $\delta\alpha_\mu$ and $\delta\beta_{\rho\sigma}$ represent the change of the Coulomb and the resonance integrals, respectively, caused by the perturbation: $\delta\alpha_\mu = \alpha_\mu^P - \alpha_\mu^O$, $\delta\beta_{\rho\sigma} = \beta_{\rho\sigma}^P - \beta_{\rho\sigma}^O$, where the exponents P and O denote the perturbed and the original systems, respectively, and $c_{m\tau}$ denotes the expansion coefficient in the mth MO with the τth AO.

more this may be accomplished on the basis of very scant data. (iii) The perturbation method permits a quantitative estimate of the characteristics of the derived system on the basis of the characteristics of the parent system. This is conditioned, of course, by our capability of making a quantitative evaluation of the magnitude of the perturbation. It is very easy to estimate relative changes in the energy parameters (π-electronic energy, W, ionization potential, electron affinity, $N \to V_1$ transition energy). The introduction of a perturbation at position μ and into a bond $\rho\sigma$ is visualized schematically in Table X.

Expressions for the perturbation calculation of the relevant energy character-istics are also summarized.

Current HMO parameters may be applied to the quantities $\delta\alpha$ and $\delta\beta$. For estimating the absolute changes $\delta E(N \to V_1)$, the spectroscopic values $\delta\alpha$ recommended by Murrell[6] can be used to advantage. Obviously δW and other values dependent on $\delta\alpha$ and $\delta\beta$ changes are summed if these perturbations occur simultaneously. If a Coulomb integral variation takes place simul-taneously at several positions, summation occurs in the resulting expressions. (iv) For the sake of illustration the expressions of Table X will now be applied to fluoranthene (78) and to five azafluoranthenes ($\alpha_N = \alpha + 0.5\beta$). The following HMO energy characteristics are appropriate to 78:

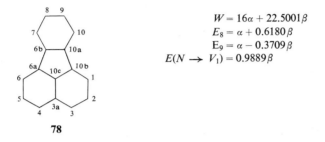

$$W = 16\alpha + 22.5001\beta$$
$$E_8 = \alpha + 0.6180\beta$$
$$E_9 = \alpha - 0.3709\beta$$
$$E(N \to V_1) = 0.9889\beta$$

78

The squares of the required expansion coefficients and π-electron densities are given in Table XI. By substituting these data into the expressions given in

TABLE XI

EXPANSION COEFFICIENTS AND π-ELECTRONIC DENSITIES FOR THE
FLUORANTHENE (78) PERTURBATION CALCULATION

	μ				
	1	2	3	7	8
$c_{8\mu}^2$	0.0461	0.0461	0.1206	0.1206	0.0461
$c_{9\mu}^2$	0.1214	0.0224	0.1632	0.0156	0.0394
$-(c_{9\mu}^2-c_{8\mu}^2)$	−0.0753	0.0137	−0.0426	0.1050	0.0067
q_μ	0.947	1.005	0.959	1.008	0.997

Table X, perturbation values are obtained. The latter are presented and compared with the exact HMO data in Table XII.

Although agreement of the HMO and the perturbation data is only semi-quantitative, it is clear at a glance that the perturbation method provides a useful evaluation of the effect of introducing a heteroatom.

TABLE XII

HMO AND PERTURBATION (PERT.) CHARACTERISTICS FOR X-AZAFLUORANTHENES (IN β UNITS)
$$(\alpha_N = \alpha + 0.5\beta)$$

		X				
		1	2	3	7	8
E_8	Pert.	0.641	0.641	0.678	0.678	0.641
	HMO	0.6354	0.6360	0.6631	0.6686	0.6325
E_9	Pert.	−0.310	−0.360	−0.289	−0.363	−0.351
	HMO	−0.3070	−0.3583	−0.2909	−0.3615	−0.3495
$E(N \rightarrow V_1)$	Pert.	0.951	0.982	0.946	1.094	0.996
	HMO	0.9424	0.9943	0.9540	1.0301	0.9820
W	Pert.	22.974	23.003	22.979	23.004	22.999
	HMO	23.0284	23.0521	23.0368	23.0512	23.0548

Ground State

The oxygen and sulfur analogs of the tropylium cation have been discussed[64, 185-187] in terms of MO theory. For a series of polynuclear thiopyrylium derivatives it has been found[187] that there is no quantitative correlation between NMR data and the HMO π-electron densities. It was shown, however, that the HMO values can nevertheless provide qualitative guidance. Another study has dealt with an aza analog of N-pyridinium cyclopentadienide.[188]

The fulvene and fulvalene heteroanalogs have also been investigated.[189-193] Good agreement was found between the π-components and the experimental dipole moments for azafulvalene **79** and for the N-benzyl of **80**. These substances are isoelectronic with sesquifulvalene. The presence of substituents was not considered in the calculation (Table XIII). For the five methyl derivatives of 2-cyclopentadienylidene-1-methyl-1,2-dihydropyridine a correlation has been discerned between the chemical shifts of the C-methyl protons and the π-electron densities calculated by the modified technique. Pseudoazulenes have been the subject of investigation as well.[64, 192]

[185] J. Fabian, A. Mehlhorn, and R. Zahradník, *J. Phys. Chem.* **72**, 3975 (1968).
[186] J. Fabian, A. Mehlhorn, and R. Zahradník, *Theoret. Chim. Acta* **12**, 247 (1968).
[187] T. E. Young and C. J. Ohnmacht, *J. Org. Chem.* **32**, 444 and 1558 (1967).
[188] G. V. Boyd, *Tetrahedron Letters* p. 3369 (1966).
[189] J. A. Berson, E. M. Evleth, Jr., and S. L. Manatt, *J. Am. Chem. Soc.* **87**, 2901 (1965).
[190] E. M. Evleth, Jr., J. A. Berson, and S. L. Manatt, *J. Am. Chem. Soc.* **87**, 2908 (1965).
[191] G. V. Boyd and N. Singer, *J. Chem. Soc.* p. 1017 (1966).
[192] G. V. Boyd, *Tetrahedron Letters* p. 1421 (1965).
[193] W. D. Kumler, *J. Org. Chem.* **28**, 1731 (1963).

TABLE XIII

EXPERIMENTAL AND CALCULATED DIPOLE MOMENTS (IN D)

System	Exp.[a]	HMO[b]	HMO–SC[b]	LCI–SCF–SC[c]
N-Methyl derivative of **79**	5.1	9.8	7.9	6.2
N-Benzyl derivative of **80**	8.9	12.9	10.5	8.2

[a] T. E. Young and C. J. Ohnmacht, *J. Org. Chem.* **32**, 444, 1558 (1967).
[b] J. A. Berson, E. M. Evleth, Jr., and S. L. Manatt, *J. Am. Chem. Soc.* **87**, 2901 (1965).
[c] R. Zahradník, unpublished calculations (1968).

The frequencies and the relative intensities of the symmetric and anti-symmetric N–H and N–O stretching vibrations of the amino- and nitro-fluoranthenes and of two nitroacenaphthylenes have been reported.[194] The

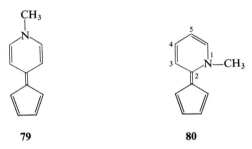

79   **80**

same order of conjugative power with respect to the amino group was found as that determined from the basicity measurements.[86]

B. ELECTRONIC SPECTRA

Most semiempirical calculations of electronic spectra of heterocycles and other derivatives of nonalternant hydrocarbons are of recent vintage. Agreement of calculated data with the experimental absorption curves is just as good as for hydrocarbons. This is not surprising, but only supports the fact that selection of empirical parameters was reasonable.

Attention has been devoted to the heteroanalogs of the tropylium cation[185, 186] (Fig. 36) of fluorenone and its benzo derivatives,[106] of thio-fluorenone[195] (the first band belongs to an $n \to \pi^*$ transition), and of

[194] J. Michl, K. Boček, and R. Zahradník, *Collection Czech. Chem. Commun.* **31**, 3471 (1966).
[195] J. Fabian and A. Mehlhorn, *Tetrahedron Letters* p. 2049 (1967).

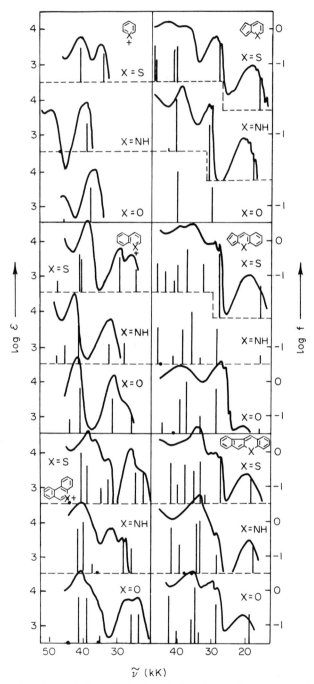

Fig. 36. Experimental and theoretical (LCI–SCF) electronic spectroscopic data for heteroanalogs of mono- and bicyclic nonalternant systems and their benzo derivatives. [J. Fabian, A. Mehlhorn, and R. Zahradník, *Theoret. Chim. Acta* **12**, 247 (1968); reproduced by permission].

tropone.[196-198] It has been demonstrated experimentally and theoretically that the first absorption band of these substances is a result of two $\pi \to \pi^*$ transitions polarized in mutually perpendicular directions. The necessity of respecting bond alternation is worth mentioning. On the other hand, Hosoya and Nagakura,[199] on the basis of a thorough LCI–SCF study of the parameters for oxygen, solvent effects, and comparison with spectra of related substances, argue that the first absorption band arises from a combination of one $n \to \pi^*$ and one $\pi \to \pi^*$ transition.

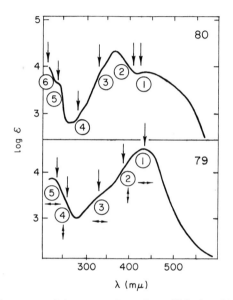

FIG. 37. Absorption curves of nitrogen analogs of sesquifulvalene **79** and **80** [E. M. Evleth, Jr., J. A. Berson, and S. L. Manatt, *J. Am. Chem. Soc.* **87**, 2908 (1965); reproduced by permission] and LCI–SCF–SC prediction of absorption band positions [$I_N = 23.13$ eV; $A_N = 10.15$ eV; $\beta_{CN}^c = -2.55$ eV; R. Zahradník, unpublished calculations (1967)]. In the case of the symmetrical compound **79** the directions of polarizations are indicated by double arrows. The encircled numbers indicate electronic states.

The *N*-heteroanalogs of sesquifulvalene, **79** and **80**, and their benzo derivatives have been studied intensively.[189, 190] Observed electronic spectra have been discussed in terms of HMO and the modified HMO theory. The authors have stated that by means of these methods it is not possible to explain why the 2-isomer **80** displays two bands in the long-wave region, whereas the 4-isomer **79** shows only one. The LCI–SCF calculation actually indicates that

[196] E. Weltin, E. Heilbronner, and H. Labhart, *Helv. Chim. Acta* **46**, 2041 (1963).
[197] H. Yamaguchi and Y. Amako, *Tetrahedron* **24**, 267 (1968).
[198] K. Inozuka and T. Yokota, *J. Chem. Phys.* **44**, 911 (1966).
[199] H. Hosoya and S. Nagakura, *Theoret. Chim. Acta* **8**, 319 (1967).

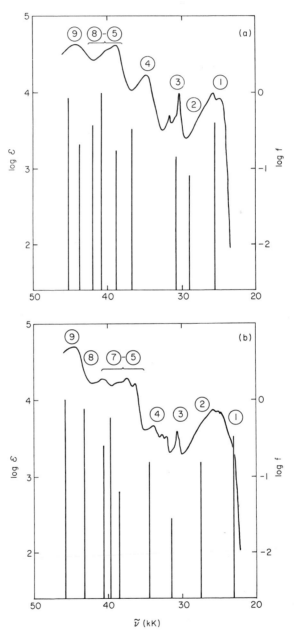

FIG. 38. Absorption curves (in cyclohexane) and LCI–SCF data for 1- and 7-amino-fluoranthenes, (*a*) and (*b*); respectively [J. Michl, personal communication (1968)].

for both isomers the long-wave band is conditioned by three transitions
(Fig. 37). It is necessary to mention that the intensity values calculated do not

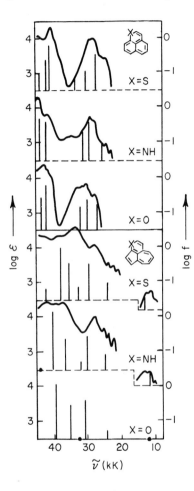

FIG. 39. Experimental and theoretical (LCI–SCF) electronic spectroscopic data for
heteroanalogs of tricyclic nonalternant systems. [J. Fabian, A. Mehlhorn, and R. Zahradník,
Theoret. Chim. Acta **12**, 247 (1968); reproduced by permission.]

express the shape of the experimental curves well. Moreover, the theory does
not reproduce well the position of the first two bands of the 2-isomer. Figure 37
also indicates the theoretical polarization directions of the individual transitions
for the 4-isomer. It would be of interest to verify, for example by dichroism
measurements, whether the main band is really composed of two transitions.

The spectra of the heteroanalogs of azulenes (Fig. 36) and aminodiaza-
azulene[200] and of five isomeric aminofluoranthenes[201] have also been studied.
The absorption curves of two of them are shown in Fig. 38. Of the other
tricyclic systems the *N, O,* and *S* heteroanalogs of pleiadiene and cycloheptazulene (Fig. 39) and two recently synthesized[202] perinaphthenone

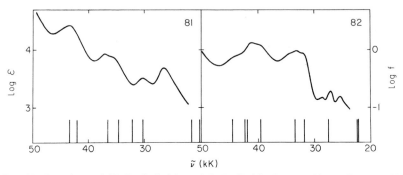

FIG. 40. Experimental [V. Boekelheide and C. D. Smith, *J. Am. Chem. Soc.* **88**, 3950
(1966); reproduced by permission] and theoretical (LCI–SCF) electronic spectroscopic data
for oxygen derivatives of the peri-condensed tricyclic system **81** and **82**.

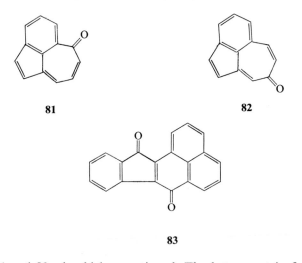

isomers, **81** and **82**, should be mentioned. The latter contain five, six, and
seven-membered rings in the same molecules (Fig. 40). Semiempirical calculations as well as HMO calculations indicate that the quinone formed by the
CrO_3 oxidation of the Reid hydrocarbon **67** possess the structure **83**.

[200] K. Hafner and U. Müller-Westerhof, *Tetrahedron Letters* p. 4341 (1967).
[201] J. Michl, personal communication (1968).
[202] V. Boekelheide and C. D. Smith, *J. Am. Chem. Soc.* **88**, 3950 (1966).

2

ESR Spectra of Radical Ions of

Nonbenzenoid Aromatics

F. Gerson and J. H. Hammons

I. Introduction

In the last few years electron spin resonance (ESR) has become a widely used spectroscopic method in organic chemistry. Although this method is restricted to paramagnetic substances, standard techniques have been developed for the conversion of neutral diamagnetic compounds to their radical ions so that such species are available for study by ESR spectroscopy.

In particular, the radical ions of aromatic compounds in solution have been intensively investigated. With the exception of a few cases—cyclooctatetraene and its derivatives are the best known—only minor changes in the π-electron structure result when an electron is added to or removed from an extended aromatic system. Therefore, both the parent compound and its radical ion may be described in terms of theoretical models of the same type, and the information extracted from the ESR spectrum of the radical ion may also be considered to be characteristic of the parent compound.

Because of the availability of benzenoid aromatic compounds and the stability of many of the radical ions derived from them, numerous studies of such radicals have been reported. In contrast, the number of radical ions generated from nonbenzenoid compounds has been relatively modest. Nevertheless, a few dozen of the latter have been investigated since 1962. The ESR spectra of these radical ions in solution will be the subject of this chapter.

Among the systems which will be considered here are alternants such as biphenylene and cyclooctatetraene (Section II); nonalternants like azulene, acenaphthylene, and acepleiadylene (Section III); and bridged [10]-, [14]-, and [18]annulenes (Section IV). Except for a few bridged annulenes containing a heteroatom in the bridging group, like 1,6-oxido[10]annulene, the compounds discussed are hydrocarbons. The emphasis will be on the spin distributions determined by means of the hyperfine splittings, and on their relation to the π-electron structures of the systems. Observed spin populations can be related in a simple way to theoretically calculated electron distribution within a single π-orbital, and thus provide a unique method for testing the correctness of eigenfunctions given by our common models of π-electron systems. Other important data, like g values, may be obtained from ESR spectra, but these are less informative for radical ions in solution than for other paramagnetic species and will not be treated here. Line widths will be mentioned only in cases of special importance.

Although this chapter is written for organic chemists, the reader is assumed to have an elementary knowledge of the field. An introduction to the principles of ESR spectroscopy and their application to organic radical ions in solution may be found in any of several sources.[1] However, to facilitate the reading of this chapter, we begin with a brief summary of the concepts used throughout.

A. HYPERFINE STRUCTURE FROM RING PROTONS. THE MCCONNELL RELATION

The hyperfine structure of an ESR spectrum is due to the interaction of the unpaired electron with the magnetic nuclei of the radical. In most cases involving hydrocarbon radicals, all of the readily observed hyperfine splittings are caused by protons, since the only abundant isotope of carbon is nonmagnetic. Of the protons in aromatic radical ions, by far the most important are those bonded directly to a π-electron center, and referred to here as ring protons. The coupling constant $a_{\mathrm{H}\mu}$ of a ring proton (i.e., the splitting due to this proton) is proportional to the π-spin population ρ_μ on the neighboring carbon[2]:

$$a_{\mathrm{H}\mu} = Q \cdot \rho_\mu \tag{1}$$

The proportionality factor Q has a value of 23–30 gauss, and a negative sign, as shown both by theory and experiment. The absolute magnitude of Q is expected to vary slightly from one type of aromatic system to another, and is commonly thought to be slightly larger for radical cations than for radical anions.[3]

A reasonable theoretical estimate for the experimental spin population is given by the square of the molecular orbital coefficient $c_{j,\mu}$, where j and μ

[1] M. Bersohn and J. C. Baird, "An Introduction to Electron Paramagnetic Resonance." Benjamin, New York, 1966; A. Carrington and A. D. McLachlan, "Introduction to Magnetic Resonance." Harper, New York, 1967; D. B. Ayscough, "Electron Spin Resonance in Chemistry." Methuen, London, 1967; F. Gerson, "Hochauflösende ESR Spektroskopie dargestellt anhand aromatischer Radikal-Ionen." Verlag Chemie, Weinheim, 1967 (English edition in press); in "Radical Ions" (E. T. Kaiser and L. Kevan, eds.), pp. 1–320. Wiley (Interscience), New York, 1968.

[2] H. M. McConnell, *J. Chem. Phys.* **24**, 764 (1956).

[3] See, however, M. T. Melchior, *J. Chem. Phys.* **50**, 511 (1969).

stand for the singly occupied orbital ψ_j and the center μ. Equation (1), known as the McConnell relation, can thus be approximated by

$$a_{H\mu} = Q \cdot c_{j,\mu}^2 \tag{2}$$

The obvious requirement that

$$\sum_\mu \rho_\mu = 1 \tag{3}$$

is satisfied by the normalization condition for ψ_j:

$$\sum_\mu c_{j,\mu}^2 = 1 \tag{4}$$

The orbital ψ_j is usually identified with the lowest antibonding orbital $\psi_a(j = a)$ in a radical anion or with the highest bonding orbital $\psi_b(j = b)$ in a radical cation, the relevant coefficients thus becoming $c_{a,\mu}$ or $c_{b,\mu}$, respectively.

B. THE HMO MODEL. ALTERNANT AND NONALTERNANT SYSTEMS

The Hückel MO (HMO) model is the most popular method for the calculation of $c_{j,\mu}^2$. Good correlation exists between these values and the coupling constants of ring protons, as is illustrated by Fig. 1, in which $a_{H\mu}$ is plotted vs $c_{j,\mu}^2$. The agreement between the experimental and calculated values is such that in cases for which unequivocal assignments cannot be made directly from experimental evidence, coupling constants are often assigned to specific protons on the basis of these theoretical values.

Hückel orbitals (HMO's) of alternant systems are known to exhibit "pairing" properties, the squares of the coefficients of two "paired" HMO's being equal. Of special importance in ESR spectroscopy is the fact that the lowest antibonding HMO, ψ_a, and the highest bonding HMO, ψ_b, represent an orbital pair of this kind, so that

$$c_{a,\mu}^2 = c_{b,\mu}^2 \quad \text{(alternants)} \tag{5}$$

Equation (5) predicts that the coupling constants of the ring protons in a radical anion derived from an alternant hydrocarbon should be nearly equal to those of the corresponding radical cation. In principle this prediction has been confirmed experimentally.

Because HMO's of nonalternant hydrocarbons in general, and ψ_a and ψ_b in particular, do not exhibit these "pairing" properties, Eq. (5) is no longer valid:

$$c_{a,\mu}^2 \neq c_{b,\mu}^2 \quad \text{(nonalternants)} \tag{6}$$

The coupling constants of the ring protons in a nonalternant radical anion are therefore expected to be different from those of the corresponding radical cation. There are only a few nonalternant hydrocarbons from which both radical ions have been produced. These cases will be dealt with in detail in Section III.

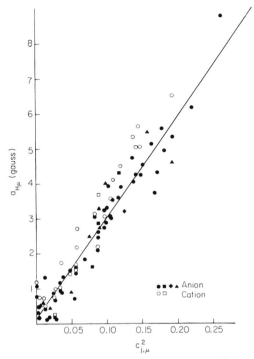

FIG. 1. Coupling constants of ring protons in aromatic radical ions vs HMO values $c_{j,\mu}^2$.

C. π-ELECTRON PERIMETERS

In monocyclic π-electron systems, here referred to as π-electron perimeters, all the HMO's except for the one lowest in energy—and in even-membered rings, also the one of highest energy—are doubly degenerate. Two degenerate orbitals can be specified as plus and minus, where + signifies symmetric and − antisymmetric relative to a mirror plane perpendicular to the molecular plane. The relevant singly occupied orbital in the radical ions can now be either ψ_{j+} or ψ_{j-}, these orbitals exhibiting a mutual degeneracy. With one exception (tropylium dianion), we will be concerned with even-membered perimeters containing $n = 4r$ or $4r + 2$ centers. In $(4r+2)$-membered perimeters the degenerate orbitals are as usual antibonding for radical anions and bonding for radical cations, and are accordingly denoted as ψ_{a+} and ψ_{a-} or as ψ_{b+} and ψ_{b-}. In perimeters with $4r$ centers the corresponding orbitals are nonbonding, and are designated ψ_{n+} and ψ_{n-}. For radical ions with unsubstituted perimeters, like benzene, planar cycloocta-tetraene, and [n]annulenes of highest symmetry D_{nh}, there is equal probability that the unpaired electron will occupy either a symmetric orbital ψ_{j+} or an antisymmetric orbital ψ_{j-} ($j = a$, b, or n). The ground state of these radical ions is therefore doubly degenerate. As a result, the spin population is evenly distributed among the n π-centers of the system ($\rho_\mu = 1/n$). However, the situation is quite different for those substituted perimeters in which the substi-tution lifts the degeneracy, like toluene, methylcyclooctatetraene, and bridged annulenes. In the radical ions of these compounds, the unpaired electron occupies preferentially one of the two relevant orbitals, which are now no longer degenerate. Whether the electron goes into ψ_{j+} or ψ_{j-} depends on the relative energy of the two orbitals, which in turn is determined by the kind and position of the substituent. Since the symmetry of the singly occupied orbital in the radical ion can be derived from the spin distribution measured by means of ESR spectroscopy, conclusions can be drawn about the relative energy of ψ_{j+} and ψ_{j-}, and thus also about the character and direction of the substituent effect.

The kind of arguments involved will be exemplified by the ESR spectrum of toluene radical anion. The unsubstituted parent compound, benzene, is a six-membered perimeter of symmetry D_{6h}, and its radical anion has an evenly distributed spin population: $\rho_\mu = 1/6$. The ground state of this radical anion is doubly degenerate, since there are two equienergetic configurations corresponding to equally probable locations of the unpaired electron in either one of the two degenerate antibonding orbitals:

$$\psi_{a+} = 0.577(\phi_1 + \phi_4) - 0.289(\phi_2 + \phi_3 + \phi_5 + \phi_6)$$
$$\psi_{a-} = 0.500(\phi_2 - \phi_3 + \phi_5 - \phi_6)$$

Substitution of methyl for hydrogen on the benzene ring to yield toluene is considered to be essentially a weak inductive perturbation, the amount of which—according to first-order treatment—should be proportional to the square of the coefficient at the perturbed center. In monosubstituted derivatives of benzene, this center $\bar\mu = 1$, as required by the symmetry of the degenerate orbitals ψ_{a+} and ψ_{a-}. (The latter are classified relative to the mirror plane passing through centers 1 and 4, as this is the only symmetry plane perpendicu-lar to the ring which remains in the monosubstituted compound.) For the symmetric orbital ψ_{a+}, the value of $c_{a+,1}{}^2$ is large (0.333), whereas in the anti-symmetric orbital ψ_{a-}, there is a nodal plane through center 1, and hence $c_{a-,1}{}^2 = 0$. The orbital ψ_{a+} should therefore be strongly perturbed by mono-substitution, but, according to this model, ψ_{a-} should be unaffected (Fig. 2). A substituent which is electron-repelling with respect to the π-electron system will destabilize ψ_{a+}, while an electron-attracting substituent will have the opposite effect. Thus, for a monosubstituted benzene in which the substituent exerts an electron-repelling inductive effect, ψ_{a+} will have higher energy than ψ_{a-}, and the odd electron in the corresponding radical anion will go into the

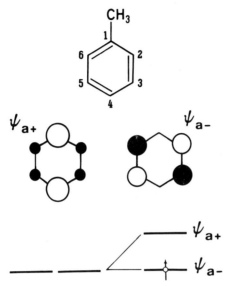

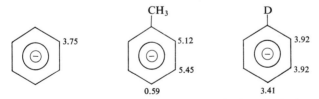

FIG. 2. Degenerate lowest antibonding orbitals of benzene. Splitting and occupancy in toluene radical anion. The radii of the circles are proportional to the absolute values of the coefficients at the respective centers. The blank and filled surfaces symbolize different signs of the coefficients.

antisymmetric orbital. Conversely, if the substituent is an electron-attractor, ψ_{a+} will be the orbital of lower energy, and will contain the unpaired electron. Comparison of the proton coupling constants in the radical anions of benzene[4]

FIG. 3. Coupling constants (in gauss) of the ring protons in the radical anions of benzene, toluene, and monodeuteriobenzene.

and toluene[5] clearly indicates that, as a consequence of the methyl substitution, the spin population has been completely redistributed and the odd electron occupies ψ_{a-} preferentially (Fig. 3). The inductive effect of the methyl group

[4] T. R. Tuttle, Jr., and S. I. Weissman, *J. Am. Chem. Soc.* **80**, 5342 (1958).
[5] J. R. Bolton and A. Carrington, *Mol. Phys.* **4**, 497 (1961).

must therefore be electron-repelling, as one would anticipate from other chemical experience. The splitting of the energy levels of ψ_{a+} and ψ_{a-} in the radical anion of toluene has been estimated to be ca. 400 cm^{-1} (ca. 0.05 eV).[6,7] This small energy difference gives rise to appreciable mixing of the two nearly degenerate orbitals, the mixing mechanism being thermal (by Boltzmann distribution) and/or vibronic (coupling by vibrations). As a result, the orbital of higher energy makes a significant contribution to the spin population of the radical ion.[6-8] In this case, it has been calculated that the minor contributor, ψ_{a+}, provides 10–25% of the spin population, and the major contributor, ψ_{a-}, accounts for 90–75%.[6,7]

Since the amount of mixing of the nearly degenerate orbitals depends inversely on the difference in their energy, which is determined in turn by the magnitude of the perturbation, substituents with a very weak perturbing effect should lead to a spin distribution in the radical anion which reflects nearly equal participation of the two orbitals. Deuterium is a substituent of this kind. Its secondary isotope effect on the π-electron system is thought to be electron-repelling, like that of alkyl substituents, but much weaker.[9] ESR studies of the radical anion of monodeuteriobenzene suggest such an effect, the spin distribution differing significantly from that of benzene radical anion and pointing to a slightly larger participation of the antisymmetric orbital ψ_{a-} (see Fig. 3).[10]

D. Refined MO Methods

1. *Negative Spin Populations*

Although the HMO approach satisfactorily accounts for the main features of the ESR hyperfine structure of aromatic ions, there are some finer details which cannot be rationalized on the basis of this simple model. As the spin populations ρ_μ are represented by the square of the coefficient $c_{j,\mu}$, they can have only a positive sign. In direct conflict with this theoretical result, there is strong evidence from proton magnetic resonance spectroscopy[11] and ESR studies of liquid crystals[12] that spin populations are negative for centers μ at

[6] E. de Boer and J. P. Colpa, *J. Phys. Chem.* **71**, 21 (1967).

[7] D. Purins and M. Karplus, *J. Chem. Phys.* **50**, 214 (1969).

[8] J. R. Bolton, A. Carrington, A. Forman, and L. E. Orgel, *Mol. Phys.* **5**, 43 (1962); A. Carrington, "Orbital Degeneracy and Spin Resonance in Free Radical Ions," Lect. Ser. Roy. Inst. Chem., London, 1965.

[9] E. A. Halevi, *Progr. Phys. Org. Chem.* **1**, 109 (1963).

[10] R. G. Lawler, J. R. Bolton, G. K. Fraenkel, and T. H. Brown, *J. Am. Chem. Soc.* **86**, 520 (1964); R. G. Lawler and G. K. Fraenkel, *J. Chem. Phys.* **49**, 1126 (1968).

[11] See, e.g., M. E. Anderson, P. J. Zandstra, and T. R. Tuttle, Jr., *J. Chem. Phys.* **33**, 1591 (1960).

[12] H. R. Falle and G. R. Luckhurst, *Mol. Phys.* **11**, 299 (1966); S. H. Glarum and J. H. Marshall, *J. Chem. Phys.* **44**, 2884 (1966).

which the HMO values $c_{j,\mu}{}^2$ are either zero or very small. The inability of the HMO model to predict negative spin populations stems chiefly from its treatment of the π-electrons as independent of one another. A refinement which allows for electron interdependence is the method of Configuration Interaction (CI).[13] It leads to spin distributions in aromatic radical ions which frequently differ in significant ways from those calculated by means of the HMO approach, and in particular correctly predicts negative spin populations in many cases.[14, 15]

McLachlan has developed a convenient perturbation procedure which is formally equivalent to the CI method, but which uses only HMO quantities, plus one additional parameter λ.[16] The formula is

$$\rho_\mu = c_{j,\mu}{}^2 + \lambda \sum_\nu \pi_{\mu,\nu} c_{j,\nu}{}^2 \tag{7}$$

where j stands for the singly occupied HMO ψ_j, and μ and ν refer to the given center and all the remaining centers, respectively. The quantities $\pi_{\mu,\nu}$ are the HMO atom–atom polarizabilities of the neutral π-electron system.[17] The additional parameter λ is usually taken as 1.2, although the absolute magnitudes of negative spin populations calculated with this value are frequently too large.

Clearly the existence of negative spin populations ρ_μ at the centers μ of vanishing HMO values $c_{j,\mu}{}^2$ is accompanied by significant increases in positive spin populations at the centers having $c_{j,\mu}{}^2$'s of appreciable magnitude. In this way the condition [Eq. (3)]

$$\sum_\mu \rho_\mu = 1$$

is still satisfied. On the other hand, the sum of the absolute magnitudes of the spin populations can by far exceed unity:

$$\sum_\mu |\rho_\mu| \geqslant 1 \tag{8}$$

The negative sign of the proportionality factor Q of the McConnell relation [Eq. (1)] means that the sign of the coupling constant $a_{H\mu}$ is opposite to that of ρ_μ. As most ρ_μ's, and in particular the relatively large ones, are positive, the large majority of $a_{H\mu}$'s are thus negative. Accordingly, positive coupling constants are both infrequent and small in absolute magnitude. In general, only

[13] R. G. Parr, "Quantum Theory of Molecular Electronic Structure." Benjamin, New York, 1963.

[14] G. J. Hoijtink, J. Townsend, and S. I. Weissman, J. Chem. Phys. 34, 507 (1961).

[15] N. M. Atherton, F. Gerson, and J. N. Murrell, Mol. Phys. 6, 265 (1963).

[16] A. D. McLachlan, Mol. Phys. 3, 233 (1960).

[17] A. Streitwieser, Jr., "Molecular Orbital Theory for Organic Chemists." Wiley, New York, 1961.

the absolute values of $a_{H\mu}$ can be determined from an ESR spectrum, and consequently we will use the coupling constants without specifying the sign.

2. "Pairing" Properties. π-Electron Perimeters

The orbitals of alternant hydrocarbons preserve their "pairing" properties in the refined MO methods like CI, which are commonly used for π-electron systems (Zero Differential Overlap approximation). As the McLachlan procedure is of the ZDO type, it also predicts similarity between the coupling constants of an alternant radical anion and those of the corresponding cation.

Refinement of the HMO model has no effect either on the degeneracy or on the coefficients of the orbitals of monocyclic π-systems, provided the highest possible symmetry (D_{nh}) of the n-membered ring is preserved, because the form of the orbitals is determined by the symmetry alone. If n is sufficiently small, as in the case of benzene and planar cyclooctatetraene, D_{nh} symmetry can be realized, but in perimeters with n greater than nine, the highest symmetry is no longer possible because of angle strain. In the derivatives of annulenes with $n = 10$ or 14 which will be dealt with in Section IV, the π-perimeters have lower symmetry, but the deviations from coplanarity do not seriously inhibit the cyclic conjugation of π-electrons, and the changes in geometry may be regarded as being mainly in-plane deformations of the perimeter. Since the HMO model is insensitive to such deformations, the relevant orbitals ψ_{j+} and ψ_{j-} ($j = a$ or b) will remain degenerate unless an electronic perturbation is introduced. Incorporation of a substituent or a triple bond causes a perturbation of this kind, which can be simulated in the HMO model by a change in the Coulomb integral $\alpha_{\bar{\mu}}$ of the perturbed center(s) $\bar{\mu}$, or in the bond integral $\beta_{\mu,\nu}$ of the linkage(s) $\mu-\nu$, respectively. On the other hand, in refined MO treatments, like CI, which allow explicitly for electron–electron interaction, both in-plane deformation of the perimeter and electronic perturbations will in general split the degeneracy of ψ_{j+} and ψ_{j-}. However, the two effects cannot be separated in the derivatives of annulenes discussed here, because in these structures the in-plane deformation of the perimeter is for steric reasons accompanied by the introduction of either bridges or triple bonds.

E. Hyperfine Structure from Nuclei Other than Ring Protons

Besides the ring protons, only those protons of an alkyl substituent which are separated by two σ-bonds from a substituted π-electron center $\bar{\mu}$, give rise to appreciable hyperfine splitting. In ESR spectroscopy one usually denotes these protons as β, whereas the ring protons are called α (Fig. 4). The alkyl protons, which are three σ-bonds away from the π-center (γ-protons) have in general very small coupling constants.

The three β-protons of a freely rotating methyl substituent at a π-electron center $\bar{\mu}$ commonly have coupling constants of absolute value similar to that of a ring proton bound to an unsubstituted center μ, provided the corresponding π-spin populations are comparable: $\rho_{\bar{\mu}} \approx \rho_{\mu}$. However, the sign of the coupling constant of a β-proton is the same as that of the spin population $\rho_{\bar{\mu}}$ at the substituted center $\bar{\mu}$, whereas, as previously indicated, the sign of the coupling

FIG. 4. α-, β-, and γ-Protons in π-electron systems.

constant a_{H_μ} of an α-proton is opposite to that of the spin population ρ_{μ}. This contrasting behavior is thought to be due to the operation of two different mechanisms for the transfer of electron spin densities from the π-MO into the hydrogen $1s$-AO: an indirect mechanism, spin-polarization, for the α-protons, and a direct mechanism, hyperconjugation, for the β-protons.[18] In accord

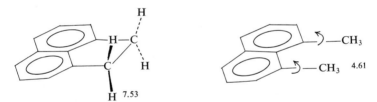

FIG. 5. Coupling constants (in gauss) of the β-protons in the radical anions of acenaphthene and 1,8-dimethylnaphthalene.

with the hyperconjugation mechanism, the coupling constants of the β-protons strongly depend on their position relative to the plane of the ring.[19] For instance the two β-protons of a methylene group in acenaphthene are situated in a position which favors hyperconjugation. The coupling constant of these protons in the radical anion[18] is therefore about 50% larger than that of the three β-protons of a freely rotating methyl group in 1,8-dimethylnaphthalene radical anion,[20] as might be expected on the basis of the simplest qualitative ideas (Fig. 5).

[18] J. P. Colpa and E. de Boer, *Mol. Phys.* **7**, 333 (1963–1964).

[19] D. Pooley and D. H. Whiffen, *Mol. Phys.* **4**, 81 (1961); C. Heller and H. M. McConnell, *J. Chem. Phys.* **32**, 1535 (1960); A. Horsfield, J. R. Morton, and D. H. Whiffen, *Mol. Phys.* **4**, 425 (1961).

[20] F. Gerson, B. Weidmann, and E. Heilbronner, *Helv. Chim. Acta* **47**, 1951 (1964).

The relation between the coupling constants of ^{14}N and ^{13}C (natural abundance 1.1%) nuclei and the π-electron spin population ρ_μ will not be dealt with, since these coupling constants will be mentioned only in isolated instances. These relations are more complex than in the case of ring protons, because they involve spin populations on more than one center.[21,22]

F. NOTATION

The following notation, which has been introduced in this section, will be used throughout the chapter.

μ	Unsubstituted π-electron center
$\bar{\mu}$	Substituted π-electron center
$a_{H\mu}$	Coupling constant of a ring proton (α-proton) bound to a carbon center μ
a_H^R	Coupling constant of a proton in the substituent R; in particular $a_H^{CH_n}$
$a_H^{CH_n}$	Coupling constant of an alkyl proton (β- or γ-proton; $n = 1, 2,$ or 3)
$a_{C\mu}$	Coupling constant of a ^{13}C nucleus in a carbon center μ
a_X	Coupling constant of a nucleus other than proton and ^{13}C; in particular a_N and a_M
a_N	Coupling constant of a ^{14}N nucleus,
a_M	Coupling constant of an alkali metal nucleus of a gegenion associated with the radical anion (M = ^7Li, ^{23}Na, or ^{39}K)
$c_{j,\mu}^2$	HMO value for the π-spin population = square of the LCAO coefficient of the center μ for the singly occupied orbital ψ_j; in particular: $c_{a,\mu}^2$, if ψ_j = lowest antibonding orbital = ψ_a $c_{b,\mu}^2$, if ψ_j = highest bonding orbital = ψ_b $c_{n,\mu}^2$, if ψ_j = nonbonding orbital = ψ_n
ρ_μ	π-Spin population as calculated from the HMO values $c_{j,\mu}^2$ by the McLachlan procedure with $\lambda = 1.2$
$c_{j+,\mu}^2$ and $c_{j-,\mu}^2$	HMO values for π-spin populations in the case of degenerate perimeter orbitals ψ_{j+} and ψ_{j-} ($j = a, b,$ or n); see definition, p. 85
$(\rho_\mu)_+$ and $(\rho_\mu)_-$	McLachlan π-spin populations calculated from $c_{j,\mu}^2$ and $c_{j-,\mu}^2$, respectively (parameter $\lambda = 1.2$)

The superscript symbols $\ominus$ and $\oplus$ are used to distinguish between radical anions and radical cations. (Not to be confused with the subscript symbols $+$ $-$ mentioned above and referring to degenerate symmetric and antisymmetric orbitals.) A superscript $\odot$ denotes a neutral (odd) radical. All coupling constants are given in gauss, with their signs unspecified as stated before. The HMO values $c_{j,\mu}^2$ were calculated with all Coulomb integrals equal to α and all bond integrals equal to β, unless otherwise stated. The abbreviations THF,

[21] M. Karplus and G. K. Fraenkel, *J. Chem. Phys.* **35**, 1312 (1961).

[22] See also J. C. M. Henning, *J. Chem. Phys.* **44**, 2139 (1966); C. L. Talcott and R. J. Myers, *Mol. Phys.* **12**, 549 (1967).

DME, and DMF stand for tetrahydrofuran, 1,2-dimethoxyethane, and
N,N-dimethylformamide, respectively.

II. Derivatives of Cyclobutadiene and Cyclooctatetraene

A. BIPHENYLENE AND BINAPHTHYLENE

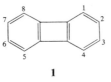

1

HMO calculations indicate that biphenylene, **1**, can only formally be
regarded as a dibenzo derivative of cyclobutadiene, since the two bonds
linking the benzene rings should have very little double bond character.[23]

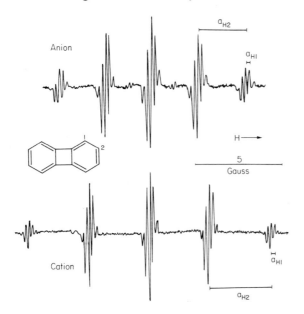

FIG. 6. ESR spectra of biphenylene radical ions. Anion; solvent: DME; gegenion: $K^{\oplus}$;
room temperature. Cation; solvent: concentrated H_2SO_4; room temperature. The spectra
are reproduced from a paper by A. Carrington and J. dos Santos-Veiga, *Mol. Phys.* **5**, 286
(1962).

This theoretical prediction has been verified by X-ray determination of the
structure, which gave a length of 1.52 Å for these two bonds, a value typical

[23] M. A. Ali and C. A. Coulson, *Tetrahedron* **10**, 41 (1960).

of a pure single bond joining two sp^2-hybridized carbons.[24] The radical anion $1^{\ominus}$ and cation $1^{\oplus}$ are both readily obtained from the neutral hydrocarbon by standard methods, like alkali metal reduction in DME for the former and sulfuric acid oxidation for the latter.[25,26] Figure 6 shows the ESR spectra of the two radical ions. The coupling constants of the ring protons[27] are given in Table I, together with the calculated spin populations. The original assignment, which is shown in Table I, has been supported by subsequent investigations of the radical anion and cation of 2,6-dimethylbiphenylene.[28]

TABLE I

PROTON COUPLING CONSTANTS AND SPIN POPULATIONS OF
BIPHENYLENE RADICAL IONS

μ	$a_{H_\mu}{}^{\ominus}$	$a_{H_\mu}{}^{\oplus}$	$c_{a,\mu}{}^2 = c_{b,\mu}{}^2$	ρ_μ
1,4,5,8	0.19	0.18	0.027	−0.009
2,3,6,7	2.74	3.58	0.087	0.091

The moderately good agreement between the experimental data, a_{H_μ}, and the computed values for $1^{\ominus}$ and $1^{\oplus}$ (Table I) may be improved slightly if the integral for the essentially single bonds between the benzene rings is decreased to $\beta' = 0.7\beta$.

Since biphenylene, **1**, is an alternant system, the theoretical spin populations are equal for the two radical ions, and the coupling constants should be very similar. The smaller $a_{H_\mu}{}^{\ominus}$ and $a_{H_\mu}{}^{\oplus}$ values ($\mu = 1,4,5,8$) are in fact almost equal, but there is a substantial difference between the larger ones ($\mu = 2,3,6,7$). Radical cations usually have somewhat larger coupling constants than the corresponding anions. One way to account for this fact in calculations of a_{H_μ} is to use slightly different Q values for the two species in the McConnell relation. However, in the case of positions $\mu = 2,3,6,7$ of **1**, the coupling constant of the cation is so much larger than that of the anion that the results can hardly be considered a verification of the "pairing" theorem. In order to test further whether the "pairing" properties do hold for the radical ions $1^{\ominus}$ and $1^{\oplus}$, Hindle et al.[27] measured the coupling constants of the ^{13}C nuclei. Two of the three ^{13}C (natural abundance) splittings could be observed for both radical ions, and assignments could be made on the basis of line-width variations. The values are given in Table II. The close similarity of the coupling constants

[24] T. C. W. Mak and J. Trotter, *Proc. Chem. Soc.* p. 163 (1961).
[25] C. A. McDowell and J. R. Rowlands, *Can. J. Chem.* **38**, 503 (1960).
[26] A. Carrington and J. dos Santos-Veiga, *Mol. Phys.* **5**, 285 (1962).
[27] P. R. Hindle, J. dos Santos-Veiga, and J. R. Bolton, *J. Chem. Phys.* **48**, 4703 (1968).
[28] J. dos Santos-Veiga, *Rev. Port. Quim.* **6**, 1 (1964); *Chem. Abstr.* **63**, 6819g (1965).

TABLE II

^{13}C COUPLING CONSTANTS OF
BIPHENYLENE RADICAL IONS

μ	$a_{C_\mu}{}^{\ominus}$	$a_{C_\mu}{}^{\oplus}$
1,4,5,8	3.48	3.20
2,3,6,7	2.36	2.48
9,10,11,12	a	a

^{a} Corresponding ^{13}C satellites not detected.

$a_{C_\mu}{}^{\ominus}$ and $a_{C_\mu}{}^{\oplus}$ in the corresponding radical ions argues in favor of the "pairing" theorem, and suggests that the unusually large difference found for the proton coupling constants is due to an anomalous Q value rather than to a redistribution of spin populations.[3]

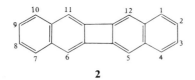

2

Binaphthylene (**2**) gave no radical cation in sulfuric acid solution, but the radical anion was readily produced by alkali metal reduction in DME.[25,26] The coupling constants of the protons, which are listed in Table III, were assigned on the basis of the calculated spin populations. The HMO bond orders

TABLE III

PROTON COUPLING CONSTANTS AND SPIN
POPULATIONS OF BINAPHTHYLENE RADICAL ANION

μ	a_{H_μ}	$c_{a,\mu}{}^{2}$	ρ_μ
1,4,7,10	1.62	0.079	0.102
2,3,8,9	0.93	0.034	0.024
5,6,11,12	4.31	0.117	0.158

indicate that the bonds between the naphthalene fragments in **2** are long,[23] and imply that—by analogy to biphenylene (**1**)—binaphthylene should be regarded as two weakly coupled benzenoid systems rather than as the dinaphtho derivative of cyclobutadiene. Thus the use of a reduced integral β'

for the two long bonds would be expected to lead to better agreement between experimental and theoretical spin populations, as in the case of **1**. Nonetheless, the values $c_{a,\mu}^{2}$ and ρ_{μ} given in Table III (which were calculated with all bond integrals equal) were found actually to be slightly better than those computed with $\beta' = 0.7\beta$. The reasons for this unexpected result are as yet unknown.

B. Cyclooctatetraene

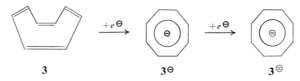

3   **3⊖**   **3⊖**

The first member of the series of cyclic $4r$ π-electron systems, cyclobutadiene ($r = 1$), has not yet been isolated in pure form,[28a] and its electronic structure is still the subject of controversy. In contrast, cyclooctatetraene (**3**) the second member ($r = 2$), has been known for over 60 years. The physical and chemical investigations of **3** clearly establish its nonplanar tub form[29] and its olefinic character.[30] On the other hand, experimental evidence indicates that in the diamagnetic dianion **3⊖**, which is formed on addition of two electrons to the neutral hydrocarbon, there is substantial cyclic conjugation of the π-electrons, and probably a planar perimeter with the full eightfold symmetry.[31] This finding is in accord with the Hückel rule, which predicts a closed shell structure and additional stability for perimeters containing $4r + 2$ π-electrons.

In solution, the cyclooctatetraene radical anion (**3⊖**) exists in equilibrium with the neutral compound **3** and the dianion **3⊖**.[32,33] ESR line-width studies of these solutions indicate the occurrence of rapid electron exchange between **3⊖** and **3⊖**, suggesting that the two ions have similar geometries; i.e., **3⊖** is believed also to be planar, with D_{8h} symmetry and a fully delocalized system of nine π-electrons.[34] The ESR hyperfine structure of this radical anion is due to eight equivalent protons having a coupling constant of 3.21 gauss. This result is consistent with an eight-membered perimeter in which the spin population at each center amounts to $1/8$.

[28a] L. Watts, J. D. Fitzpatrick, and R. Pettit, *J. Am. Chem. Soc.* **87**, 3253 (1965).

[29] O. Bastiansen and O. Hassel, *Acta. Chem. Scand.* **3**, 209 (1949); I. L. Karle, *J. Chem. Phys.* **20**, 65 (1952); H. Narain and B. D. Saksena, *Nature* **165**, 723 (1950).

[30] R. A. Raphael, *in* "Non-Benzenoid Aromatic Compounds" (D. Ginsburg, ed.), p. 465. Wiley (Interscience), New York, 1959.

[31] T. J. Katz, *J. Am. Chem. Soc.* **82**, 3784 (1960).

[32] T. J. Katz, *J. Am. Chem. Soc.* **82**, 3785 (1960).

[33] T. J. Katz and H. L. Strauss, *J. Chem. Phys.* **32**, 1873 (1960); H. L. Strauss, T. J. Katz, and G. K. Fraenkel, *J. Am. Chem. Soc.* **85**, 2360 (1963).

[34] F. J. Smentowski and G. R. Stevenson, *J. Am. Chem. Soc.* **89**, 5120 (1967).

The overall spectral width for $3^{\ominus}$ is $8 \times 3.21 = 25.68$ gauss. Since in $3^{\ominus}$, as in benzene radical anion, all centers carry protons and must, by symmetry, exhibit positive spin populations, this width has the special significance of being in itself an unequivocal value of $|Q|$. Relative to the corresponding value for benzene radical anion (22.5 gauss), there is an appreciable increase, contrary to a theoretical prediction based on a previous calculation (see Section III, B).

The coupling constant $a_{C_{\mu}}$ of a ^{13}C nucleus (natural abundance) in one of the eight equivalent carbon centers of $3^{\ominus}$ amounts to 1.28 gauss.[35]

Ion pairing of $3^{\ominus}$ with alkali metal cations $Li^{\oplus}$, $Na^{\oplus}$, and $K^{\oplus}$ has also been extensively studied. It was found that the gegenion has a large influence on the electron-transfer between $3^{\ominus}$ and the diamagnetic dianion $3^{\ominus\ominus}$, this transfer being the main cause of line-broadening in the ESR spectrum. Additional hyperfine splitting from alkali metal nuclei of the gegenion associated with $3^{\ominus}$ was observed for ^7Li and ^{23}Na in THF at room temperature: $a_{Li} = 0.2$ and $a_{Na} = 0.9$ gauss.[34]

A point of theoretical importance is that the ground state of fully symmetrical (D_{8h}) cyclooctatetraene radical anion—like that of benzene radical anion, which was discussed in Section I,C—should be doubly degenerate. This degeneracy is predicted by any MO model provided that the eight-membered perimeter has the full D_{8h} symmetry. (As was stated in Section I,D, the orbitals of a D_{8h} perimeter are completely determined by the symmetry, and thus are independent of the MO approximation used.) In a radical anion having this kind of perimeter, three electrons occupy the two degenerate nonbonding orbitals

$$\psi_{n+} = 0.500(\phi_1 - \phi_3 + \phi_5 - \phi_7)$$

and

$$\psi_{n-} = 0.500(\phi_2 - \phi_4 + \phi_6 - \phi_8)$$

As shown in Fig. 7, this situation gives rise to two equienergetic configurations (ψ_{n+} being doubly and ψ_{n-} singly occupied, or the reverse), and a doubly degenerate ground state. By analogy to the case of the benzene radical anion, a perturbation which lifts the degeneracy of the two relevant orbitals and makes one of the two electronic configurations more stable than the other, is consequently expected to have a profound effect on the spin distribution. In fact, such an effect has been observed for monosubstituted alkyl derivatives of $3^{\ominus}$ (see Section II,C). Introduction of a deuteron into $3^{\ominus}$, however, does not cause any measurable redistribution of the spin population. The hyperfine structure of the radical anion of monodeuteriocyclooctatetraene resembles that of the parent undeuterated species, the only difference being the replacement of one proton splitting of 3.21 gauss by a corresponding deuteron splitting[36] (Fig. 8). This

[35] H. L. Strauss and G. K. Fraenkel, *J. Chem. Phys.* **35**, 1738 (1961).

[36] A. Carrington, H. C. Longuet-Higgins, R. E. Moss, and P. F. Todd, *Mol. Phys.* **9**, 187 (1965).

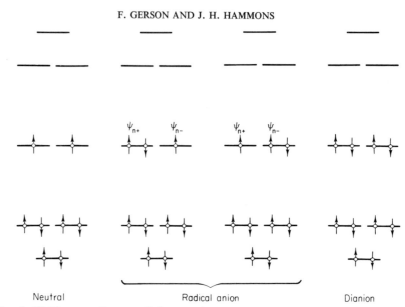

FIG. 7. HMO energy diagram of planar cyclooctatetraene. Occupancy in neutral compound, radical anion, and dianion.

behavior contrasts with that of benzene radical anion, for which mono-deuteration results in an appreciable change in the spin population (Section I,C).

A simple HMO model which allows for the weak electron-repelling effect of the deuterium cannot explain the different consequences of deuteration on the ESR spectra of the benzene and cyclooctatetraene radical anions. According to this model, the effect of the deuterium substitution is simulated by

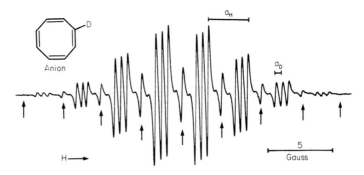

FIG. 8. ESR spectrum of monodeuteriocyclooctatetraene radical anion. Solvent: DME; gegenion: Li$^{\oplus}$; room temperature. The lines marked by arrows stem from undeuterated species present as isotopic impurity. The spectrum is reproduced from a paper by A. Carrington, H. C. Longuet-Higgins, R. E. Moss, and P. F. Todd, *Mol. Phys.* 9, 187 (1965).

alteration of the Coulomb integral $\alpha_{\bar{\mu}}$ of the substituted center, and the change in the energy of an orbital is—in first-order approximation—proportional to the square of the coefficient at the perturbed center. By convention this center $\bar{\mu}$ must be 1 in cyclooctatetraene, the relevant squared coefficients therefore being $c_{n+,1}^2 = 0.250$ and $c_{n-,1}^2 = 0$. The splitting of the degenerate orbitals ψ_{n+} and ψ_{n-} in deuteriocyclooctatetraene should thus be 2/3 of that found for the degenerate orbitals ψ_{a+} and ψ_{a-} in deuteriobenzene, in pronounced disagreement with experiment.

In view of the failure of the inductive model, it was instead suggested that the changes of the integral $\beta_{\bar{\mu},\nu}$ of the C—C bonds adjacent to the deuterated center might be more important than the change in $\alpha_{\bar{\mu}}$.[37,10] Out-of-plane vibrations would be expected to weaken the π-bonding, and thus to decrease the effective β value. As these vibrations are of smaller amplitude for centers bearing deuterons than for those bearing protons, the $\beta_{1,2}$ and $\beta_{1,n}$ integrals can be taken as larger than those of the remaining bonds. An increase in $\beta_{1,2} = \beta_{1,n}$ produces a change in the energy of the orbital ψ_j which is proportional to the two products of the corresponding coefficients, $c_{j,1} \times c_{j,2}$ and $c_{j,1} \times c_{j,n}$. Since these products are zero for both nonbonding orbitals in cyclooctatetraene, the perturbation will not split the degeneracy. In contrast, only the antisymmetric orbital ψ_{a-} of benzene has a vanishing product of the coefficients ($c_{a-,1} \times c_{a-,2} = c_{a-,1} \times c_{a-,6} = 0$; see Section I,C), whereas this product for ψ_{a+} is negative ($c_{a+,1} \times c_{a+,2} = c_{a+,1} \times c_{a+,6} = -0.167$), indicating that ψ_{a+} is destabilized by the perturbation. The unpaired electron should occupy ψ_{a-} preferentially, and the experimental results show that it does so. Thus the model based on changes in β succeeds in accounting for the ESR spectra of the monodeuterio derivatives of both benzene and cyclooctatetraene radical anions.

C. ALKYLCYCLOOCTATETRAENES

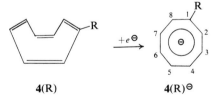

4(R)   4(R)$^{\ominus}$

As mentioned in Section II,B, alkyl substitution in cyclooctatetraene leads to the expected redistribution of the spin population in the radical anion. The ESR spectra of the radical anions of alkylcyclooctatetraenes, 4(R), with

[37] M. Karplus, R. G. Lawler, and G. K. Fraenkel, *J. Am. Chem. Soc.* **87**, 5260 (1965); R. G. Lawler, J. R. Bolton, M. Karplus, and G. K. Fraenkel, *J. Chem. Phys.* **47**, 2149 (1967).

TABLE IV

PROTON COUPLING CONSTANTS OF ALKYL-MONOSUBSTITUTED
CYCLOOCTATETRAENE RADICAL ANIONS

μ	R = CH₃		R = C_2H_5, n-C_3H_7, n-C_4H_9	
	$a_{H\mu}$	$a_H^{CH_3}$	$a_{H\mu}$	$a_H^{CH_2}$ (β-Protons)
1	—	5.1	—	2.5–2.6
2,4,6,8	1.6	—	1.9	—
3,5,7	4.8	—	4.5	—

R = methyl, ethyl, isopropyl, or n-butyl, exhibit a large coupling constant for three equivalent ring protons and a small splitting from the remaining four (Table IV).[38] The assignment of the larger value to the protons at the centers

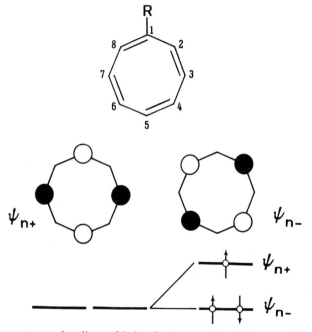

FIG. 9. Degenerate nonbonding orbitals of planar cyclooctatetraene. Splitting and occupancy in monoalkylcyclooctatetraene radical anions. Schematic representation as in Fig. 2.

3,5,7 and the smaller to those at the centers 2,4,6,8 is consistent with the assumption that the singly occupied orbital in $4(R)^{\ominus}$ is essentially ψ_{n+} (Fig. 9).

[38] A. Carrington and P. F. Todd, *Mol. Phys.* **7**, 533 (1963–1964); **8**, 299 (1964).

The occupancy of ψ_{n+} by the odd electron is predicted by the simple HMO model. As was pointed out before, the quantity $c_{n+,1}^2$ at the substituted center is 0.250 for the symmetric orbital, ψ_{n+}, whereas in the antisymmetric orbital ψ_{n-} this center lies in a nodal plane, and the corresponding value, $c_{n-,1}^2$, is zero. It is thus evident that ψ_{n+} should be strongly destabilized by the inductive electron-repelling effect of the alkyl substitutent, but ψ_{n-} in the first-order approximation should remain unaffected (see Fig. 9). As a result, ψ_{n-} will be occupied by two electrons, and the odd electron will go into ψ_{n+}.

The simple model indicates that the large coupling constant in **4(R)**$^\ominus$ should be about 6.4 gauss, i.e., twice as big as that of the protons in the parent radical anion **3**$^\ominus$, the spin population now being distributed over four centers instead of eight. As can be seen from Table IV, the experimental results[38] deviate markedly from this expectation. Allowance for negative spin populations by the use of the McLachlan refinement can account for the nonzero value of the smaller coupling constants, but would predict for the larger coupling constants a value substantially greater than 6.4 gauss rather than the decreased value of 4.5–4.8 which is actually observed. The only theoretical approach which can satisfactorily interpret the results is a mixing between the orbitals ψ_{n+} and ψ_{n-}. This mixing, the mechanism of which was indicated in Section I,C, signifies 'hat the odd electron does not occupy a "pure" orbital ψ_{n+}, but rather one which is expressed as a combination of ψ_{n+} and ψ_{n-}, with greater participation of the former.

D. Orthotetraphenylene

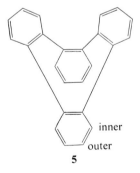

inner
outer

5

According to X-ray analysis,[39] the centers of the four benzene rings of orthotetraphenylene (**5**) lie approximately at the vertices of a regular tetrahedron. The uptake of one or of two electrons is not likely to lead to a planar geometry because of the extreme steric strain that would result. Also, in contrast to the cyclooctatetraene radical anion (**3**$^\ominus$), HMO calculations

[39] I. L. Karle and L. O. Brockway, *J. Am. Chem. Soc.* **66**, 1974 (1944).

suggest that there would be only a modest amount of electronic stabilization of the radical anion of orthotetraphenylene ($5^{\ominus}$) if the eight-membered ring were to become a regular octagon. A model of four weakly coupled benzene

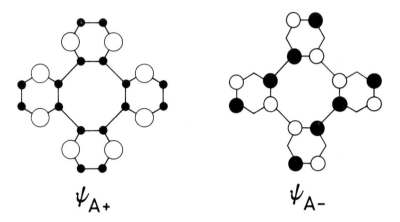

FIG. 10. Lowest antibonding orbitals of orthotetraphenylene as linear combinations of corresponding benzene orbitals. Schematic representation as in Fig. 2.

fragments should therefore apply to $5^{\ominus}$, and it should be possible to represent the singly occupied orbital in this radical anion as a linear combination of four degenerate antibonding benzene orbitals.[40] The two combinations of this kind

TABLE V

PROTON COUPLING CONSTANTS AND SPIN
POPULATIONS OF ORTHOTETRAPHENYLENE
RADICAL ANION

μ	$a_{H\mu}$	$c_{A+,\mu}^2$	$c_{A-,\mu}^2$
Inner	0.2^a	0.083	0.000
Outer	1.4^a	0.021	0.063

^{a} Each value for eight equivalent protons; assignment based on correlation with $c_{A-,\mu}^2$.

which have to be considered are: ψ_{A+}, the linear combination of the symmetric benzene orbitals ψ_{a+}, and ψ_{A-}, the corresponding combination of antisymmetric orbitals ψ_{a-}. By inspection of Fig. 10 one can readily decide whether ψ_{A+} or

[40] A. Carrington, H. C. Longuet-Higgins, and P. F. Todd, *Mol. Phys.* **8**, 45 (1964).

ψ_{A-} will be energetically favored and thus house the unpaired electron. Since the coupling of the fragments results in stabilization when the coefficients at the coupled centers μ and ν are of the same sign, and since the magnitude of the stabilization is proportional to the product of these coefficients, ψ_{A-} must have lower energy than ψ_{A+}. Relative to the degenerate orbitals of the uncoupled fragments, the first-order stabilization energies amount to $4 \times 2 \times 0.25 \times c_{a+,\mu} \times c_{a+,\nu} \times \beta' = +0.167\beta'$ for ψ_{A+}, and $4 \times 2 \times 0.25 \times c_{a-,\mu} \times c_{a-,\nu} \times \beta' = +0.500\beta'$ for ψ_{A-}, where β' is the integral for the weak bond, joining the fragments, and $c_{j,\mu}$ and $c_{j,\nu}$ ($j = a+$ or $a-$) are the relevant coefficients of the centers μ and ν flanking this bond.

The coupling constants of the protons[40] are given in Table V, together with the spin populations expected for occupancy of ψ_{A+} and ψ_{A-}. The experimental results are consistent with the model of weakly linked fragments, but the correlation with the spin populations $c_{A-,\mu}^2$ is only slightly better than with $c_{A+,\mu}^2$.

E. 1,2:5,6-DIBENZOCYCLOOCTATETRAENE

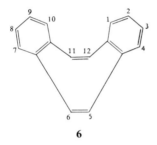

6

The 1,2:5,6-dibenzo derivative (**6**) of cyclooctatetraene may be considered to be an intermediate case between the parent compound **3** and orthotetraphenylene (**5**). In contrast to **5**, steric strain in the planar form is not prohibitively great, but the energy gain resulting from π-electron delocalization in the eight-membered ring of **6** should be less pronounced than in the parent monocyclic system **3**.

There has been some controversy in the literature about the structure of the radical anion **6**⊖ of 1,2:5,6-dibenzocyclooctatetraene. Carrington *et al.*[40] assumed the structure to be nonplanar, and adopted a model of weakly coupled benzene and ethylene fragments similar to that used for the radical anion **5**⊖ of orthotetraphenylene. In a subsequent paper, however, Katz *et al.*[41] showed that on reduction the dibenzo derivative **6** behaves in several

[41] T. J. Katz, M. Yoshida, and L. C. Siew, *J. Am. Chem. Soc.* **87**, 4516 (1965).

respects very much like the parent hydrocarbon **3**. In particular, the polarographic reduction potential is far lower than would be expected for non-interacting fragments ($\beta' = 0$), and is closer to that of a fully delocalized system ($\beta' = \beta$). Therefore, the use of a model of weakly coupled fragments is at best questionable. Moreover, the compound **6** readily undergoes a two-electron reduction to the dianion **6**$^{\ominus}$, the four "vinyl" protons of which, like the protons of the dianion **3**$^{\ominus}$ of cyclooctatetraene, exhibit a substantial ring current effect in the NMR spectrum. This observation argues in favor of a planar (or nearly planar) structure for **6**$^{\ominus}$, and it is not unlikely that the radical anion **6**$^{\ominus}$ has a similar geometry, with appreciable delocalization of the π-electrons about the central ring.

The coupling constants are slightly dependent on the solvent and the gegenion. Table VI lists the data obtained for **6**$^{\ominus}$ in THF solution with K$^{\oplus}$ as

TABLE VI

PROTON COUPLING CONSTANTS AND SPIN
POPULATIONS OF 1,2:5,6-DIBENZOCYCLOOCTATETRAENE
RADICAL ANION

		$c_{a,\mu}^2$	
μ	$a_{H\mu}$	$\beta' = 0^a$	$\beta' = \beta$
1,4,7,10	0.17	0.000	0.024
2,3,8,9	1.80	0.063	0.045
5,6,11,12	2.51	0.125	0.117

[a] When $\beta' = 0$, there are six degenerate antibonding MO's. The orbital chosen is a proper linear combination of two antisymmetric benzene MO's ψ_{a-} with the two ethylene antibonding MO's. This combination has the same symmetry as the lowest antibonding orbital for the whole system when β' is nonzero.

gegenion,[41] together with the $c_{a,\mu}^2$ values ($\beta' = 0$ or $\beta' = \beta$). The assignment of the coupling constants is made on the basis of calculated spin populations. Although the agreement is somewhat better for $\beta' = 0$ than for $\beta' = \beta$, the differences are too small to allow any conclusions to be drawn about the strength of the interaction between the π-fragments. In the absence of definitive evidence from the ESR spectrum to the contrary, the low polarographic reduction potential of 1,2:5,6-dibenzocyclooctatetraene must be taken to indicate that this interaction is not a weak one.

F. MONOHOMOCYCLOOCTATETRAENE

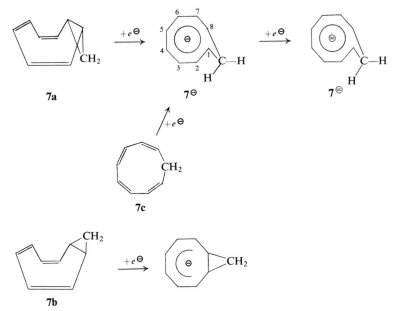

The *cis*-monomethylene adduct of cyclooctatetraene (**7a**) undergoes a two-electron reduction to a dianion, the proton resonance spectrum of which points to a homoaromatic structure $7^{\ominus}$.[42,43] The behavior of **7a** on reduction thus parallels that of cyclooctatetraene itself, in that an uptake of two electrons causes a flattening of the ring, accompanied by a dramatic increase in cyclic conjugation. The radical anion $7^{\ominus}$, which is an intermediate reduction product between the neutral compound **7a** and its diamagnetic dianion $7^{\ominus}$, has been observed by ESR spectroscopy. Its hyperfine structure suggests homoconjugation, two of the experimental results being particularly important. First, the large magnitude of the coupling constant (ca. 12 gauss) observed for one of the methylene protons rules out a classical structure having a fully closed cyclopropane ring. Second, the great difference between this coupling constant and that of the second methylene proton eliminates a planar structure, in which the two protons would be equivalent. The spectrum is consistent, however, with the formulation of $7^{\ominus}$ as homoaromatic. The coupling constants of the protons are listed in Table VII for two sets of experimental conditions: electrolysis in liquid NH_3, with $(CH_3)_4N^{\oplus}I^{\ominus}$ as supporting salt,[43] and reduction with K in

[42] M. Ogliaruso, R. Rieke, and S. Winstein, *J. Am. Chem. Soc.* **88**, 4731 (1966); R. Rieke, M. Ogliaruso, R. McClung, and S. Winstein, *ibid.* p. 4729.

[43] T. J. Katz and C. L. Talcott, *J. Am. Chem. Soc.* **88**, 4732 (1966).

TABLE VII

PROTON COUPLING CONSTANTS AND SPIN POPULATIONS[a] OF
MONOHOMOCYCLOOCTATETRAENE RADICAL ANION

| μ | DME | | NH$_3$ | | $c_{a,\mu}^{2}$ | ρ_μ |
	$a_{H\mu}$	$a_H^{CH_2}$	$a_{H\mu}$	$a_H^{CH_2}$		
1,8	5.72	$\begin{cases}12.18^b\\ 4.54^b\end{cases}$	5.24	$\begin{cases}12.00^b\\ 4.80^b\end{cases}$	0.175	0.240
2,7	0.87	—	1.02	—	0.074	0.024
3,6	5.12	—	5.24	—	0.143	0.156
4,5	1.99	—	2.04	—	0.108	0.080

[a] Calculated with $\beta' = 0.5\beta$.
[b] Each value for one proton only.

DME.[45] The assignment is based on the calculated spin populations, except for the two methylene protons, which were assigned unequivocally by deuterium substitution. In the calculation, the value of the integral β' for the weak π-bonding between the two centers 1 and 8 was taken as equal to 0.5β.

The coupling constants of the protons in the radical anion $7^\ominus$ are relatively insensitive to changes in solvent, temperature, and gegenion.[44] These results indicate that $7^\ominus$ is not in equilibrium with any other radical anion, and imply that it is the only species observed.

In contrast to the cis compound **7a**, the isomeric *trans*-monomethylene adduct of cyclooctatetraene (**7b**) yields a radical anion the ESR spectrum of which is like that expected for derivatives of 1,3,5-hexatriene.[45] This radical anion appears not to be homoaromatic. The difference between the two cases has been explained in terms of orbital symmetry arguments for electrocyclic reactions.[45,45a]

The radical anion $7^\ominus$ of monohomocyclooctatetraene has also been obtained by the reduction of *cis,cis,cis,cis*-cyclononatetraene (**7c**) with K in DME.[46]

[44] F. J. Smentowski, R. M. Owens, and B. D. Faubion, *J. Am. Chem. Soc.* **90**, 1537 (1968).
[45] G. Moshuk, G. Petrowski, and S. Winstein, *J. Am. Chem. Soc.* **90**, 2179 (1968).
[45a] R. B. Woodward and R. Hoffmann, "The Conservation of Orbital Symmetry," p. 60. Academic Press, New York, 1970.
[46] G. Moshuk and S. Winstein, to be published.

III. Even Nonalternants

A. 6,6-DIPHENYLFULVENE

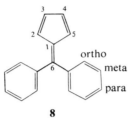

8

As yet, no radical ions having the simple fulvene π-system are known. Attempts to produce the radical anion of 6,6-dimethylfulvene were unsuccessful,[47] although the low energy calculated for the first antibonding orbital ($E_a = \alpha - 0.254\beta$) indicates that these compounds should be readily reduced. The radical anion seems to be short-lived; it immediately forms secondary products, presumably polymers. In contrast to the dimethyl derivative, the radical anion of 6,6-diphenylfulvene (**8**) is relatively stable.[47] It can be produced either by alkali metal reduction in DME or THF, or by electrolysis in DMF. The coupling constants of the electrolytically produced radical anion, which appear in Table VIII, are in reasonable agreement with the $c_{a,\mu}{}^2$ and ρ_μ values

TABLE VIII

PROTON COUPLING CONSTANTS AND SPIN POPULATIONS
OF 6,6,-DIPHENYLFULVENE RADICAL ANION

μ	$a_{H\mu}$	$c_{a,\mu}{}^2$	ρ_μ
2,5	1.86[a]	0.065	0.067
3,4	1.86[a]	0.050	0.027
Ortho	2.09	0.073	0.109
Meta	0.88	0.000	−0.036
Para	2.55[a]	0.075	0.108

[a] Assignment uncertain.

calculated. No significant improvement in the correlation occurs when the integral β for the bonds between the fulvene moiety and the phenyl substituents is slightly decreased. The tentative assignment is based on comparison of $a_{H\mu}$'s with the calculated values.

The ESR spectrum of 6,6-bis(pentadeuteriophenyl)fulvene radical anion consists of five overlapping broad bands spaced about 2 gauss apart,[47] thus

[47] F. Gerson and J. H. Hammons, unpublished work (1969).

confirming that two of the larger coupling constants must belong to the two pairs of fulvene protons.

B. AZULENE

9

The ESR spectrum of the radical anion **9**⊖ of azulene, which is shown in Fig. 11, was first reported by Fraenkel and co-workers,[48] who produced it by

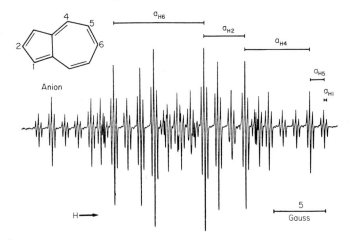

FIG. 11. ESR spectrum of azulene radical anion. Solvent: DMF; gegenion: $(n\text{-}C_3H_7)_4N^\oplus$; room temperature.[48]

electrolytic reduction. Later **9**⊖ was also obtained by reduction with alkali metals in DME.[49] The coupling constants observed in the two types of experiments differ significantly, in accord with the expectation that the azulene system, like nonalternants in general, should be more sensitive to solvent and gegenion than benzenoid hydrocarbons. Table IX lists the experimental values found for **9**⊖ at room temperature in DMF, with $(n\text{-}C_3H_7)_4N^\oplus$ as gegenion,[48] and in DME, with $Li^\oplus$ as gegenion.[49] The differences are attributable to the effect of $Li^\oplus$ in the relatively nonpolar DME, since either dilution or an increase in the polarity of the solvent (produced by a decrease in temperature) makes the coupling constants for the DME solution approach the DMF values.[49] The

[48] I. Bernal, P. H. Rieger, and G. K. Fraenkel, *J. Chem. Phys.* **37**, 1489 (1962).

[49] A. H. Reddoch, *J. Chem. Phys.* **41**, 444 (1964).

assignment of the coupling constants was established by comparison of the ESR spectrum of **9**$^\ominus$ with those of radical anions of 1,3-dideuterio- and 4,6,8-trimethylazulene.[48] The satisfactory correlation between the experimental and the calculated values is not essentially changed if the integral β of the long C-9 to C-10 bond is decreased slightly. The fact that the ratio $a_{H\mu}/c_{a,\mu}^2$ or $a_{H\mu}/\rho_\mu$ is larger for the centers in the five-membered ring than for those in the seven-membered ring, led the authors to postulate a dependence of Q on bond angle.[48] Subsequent experimental data have cast doubt on this postulate (see Section II,B).

Despite the appreciable effect of the gegenion on the ESR spectrum of **9**$^\ominus$ in DME solution, no additional splitting from alkali metal nuclei was detected.

TABLE IX

PROTON COUPLING CONSTANTS AND SPIN POPULATIONS OF
AZULENE RADICAL ANION

	$a_{H\mu}$			
μ	DME	DMF	$c_{a,\mu}^2$	ρ_μ
1,3	0.22	0.27	0.004	−0.025
2	3.80	3.95	0.100	0.109
4,8	6.46	6.22	0.221	0.304
5,7	1.46	1.34	0.011	−0.085
6	9.07	8.83	0.261	0.375

However, when DME was replaced by THF as solvent, splitting by the nucleus of the gegenion was observed.[50] The coupling constants measured at room temperature were: $a_{Li} = 0.17$; $a_{Na} = 0.54$; and $a_K = 0.2$ gauss. A simple MO model which treats the effect of the gegenion as an electrostatic perturbation of the π-electron system of the radical anion accounts qualitatively for the changes in the ESR spectrum of **9**$^\ominus$, which are caused by the increased association with the gegenion.[50]

Recently the radical anions of 1,1′- and 2,2′-biazulenyl have also been studied by ESR spectroscopy.[51] Whereas the hyperfine structure of the radical anion of the 2,2′-compound is due to the protons of both azulenyl fragments, the spectrum of the radical anion of the 1,1′-compound exhibits splitting from only one azulene moiety of the molecule. The splittings are very similar to those found for **9**$^\ominus$, and it has been assumed that the transfer of the odd electron between the two azulenyl moieties in the latter case is quite slow because steric strain makes it impossible for them to be coplanar.

[50] A. H. Reddoch, *J. Chem. Phys.* **43**, 225 (1965).
[51] Y. Ikegami and S. Seto, *Mol. Phys.* **16**, 101 (1969).

C. ACENAPHTHYLENE AND FLUORANTHENE

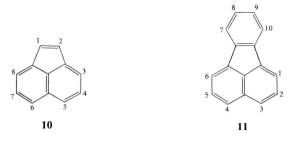

10   **11**

Consistent with the low energy calculated for the lowest antibonding orbital $\psi_a(E_a = \alpha - 0.284\beta)$, acenaphthylene (**10**) can be easily reduced to the radical anion **10**$^\ominus$ by standard methods.[52] The spin distribution in **10**$^\ominus$, such as that in the radical anion of azulene (**9**$^\ominus$) is very sensitive to environmental effects. This sensitivity is evident from Table X, which compares the coupling constants

TABLE X

PROTON COUPLING CONSTANTS AND SPIN POPULATIONS OF
ACENAPHTHYLENE RADICAL ANION

	$a_{H\mu}$			
μ	THF	DMF	$c_{a,\mu}^2$	ρ_μ
1, 2	3.10	3.07	0.104	0.100
3, 8	4.99	4.50	0.151	0.205
4, 7	0.71	0.46	0.014	−0.052
5, 6	5.92	5.60	0.178	0.261

obtained at room temperature for DMF solution [gegenion: $(C_2H_5)_4N^\oplus$] with those for THF solution (gegenion: Na$^\oplus$).[53, 54] As in the case of the radical anion **9**$^\ominus$, either dilution or a decrease in the temperature of the THF solution makes the coupling constants approach those found for the DMF solution. The difference in the coupling constants must in this case, too, be due to association with the alkali metal cation. In Fig. 12 the ESR spectra of the ion-paired and nonassociated radical anions are reproduced for comparison.[54]

[52] E. de Boer and S. I. Weissman, *J. Am. Chem. Soc.* **80**, 4549 (1958).
[53] M. Iwaizumi and T. Isobe, *Bull. Chem. Soc. Japan* **37**, 1651 (1964).
[54] F. Gerson and J. Heinzer, unpublished work (1965).

The excellent correlation between the experimental and calculated values—especially the one between a_{H_μ} (DMF) and $c_{a,\mu}^2$—allows a reliable assignment on a theoretical basis. This assignment has been confirmed by a study of the radical anions of the 1- and 5-deuterio derivatives.[55]

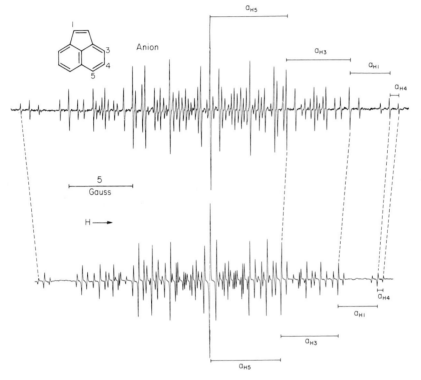

FIG. 12. ESR spectra of acenaphthylene radical anions. Upper spectrum: associated radical anion. Solvent: THF; gegenion: $Na^\oplus$; room temperature. Lower spectrum: non-associated radical anion. Solvent: DMF; gegenion: $(C_2H_5)_4N^\oplus$; room temperature. The corresponding hyperfine components are connected by dashed lines.

For THF solutions of $10^\ominus$ at room temperature, additional splitting from an alkali metal nucleus was found only in the case of $K^\oplus$ and $Cs^\oplus$ gegenions: $a_K = 0.07$ and $a_{Cs} = 1.17$ gauss. Use of dioxane as solvent resulted in the observation of splitting by the nuclei of $Li^\oplus$ and $Na^\oplus$ cations as well. Calculations based on a model which treated the association with the gegenion as an electrostatic perturbation led to the conclusion that in the most favored ion-pair configuration the cation lies above the five-membered ring of $10^\ominus$.[56]

[55] F. Gerson and B. Weidmann, *Helv. Chim. Acta.* **49**, 1837 (1966).

[56] M. Iwaizumi, M. Suzuki, T. Isobe, and H. Azumi, *Bull. Chem. Soc. Japan* **40**, 2754 (1967).

Irradiation of **10** yields two dimeric products, *cis*- and *trans*-dinaphthylene-cyclobutane.[57] Either of these stereoisomers is reduced by an alkali metal to two equivalents of **10**$^\ominus$ with no radical anion of the dimer (cis or trans) appearing.[55]

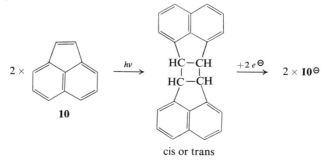

cis or trans

The driving force for the latter reaction is presumably the low energy of the orbital ψ_a of acenaphthylene (see above), which should be compared with that of the corresponding orbital of naphthalene ($E_a = \alpha - 0.618\beta$).

Fluoranthene (**11**) is formally a benzo derivative of acenaphthylene (**10**) although the ground state of the neutral compound **11** is best represented by weakly coupled naphthalene and benzene fragments.[58] (Analogously, the ground state of **10** may be regarded as weakly linked naphthalene and ethylene fragments.) This description, however, is less suitable for the radical anion, since the uptake of an additional electron in the lowest antibonding orbital ψ_a strengthens the weak bonds between the fragments; the contributions of the odd electron in ψ_a to the relevant HMO bond order are +0.074 for **10**$^\ominus$, and +0.068 for **11**$^\ominus$.

In general, the radical anions and cations of nonalternant hydrocarbons exhibit more even distribution of π-electron charge and smaller differences in π-bond orders than the neutral systems. In this sense, these radical ions may be considered to be more aromatic than the corresponding neutral compounds, and simple MO models may well be more adequate for the former than for the latter. Consistent with this viewpoint, good agreement exists between the coupling constants, a_{H_μ}[54, 59] and the calculated values $c_{a,\mu}^2$ and ρ_μ for the radical anion **11**$^\ominus$ prepared from fluoranthene by reduction with K in DME. As can be seen from Table XI, the three larger coupling constants can be

[57] K. Dziewonski and G. Rapalski, *Ber. Deut. Chem. Ges.* **45**, 2491 (1912); K. Dziewonski and C. Paschalski, *ibid.* **46**, 1986 (1913).

[58] E. Heilbronner, J.-P. Weber, J. Michl, and R. Zahradnik, *Theoret. Chim. Acta* **6**, 141 (1966).

[59] R. G. Lawler, as quoted in B. G. Segal, M. Kaplan, and G. K. Fraenkel, *J. Chem. Phys.* **43**, 4191 (1965).

TABLE XI

PROTON COUPLING CONSTANTS AND SPIN
POPULATIONS OF FLUORANTHENE RADICAL ANION

μ	$a_{H\mu}$	$c_{a,\mu}{}^2$	ρ_μ
1, 6	3.90	0.121	0.164
2, 5	0.17^a	0.022	−0.032
3, 4	5.20	0.163	0.241
7, 10	0.08^a	0.016	−0.037
8, 9	1.21	0.039	0.037

^{a} Assignment uncertain.

reliably assigned, and only the two smaller ones, both of which are quite close to zero, cannot be assigned with certainty. (These small coupling constants presumably arise from negative spin populations ρ_μ, and—as mentioned in Section I,D—such values, calculated by means of McLachlan's method with $\lambda = 1.2$, are frequently too large.)

D. ACEPLEIADYLENE

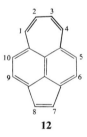

12

To date only radical anions of the nonalternant hydrocarbons **8–11** have been prepared and investigated by ESR spectroscopy; the corresponding cations are still unknown. In the case of fluoranthene (**11**) an ESR signal ascribed to the radical cation **11**$^\oplus$ was observed,[60] but lack of resolution prevented the analysis of the hyperfine structure. In none of these systems, therefore, has it been possible to test the prediction that the radical anion and cation of the same nonalternant hydrocarbon should have different coupling constants and exhibit dissimilar ESR spectra. As long ago as 1958, de Boer and Weissman[52] reported that acepleiadylene (**12**)[61] forms a radical cation

[60] I. C. Lewis and L. S. Singer, *J. Chem. Phys.* **43**, 2712 (1965); I. C. Lewis and L. S. Singer, *J. Chem. Phys.* **44**, 2082 (1966).

[61] V. Boekelheide and G. K. Vick, *J. Am. Chem. Soc.* **78**, 653 (1956).

in concentrated sulfuric acid, but they were unable to resolve the ESR spectrum. Later investigations, however, have proved that this spectrum of complex hyperfine structure and large overall range (42.9 gauss) was not due solely to the radical cation **12**⊕, but also to some other intermediate paramagnetic species.[62] When the solution of **12** in concentrated sulfuric acid is heated from

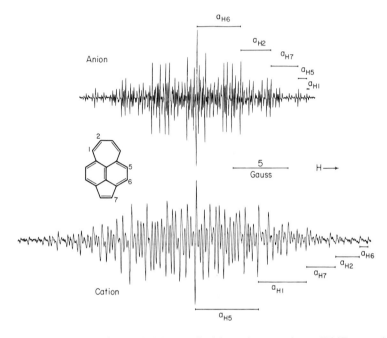

FIG. 13. ESR spectra of acepleiadylene radical ions. Anion, solvent: DME; gegenion: Na⊕; temperature: −70°C. Cation, solvent: concentrated H_2SO_4; temperature: +65°C.

room temperature to +80°C (or when potassium persulfate is added), another spectrum of smaller overall range (32.0 gauss) and simpler hyperfine structure appears. This spectrum is consistent with that expected for the radical cation **12**⊕.[63] It is shown in Fig. 13, below the spectrum of the radical anion **12**⊖

[62] F. Gerson and J. Heinzer, *Helv. Chim. Acta* **49**, 7 (1966).

[63] The reaction of **12** with sulfuric acid can be rationalized in terms of three steps:

1. $\mathbf{12} + \text{H}\oplus \xrightarrow{\text{r.t.}} \mathbf{12}\text{—H}\oplus$ Protonation

2. $\mathbf{12}\text{—H}\oplus + \mathbf{12} \xrightarrow{\text{r.t.}} \mathbf{12}\text{—H}\odot + \mathbf{12}\oplus$ Electron transfer

3. $\mathbf{12}\text{—H}\odot + 2\text{H}\oplus + \text{HSO}_4\ominus \xrightarrow{+80°C} \mathbf{12}\oplus + 2\text{H}_2\text{O} + \text{SO}_2$ Oxidation

According to this scheme the spectrum observed at room temperature (r.t.) is due to a mixture of **12**—H⊙ and **12**⊕.

generated from acepleiadylene by alkali metal reduction in DME. The two spectra can be seen to differ appreciably, the overall ranges being 20.50 and 32.04 gauss for **12**$^\ominus$ and **12**$^\oplus$, respectively. In Fig. 14 the lowest antibonding

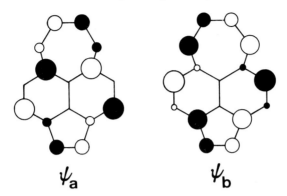

FIG. 14. Lowest antibonding (ψ_a) and highest bonding (ψ_b) orbitals of acepleiadylene. Schematic representation as in Fig. 2.

(ψ_a) and the highest bonding orbital (ψ_b) of acepleiadylene are schematically depicted. The spin distributions, as represented in Fig. 14, account remarkably well for the observed spectra.[64] The relevant data[62] are listed in Table XII.

TABLE XII

PROTON COUPLING CONSTANTS AND SPIN POPULATIONS OF ACEPLEIADYLENE
RADICAL IONS

μ	$a_{H\mu}{}^{\ominus}$	$a_{H\mu}{}^{\oplus}$	$c_{a,\,\mu}{}^2$	$c_{b,\,\mu}{}^2$	$\rho_\mu{}^{\ominus}$	$\rho_\mu{}^{\oplus}$
1, 4	0.21[a]	4.53	0.027	0.116	−0.003	0.147
2, 3	2.76	2.13[a]	0.087	0.056	0.108	0.040
5, 10	0.80[a]	5.88	0.000	0.143	−0.057	0.206
6, 9	4.04	0.78	0.136	0.007	0.195	−0.045
7, 8	2.44	2.70[a]	0.087	0.056	0.096	0.050
	20.50[b]	32.04[b]	0.674[b]	0.756[b]	0.918[b]	0.976[b]

[a] Assignment uncertain.
[b] These numbers are twice the sums of the absolute values in the columns above.

The assignment of the coupling constants, which is based on the theoretical values, is unequivocal for the large $a_{H\mu}$'s, but subject to some uncertainty for the smaller ones.

[64] F. Gerson and J. Heinzer, *Chem. Commun.* p. 488 (1965).

E. ACENAPHTH[1,2-*a*]ACENAPHTHYLENE

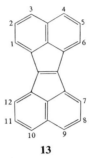

13

For another nonalternant hydrocarbon, acenaphth[1,2-*a*]acenaphthylene (**13**),[65] the agreement between the ESR results and HMO predictions is still more

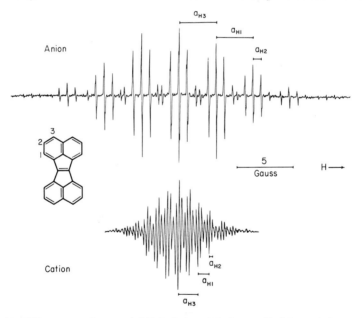

FIG. 15. ESR spectra of acenaphth[1,2-*a*]acenaphthylene radical ions. Anion, solvent: DME; gegenion: $Na^{\oplus}$; temperature: $-70°C$. Cation, solvent: concentrated H_2SO_4; temperature: $+65°C$.

spectacular.[64] Both radical ions (**13**⊖ and **13**⊕) can readily be generated, the anion by alkali metal reduction in DME, and the cation by oxidation in sulfuric acid.[62] Their ESR spectra, which are reproduced in Fig. 15, differ to an extra-

[65] R. L. Letsinger and J. A. Gilpin, *J. Org. Chem.* **29**, 243 (1964).

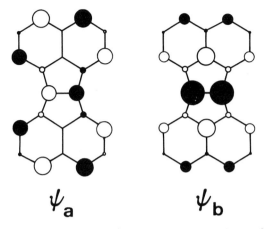

ψ_a ψ_b

FIG. 16. Lowest antibonding (ψ_a) and highest bonding (ψ_b) orbitals of acenaphth[1,2-a]-acenaphthylene. Schematic representation as in Fig. 2.

ordinary degree in their overall ranges: 29.44 gauss for $13^{\ominus}$, and 12.00 gauss for $13^{\oplus}$. This difference can easily be explained by consideration of the electron distributions in the singly occupied orbitals ψ_a and ψ_b, as depicted by the schemes in Fig. 16. It can be seen that in ψ_a the unpaired electron resides mainly at the

TABLE XIII

PROTON COUPLING CONSTANTS AND SPIN POPULATIONS OF
ACENAPHTH[1,2-a]ACENAPHTHYLENE RADICAL IONS

μ	$a_{H\mu}{}^{\ominus}$	$a_{H\mu}{}^{\oplus}$	$c_{a,\mu}{}^2$	$c_{b,\mu}{}^2$	$\rho_\mu{}^{\ominus}$	$\rho_\mu{}^{\oplus}$
1,6,7,12	3.30^a	1.00	0.093	0.030	0.132	0.035
2,5,8,11	0.71	0.24	0.003	0.001	−0.038	−0.021
3,4,9,10	3.35^a	1.76	0.100	0.036	0.149	0.045
	29.44^b	12.00^b	0.784^b	0.268^b	1.276^b	0.404^b

^{a} Assignment uncertain.
^{b} These numbers are four times the sums of absolute values in the columns above.

outer, proton-carrying centers, whereas in ψ_b it favors the inner, "blind" centers which are devoid of protons. The sum of the values $c_{a,\mu}{}^2$ for the spin populations at the proton-carrying centers is thus three times as large as the corresponding sum for the values $c_{b,\mu}{}^2$. In Table XIII the coupling constants of the ring protons in $13^{\ominus}$ and $13^{\oplus62}$ are correlated with $c_{a,\mu}{}^2$ and $c_{b,\mu}{}^2$, respectively, as well as with the spin populations $\rho_\mu{}^\ominus$ and $\rho_\mu{}^\oplus$. The assignment is well established by this correlation, except perhaps for the two large coupling constants $a_{H_\mu}{}^\ominus$ in the anion, which are nearly equal. The excellent agreement between the experimental and theoretical values provides additional evidence that the HMO model is adequate not just for alternant systems, but for non-alternants as well. In fact, for reasons already mentioned this model seems to work better for nonalternant radical ions than for the parent neutral hydro-carbons.

F. ACEPLEIADIENE

14

Acepleiadiene **(14)**[66] which differs by two hydrogens from acepleiadylene **(12)** is a dimethylene derivative of pleiadiene, a nonalternant hydrocarbon. The formation of a radical cation from acepleiadiene in sulfuric acid was first reported by de Boer and Weissman in their pioneering paper[52] mentioned in Section III,D. The radical cation **14**⊕ obtained by this method is relatively unstable,[67] and after 30 min the spectrum changes to that observed when acepleiadylene **(12)** is dissolved in sulfuric acid (see above).[62, 68] One can instead obtain a more stable solution of the radical cation **14**⊕ of acepleiadiene by dissolving the neutral compound **14** in unpurified moist $SbCl_3$ at $+80°C$. The oxidizing agent is thought to be $SbCl_5$, present in trace amounts.[69]

The spectra of the radical cation **14**⊕ in concentrated sulfuric acid and radical anion **14**⊖ generated by reduction with sodium in DME are reproduced in Fig. 17. The dissimilarity of the two spectra is again apparent. The spectra can be rationalized on the basis of the lowest antibonding and highest bonding orbitals, ψ_a and ψ_b, of pleiadiene, which are depicted in Fig. 18. The two substi-tuted centers $\bar\mu$ are marked by arrows. Again, pronounced differences between

[66] V. Boekelheide, W. E. Langeland, and Chu-Tsin Liu, *J. Am. Chem. Soc.* **73**, 2432 (1951).

[67] F. Gerson and J. Heinzer, *Helv. Chim. Acta* **50**, 1852 (1967).

[68] This observation supports the scheme of reaction of **12** with sulfuric acid (see footnote 63), since acepleiadiene, **14**≡**12**—H_2, has one more hydrogen than the radical **12**—$H^\ominus$ postulated as an intermediate in this scheme.

[69] E. C. Baughan, T. P. Jones, and L. G. Stoodley, *Proc. Chem. Soc.* p. 274 (1963).

the electron distributions of the two orbitals can be seen. In particular, the values for the substituted centers $\bar{\mu}$ differ considerably: $c_{b,\bar{\mu}}^2 \gg c_{a,\bar{\mu}}^2$. It is this inequality which causes the overall spectral range of the radical cation (56.66 gauss) to be so much greater than that of the corresponding anion (31.64 gauss).

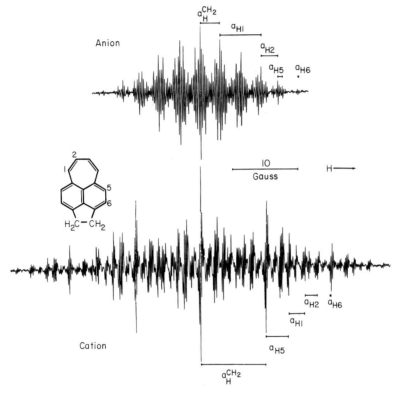

FIG. 17. ESR spectra of acepleiadiene radical ions. Anion, solvent: DME; gegenion: Na⊕; temperature: −80°C. Cation, solvent: concentrated H_2SO_4; room temperature.

The overall range is especially sensitive to the value of $c_{j,\bar{\mu}}^2$ ($j = a$ or b) for two reasons. First, in the special case of acepleiadiene the position of the methylene protons is favorable for hyperconjugation (see Section I,E), so that the proportionality factor between the coupling constant $a_H^{CH_2}$ and $\rho_{\bar{\mu}}$ is significantly higher than the $|Q|$ value for ring protons in the McConnell relation [Eq. (1)]. Second, there are two methylene protons for each center $\bar{\mu}$, against one ring proton at each center μ. The coupling constants of the protons in **14**⊖ and **14**⊕[67] are given in Table XIV, along with the corresponding theoretical values calculated for the unsubstituted pleiadiene system. If one allows for the inductive effect of the methylene chain by setting the Coulomb integral $\alpha_{\bar{\mu}} = \alpha - 0.3\beta$ for the substituted center $\bar{\mu}$,[67] these values are changed only slightly.

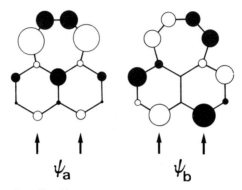

FIG. 18. Lowest antibonding (ψ_a) and highest bonding (ψ_b) orbitals of pleiadiene. The centers which are joined by the dimethylene chain in acepleiadiene have been marked by arrows. Schematic representation as in Fig. 2.

The assignment of the coupling constant of the methylene protons is made directly from the ESR spectra, whereas the coupling constants of the ring

TABLE XIV

PROTON COUPLING CONSTANTS AND SPIN POPULATIONS OF ACEPLEIADIENE
RADICAL IONS

μ	$a_{H\mu}{}^{\ominus}$	$a_{H\mu}{}^{\oplus}$	$c_{a,\mu}{}^2$	$c_{b,\mu}{}^2$	$\rho_\mu{}^{\ominus}$	$\rho_\mu{}^{\oplus}$
1,4	6.33	2.44	0.238	0.104	0.301	0.117
2,3	2.56	2.10	0.112	0.068	0.088	0.046
5,10	0.71	3.50	0.032	0.135	0.032	0.191
6,9	0.20	0.17	0.001	0.009	−0.022	−0.051
	19.60[a]	16.42[a]	0.766[a]	0.632[a]	0.886[a]	0.810[a]

$\bar{\mu}$	$a_H^{CH_2\ominus}$	$a_H^{CH_2\oplus}$	$c_{a,\bar{\mu}}{}^2$	$c_{b,\bar{\mu}}{}^2$	$\rho_{\bar{\mu}}{}^{\ominus}$	$\rho_{\bar{\mu}}{}^{\oplus}$
7,8	3.05	10.06	0.037	0.152	0.040	0.228

[a] These numbers are twice the sums of the absolute values in the columns above.

protons are assigned on the basis of the theoretical values. This assignment is reliable for the large values of $a_{H\mu}$, but may be subject to some doubt in the case of the smaller values.

G. 3,5,8,10-Tetramethylcyclopenta[ef]heptalene

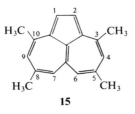

15

This compound **(15)** is one of the so-called Hafner hydrocarbons,[70] and, like azulene **(9)**, it contains only fused odd-membered rings. The parent unsubstituted hydrocarbon is still unknown. Relative to Hammett's H_0 function in a solvent system of ethanol (20%)–H$_2$SO$_4$, the pK^* value of **15**

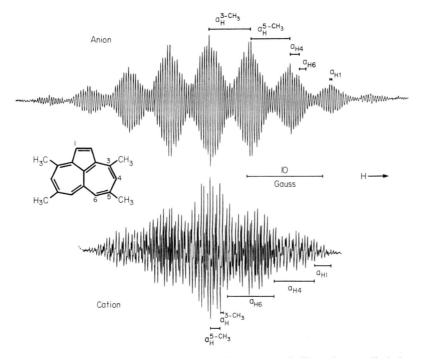

FIG. 19. ESR spectra of 3,5,8,10-tetramethylcyclopenta[ef]heptalene radical ions. Anion, solvent: DME; gegenion: Na$^\oplus$; temperature: −80°C. Cation, solvent: unpurified SbCl$_3$; temperature: +100°C.

[70] See, e.g., K. Hafner and G. K. Schneider, *Ann.* **672**, 194 (1964); K. Hafner and co-workers, unpublished work (1964–66).

is $-0.9.$[71] It is therefore not surprising that, as in the case of azulene, the compound **15** is almost completely converted to its diamagnetic proton adduct in concentrated sulfuric acid, and thus the use of this medium in the preparation of the radical cation **15**$^\oplus$ is precluded. However, since the highest bonding orbital ψ_b of cyclopenta[*ef*]heptalene has the energy $E_b = \alpha + 0.241\beta$, this cation **15**$^\oplus$ should be relatively stable. In fact, the radical cation **15**$^\oplus$ is readily formed from the neutral compound **15** in moist, unpurified $SbCl_3$ at $+80°C.$[67] The spectrum of **15**$^\oplus$ is shown in Fig. 19, below that of the corresponding radical anion **15**$^\ominus$, produced by reduction with Na in DME.

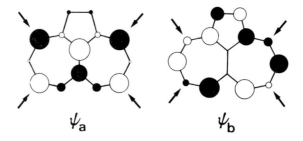

FIG. 20. Lowest antibonding (ψ_a) and highest bonding (ψ_b) orbitals of cyclopenta[*ef*]-heptalene. The centers which are substituted in the 3,5,8,10-tetramethyl derivative, have been marked by arrows. Schematic representation as in Fig. 2.

As in the case of acepleiadiene (**14**), the radical ions of which were discussed in Section III,F, the spectra of **15**$^\ominus$ and **15**$^\oplus$ are quite dissimilar. In contrast to **14**$^\ominus$ and **14**$^\oplus$, it is now the radical anion **15**$^\ominus$ which has a larger overall spectral range: 68.54 gauss, compared to the range of only 35.68 gauss for the corresponding cation **15**$^\oplus$. One can rationalize this difference in a manner similar to that used for **14**$^\ominus$ and **14**$^\oplus$, i.e., by examination of the electron distribution between the proton-carrying and alkyl-substituted centers in the lowest antibonding and highest bonding orbitals, ψ_a and ψ_b, of cyclopenta[*ef*]-heptalene (Fig. 20). It can be seen that the $c_{a,\bar{\mu}}^2$ values at the substituted centers $\bar{\mu}$ (marked by arrows) are large, while the $c_{b,\bar{\mu}}^2$ values at these centers are small. The opposite is true for $c_{a,\mu}^2$ and $c_{b,\mu}^2$ at the unsubstituted, proton-carrying centers μ. Again it is apparent that the great difference in the overall spectral ranges of **15**$^\ominus$ and **15**$^\oplus$ results primarily from the difference in the coupling constants ($a_H^{CH_3}$) of the methyl β-protons, which is in turn a consequence of the widely differing spin populations at the substituted centers $\bar{\mu}$. It is true that for the freely rotating methyl substituents the $a_H^{CH_3}$ values are expected to be smaller than the corresponding values $a_H^{CH_2}$ for methylene β-protons in compounds like **14** (see Fig. 5 in Section I,E). The effect on the overall spectral

[71] F. Gerson and E. Haselbach, unpublished results (1965).

range, however, is compensated for by the larger number (three) of methyl protons relative to that (two) of methylene protons.

There is a good correlation between the observed coupling constants of the protons in **15**$^\ominus$ and **15**$^\oplus$ and the corresponding theoretical values as calculated for cyclopenta[ef]heptalene[67] (Table XV). The different numbers of equivalent

TABLE XV

PROTON COUPLING CONSTANTS AND SPIN POPULATIONS OF
3,5,8,10-TETRAMETHYLCYCLOPENTA[ef]HEPTALENE RADICAL IONS

μ	$a_{H\mu}^{\ominus}$	$a_{H\mu}^{\oplus}$	$c_{a,\mu}^{2}$	$c_{b,\mu}^{2}$	$\rho_{\mu}^{\ominus}$	$\rho_{\mu}^{\oplus}$
1,2	0.32	2.07	0.005	0.068	−0.003	0.054
4,9	1.28[a]	4.99	0.001	0.135	−0.061	0.187
6,7	0.99[a]	6.07	0.027	0.152	−0.029	0.227
	5.18[b]	26.26[b]	0.066[b]	0.710[b]	0.186[b]	0.936[b]

$\bar{\mu}$	$a_{H}^{CH3\ominus}$	$a_{H}^{CH3\oplus}$	$c_{a,\bar{\mu}}^{2}$	$c_{b,\bar{\mu}}^{2}$	$\rho_{\bar{\mu}}^{\ominus}$	$\rho_{\bar{\mu}}^{\oplus}$
3,10	5.44	0.35[a]	0.158	0.033	0.231	−0.014
5,8	5.12	1.22[a]	0.149	0.009	0.203	−0.048

[a] Assignment uncertain.
[b] These numbers are twice the sums of the absolute values in the columns above.

protons make it possible to distinguish unequivocally between the coupling constants of the ring protons and those of the methyl protons. Some ambiguity exists, though, in the assignment of the coupling constants to particular sets of methyl or ring protons on the basis of the theoretical values.

H. $\Delta^{9,9'}$-BIFLUORENE

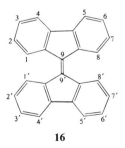

16

The radical cation of $\Delta^{9,9'}$-bifluorene (**16**) was prepared by treatment with SbCl$_5$ in methylene chloride.[60, 60a] The coupling constants derived from the ESR spectrum of **16**$^\oplus$, are listed in Table XVI, along with the corresponding

TABLE XVI

PROTON COUPLING CONSTANTS AND SPIN POPULATIONS OF $\Delta^{9,9'}$-BIFLUORENE
RADICAL IONS

μ	$a_{H\mu}^{\ominus}$	$a_{H\mu}^{\oplus}$	$c_{a,\mu}^{2}$	$c_{b,\mu}^{2}$	$\rho_{\mu}^{\ominus}$	$\rho_{\mu}^{\oplus}$
1,1′,8,8′	1.51[a]	2.14	0.034	0.067	0.038	0.092
2,2′,7,7′	0.53[b]	0.15[b]	0.017	0.002	0.005	−0.023
3,3′,6,6′	1.90	1.98	0.042	0.056	0.052	0.073
4,4′,5,5′	0.27[b]	0.46[b]	0.009	0.027	−0.007	0.017
	16.84[c]	18.92[c]	0.408[c]	0.608[c]	0.408[c]	0.820[c]

[a] Corrected value [L. S. Singer, private communication (1967)].
[b] Assignment uncertain.
[c] These numbers are four times the sums of absolute values in the columns above.

values for the radical anion **16**$^{\ominus}$, which was obtained by reduction of the parent compound with K in THF.[60a] The $a_{H\mu}$ values are rather small and are of similar magnitude in the two radical ions. Moreover, there is only moderate agreement between the calculated spin populations and the experimental data, and consequently the assignment of the coupling constants is somewhat uncertain. Experimental support for these assignments might be desirable.

I. INDENO[1,2,3-*cd*]FLUORANTHENE

17

The use of SbCl$_5$ in methylene chloride as a method for preparation of radical cations[60, 60a] has also been successful with indeno[1,2,3-*cd*]fluoranthene (**17**),[72] a dibenzo derivative of the nonalternant hydrocarbon pyracylene, which has recently been synthesized.[73] The coupling constants for **17**$^{\oplus}$ given in

[72] C. Elschenbroich and F. Gerson, *Helv. Chim. Acta* **53**, 838 (1970).
[73] B. M. Trost and G. M. Bright, *J. Am. Chem. Soc.* **89**, 4244 (1967). Indeno[1,2,3-*cd*]-fluoranthene was first synthesized by H. W. D. Stubbs and S. H. Tucker [*J. Chem. Soc.* p. 2936 (1951)].

Table XVII compare favorably with the calculated spin populations. The same is true for the corresponding values for the radical anion $17^{\ominus}$ obtained from the neutral compound **17** by electrolytic reduction in DMF. The assignment of coupling constants can thus be based on the calculated values, although the relatively small magnitudes of $a_{H\mu}$ make such an assignment

TABLE XVII

PROTON COUPLING CONSTANTS AND SPIN POPULATIONS OF
INDENO[1,2,3-*cd*]FLUORANTHENE RADICAL IONS

μ	$a_{H\mu}{}^{\ominus}$	$a_{H\mu}{}^{\oplus}$	$c_{a,\mu}{}^2$	$c_{b,\mu}{}^2$	$\rho_\mu{}^{\ominus}$	$\rho_\mu{}^{\oplus}$
1,4,7,10	0.33	0.15[a]	0.020	0.009	0.012	−0.012
2,3,8,9	0.92	1.99	0.030	0.053	0.030	0.059
5,6,11,12	1.68	0.70[a]	0.066	0.019	0.068	0.009
	11.72[b]	11.36[b]	0.464[b]	0.364[b]	0.440[b]	0.320[b]

[a] Assignment uncertain.
[b] These numbers are four times the sums of the absolute values in the columns above.

subject to some uncertainty. The spin population in $17^{\ominus}$ and $17^{\oplus}$ seems to lie at the "blind" centers even to a greater extent than in the radical ions $16^{\ominus}$ and $16^{\oplus}$ of $\Delta^{9,9'}$-bifluorene discussed before. This spin distribution accounts for the small overall ranges observed: 11.72 gauss for $17^{\ominus}$ and 11.36 gauss for $17^{\oplus}$.

J. HEPTAFULVALENE

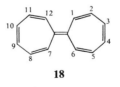

18

Recently the radical cation and anion of the nonalternant heptafulvalene (**18**) were prepared and investigated by ESR spectroscopy.[74] The cation $18^{\oplus}$ was generated by thermal cleavage of the dication dimer (from +30 to 60°C), whereas the anion was produced from **18** by the standard reaction with K in

[74] M. D. Sevilla, S. H. Flajser, G. Vincow, and H. J. Dauben, Jr., *J. Am. Chem. Soc.* **91**, 4139 (1969).

several ethereal solvents at −80°C. It was found that the radical anion **18**$^\ominus$ disappears if the temperature is raised to −50°C, but reappears at higher temperature (+30°C). This behavior is thought to be due to dimerization at intermediate temperature and subsequent cleavage on further warming. The coupling constants of the ring protons[74] are given in Table XVIII for **18**$^\ominus$ in THF solution at −90°C and for **18**$^\oplus$ in nitroethane at +50°C.

The HMO model of heptafulvalene exhibits some unusual features which have not been encountered in the nonalternant systems discussed so far. In the case of **18**, there are only six bonding HMO's available for the 14 π-electrons of the neutral compound, so that the lowest antibonding orbital ψ_a is the highest occupied one ($E_a = \alpha - 0.182\beta$). The next higher antibonding

TABLE XVIII

Proton Coupling Constants and Spin Populations[a] of Heptafulvalene Radical Ions

μ	$a_{\mathrm{H}\mu}{}^{\ominus}$	$a_{\mathrm{H}\mu}{}^{\oplus}$	$c_{a^*,\mu}{}^2$	$c_{a,\mu}{}^2$	$\rho_\mu{}^{\ominus}$	$\rho_\mu{}^{\oplus}$
1,6,7,12	4.18	0.08	0.136	0.039	0.184	0.016
2,5,8,11	<0.15	2.90	0.027	0.090	0.002	0.107
3,4,9,10	2.53	1.72	0.087	0.064	0.094	0.064
From 26.84[b]	18.80[b]	1.000[b]	0.772[b]	1.120[b]	0.748[b]
to 27.44[b]						

[a] Assuming that the unpaired electron occupies ψ_{a^*} in the anion, and ψ_a in the cation (see text).

[b] These numbers are four times the sums of the absolute values in the columns above.

HMO, denoted ψ_{a^*}, must therefore be taken as the lowest vacant orbital of the neutral heptafulvalene (**18**) ($E_{a^*} = \alpha - 0.445\beta$). This HMO belongs to a pair of accidently degenerate orbitals which can be represented as being localized on either of the two seven-membered rings. The two localized degenerate orbitals are therefore specified as $\psi_{a^*}{}^{(1)}$ and $\psi_{a^*}{}^{(2)}$.

It is assumed that ψ_a is the singly occupied orbital in the radical cation **18**$^\oplus$, and that in the radical anion **18**$^\ominus$ the odd electron occupies a delocalized orbital ψ_{a^*} to which the two degenerate localized HMO'S $\psi_{a^*}{}^{(1)}$ and $\psi_{a^*}{}^{(2)}$ contribute equally. The delocalized orbital ψ_{a^*} can thus be expressed as either $(1/\sqrt{2})(\psi_{a^*}{}^{(1)} + \psi_{a^*}{}^{(2)})$ or $(1/\sqrt{2})(\psi_{a^*}{}^{(1)} - \psi_{a^*}{}^{(2)})$. The electron distributions in $\psi_{a^*} = (1/\sqrt{2})(\psi_{a^*}{}^{(1)} + \psi_{a^*}{}^{(2)})$ and in ψ_a are schematically depicted in Fig. 21. Table XVIII shows that the assumptions made for the singly occupied orbitals of the radical ions **18**$^\ominus$ and **18**$^\oplus$ are justified by the good agreement between the coupling constants $a_{H\mu}$ and the calculated spin populations $c_{a*,\mu}^2$ and $c_{a,\mu}^2$.[74]

The good agreement between the experimental and the computed values also provides further support for the statement made previously (Section III,B and E) that the radical ions of nonalternants may be more aromatic than the parent neutral compounds. In contrast to heptafulvalene itself, which exhibits alternating single and double bonds, the radical ions $18^{\ominus}$ and $18^{\oplus}$ probably have extensive π-electron conjugation, in better accord with the HMO model. The ease with which the dimers of the two radical ions dissociate also points to this conclusion.

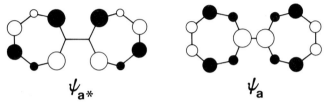

FIG. 21. Lowest antibonding orbitals ψ_{a*} and ψ_a of heptafulvalene (see text). Schematic representation as in Fig. 2.

In 1-methyl-THF solution at $-120°C$, the radical anion $18^{\ominus}$ gives rise to a spectrum which must be analyzed in terms of pairs of equivalent protons, with splittings which are approximately double those in the "normal" spectrum.[74] The spectrum of $18^{\ominus}$ in 1-methyl-THF can be ascribed to a radical anion in which the unpaired electron is localized on one of the seven-membered rings. The hyperfine structure of this spectrum is best rationalized in terms of a preferential occupancy of one of the degenerate localized HMO's $\psi_{a*}^{(1)}$ or $\psi_{a*}^{(2)}$ (see above). The localization of the odd electron on one seven-membered ring is thought to be due to association of the gegenion $K^{\oplus}$ with this particular ring, such a localization being favored by the use of a medium of low solvating power.

K. CONCLUDING REMARKS

The success of the MO predictions with respect to presence or absence of orbital pairing in π-electron systems can be dramatically illustrated by a comparison of the ratios $\sum_\mu a_{H\mu}^{\oplus}/\sum_\mu a_{H\mu}^{\ominus}$, i.e., the ratios of the overall spectral ranges due to the ring protons in the corresponding radical cations and anions. These ratios are listed in Table XIX for a number of alternant benzenoid aromatics and for the six nonalternants 12–18, together with the ratios $\sum_\mu c_{b,\mu}^2/\sum_\mu c_{a,\mu}^2$ and $\sum_\mu |\rho_\mu^{\oplus}|/\sum_\mu |\rho_\mu^{\ominus}|$ for the proton-carrying centers μ. (The relevant sums of the experimental and calculated values have been given at the bottoms of the appropriate columns in Tables XII–XVIII.) It is obvious that the ratio $\sum_\mu a_{H\mu}{}^\oplus / \sum_\mu a_{H\mu}{}^\ominus$ is nearly constant (1.09 ± 0.05) for the alternant benzenoid hydrocarbons. The deviation from the theoretical ratio of unity can be reasonably accounted for by the use of a 10% larger value of $|Q|$ for the

TABLE XIX

RADICAL CATIONS VS RADICAL ANIONS OF SOME ALTERNANT AND NONALTERNANT
HYDROCARBONS: RATIOS OF OVERALL SPECTRAL RANGES AND OF SUMMED
ABSOLUTE VALUES OF THE SPIN POPULATIONS AT PROTON-CARRYING CENTERS

Radical ions of	$\dfrac{\sum\limits_\mu a_{H\mu}{}^\oplus}{\sum\limits_\mu a_{H\mu}{}^\ominus}$	$\dfrac{\sum\limits_\mu c_{b,\mu}{}^2}{\sum\limits_\mu c_{a,\mu}{}^2}$	$\dfrac{\sum\limits_\mu |\rho_\mu{}^\oplus|}{\sum\limits_\mu |\rho_\mu{}^\ominus|}$	Table
Alternants[a]				
Anthracene	1.11	1.00	1.00	
Tetracene	1.12	1.00	1.00	
Pentacene	1.13	1.00	1.00	
Pyrene	1.09	1.00	1.00	
Perylene	1.08	1.00	1.00	
Coronene	1.04	1.00	1.00	
Nonalternants[b]				
Acepleiadylene 12	1.56	1.12	1.06	XII
Acenaphth[1,2-*a*]acenaphthylene 13	0.41	0.34	0.32	XIII
Acepleiadiene 14	0.84[c]	0.83[d]	0.91	XIV
3,5,8,10-Tetramethylcyclopenta[*ef*]-heptalene 15	5.07[c]	10.8[d]	5.03	XV
$\Delta^{9,9'}$-Bifluorene 16	1.12	1.49	2.01	XVI
Indeno[1,2,3-*cd*]fluoranthene 17	0.97	0.78	0.73	XVII
Heptafulvalene 18	0.69 (± 0.01)	0.77[e]	0.67	XVIII

[a] Data taken from F. Gerson, "Hochauflösende ESR-Spektroskopie, dargestellt anhand aromatischer Radikal-Ionen." Verlag Chemie, Weinheim, 1967 (English edition in press).

[b] Data taken from the Tables XII–XVIII of this section. The number of the relevant table is given in the last column.

[c] Coupling constants of ring protons only.

[d] Calculated for the unsubstituted systems: pleiadiene (instead of 14) and cyclopenta[*ef*]-heptalene (instead of 15).

[e] This number is the ratio $\sum_\mu c_{a,\mu}{}^2 / \sum_\mu c_{a*,\mu}{}^2$ (see Table XVIII).

cations. In striking contrast, the corresponding ratios $\sum_\mu a_{H\mu}{}^\oplus / \sum_\mu a_{H\mu}{}^\ominus$ for the nonalternant radical ions vary from 0.41 to 5.13, i.e., more than an order of magnitude. These variations are paralleled by the ratios $\sum_\mu c_{b,\mu}{}^2 / \sum_\mu c_{a,\mu}{}^2$ and $\sum_\mu |\rho_\mu{}^\oplus| / \sum_\mu |\rho_\mu{}^\ominus|$ which agree qualitatively—and in most cases even semiquantitatively—with the ratios of the experimental quantities.

IV. Compounds Related to [10]-, [14]-, and [18]Annulenes

A. 1,6-METHANO[10]ANNULENE

19

Full D_{10h} symmetry is highly unfavorable for a ten-membered cyclic π-electron system because of angle strain (144 degrees for the C—C—C angle, compared with an optimum angle of 120 degrees for sp^2 hybridization). This strain can be avoided if two trans double bonds are introduced, but only at the cost of severe van der Waals interference between the two inner hydrogens (see perimeter of symmetry D_{2h} in Fig. 22). As a consequence, cyclodecapentaene (or [10]annulene), the second representative in the series of $(4r + 2)$-membered

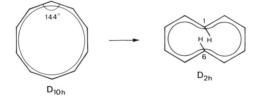

FIG. 22. Ten-membered perimeters of symmetry D_{10h} and D_{2h}.

perimeters $(r = 2)$ is known only as an unstable intermediate which cannot be isolated in pure form.[75] When a C-1 to C-6 σ-bond replaces the two inner hydrogens, the resulting compound, naphthalene, has strong π-conjugation across the ring and can hardly be considered as a ten-membered perimeter. However, if C-1 and C-6 are joined by an alkyl bridge, as in 1,6-methano[10]-annulene (**19**)[76] transannular conjugation is effectively interrupted. Although the perimeter of **19** has only C_{2v} symmetry and is no longer planar[77] (see Fig. 27) cyclic conjugation is not seriously inhibited.[76,78,79]

[75] E. E. van Tamelen and T. L. Burkroth, *J. Am. Chem. Soc.* **89**, 151 (1967); see, however, S. Masamune and R. T. Seidner, *Chem. Commun.* p. 542 (1969).

[76] E. Vogel and H. D. Roth, *Angew. Chem.* **76**, 145 (1964); *Angew. Chem. Intern. Ed. Engl.* **3**, 228 (1964).

[77] M. Dobler and J. D. Dunitz, *Helv. Chim. Acta* **48**, 1429 (1965).

[78] H.-R. Blattmann, W. A. Böll, E. Heilbronner, G. Hohlneicher, E. Vogel, and J.-P. Weber, *Helv. Chim. Acta* **49**, 2017 (1966).

[79] E. Vogel, *Chem. Soc. (London), Spec. Publ.* **21**, 113 (1967); E. Vogel and H. Günther, *Angew. Chem.* **79**, 429 (1967); *Angew. Chem. Intern. Ed. Engl.* **6**, 385 (1967).

As was pointed out in Section I,D the two lowest antibonding orbitals, ψ_{a+} and ψ_{a-}, of $(4r + 2)$-membered perimeters are degenerate, regardless of the MO model used, if the full $D_{(4r+2)h}$ symmetry is preserved. When the symmetry of the perimeter is reduced or when a bridge is introduced, the degeneracy is lifted; the two effects are inseparable experimentally. In this discussion the out-of-plane deformation of the bridged [10]annulene is neglected, and the π-system is treated as a planar ten-membered perimeter of D_{2h} symmetry.

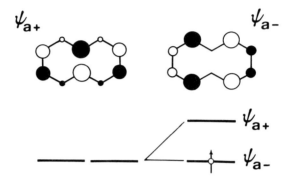

FIG. 23. Degenerate lowest antibonding orbitals of a ten-membered perimeter. Splitting and occupancy in the 1,6-methano[10]annulene radical anion. Schematic representation as in Fig. 2.

Moreover, if the HMO model, which is insensitive to in-plane deformations, is used, the degenerate D_{10h} perimeter orbitals

$$\psi_{a+} = 0.447(\phi_1 - \phi_6) - 0.138(\phi_2 - \phi_5 - \phi_7 + \phi_{10})$$
$$-0.362(\phi_3 - \phi_4 - \phi_8 + \phi_9)$$

and

$$\psi_{a-} = 0.425(\phi_2 + \phi_5 - \phi_7 - \phi_{10}) - 0.263(\phi_3 + \phi_4 - \phi_8 - \phi_9)$$

are not affected by the change of the perimeter to D_{2h} symmetry until a perturbation due to the bridging methylene group is introduced. As this group is electron-repelling, the energy of a π-orbital—to a first-order approximation —will be increased by an amount proportional to the squares of the coefficients at the bridged centers 1 and 6. It is evident from Fig. 23, in which the orbitals ψ_{a+} and ψ_{a-} are schematically represented, that ψ_{a+} should be significantly destabilized $(c_{a+,1}{}^2 = c_{a+,6}{}^2 = 0.200)$, while ψ_{a-}, which has a nodal plane passing through C-1 and C-6 $(c_{a-,1}{}^2 = c_{a-,6}{}^2 = 0)$, should be unchanged. Thus the HMO model predicts that the unpaired electron will occupy ψ_{a-}. This prediction is confirmed by the ESR spectra of the radical anions of **19** and its 2,5,7,10-tetradeuterio derivative[80] (see Fig. 24).

[80] F. Gerson, E. Heilbronner, W. A. Böll, and E. Vogel, *Helv. Chim. Acta* **48**, 1494 (1965).

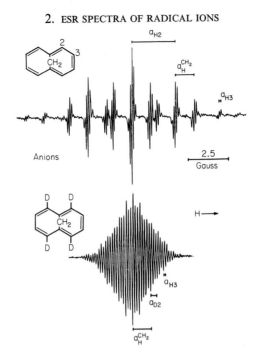

FIG. 24. ESR spectra of the radical anions of 1,6-methano[10] annulene and its 2,5,7,10-tetradeuterio derivative. Solvent: DME; gegenion: Na⊕; temperature: −70°C.

The experimental data are given in Table XX, where they are compared with the values $c_{a+,\mu}^2$ and $c_{a-,\mu}^2$, as well as with the corresponding McLachlan spin populations $(\rho_\mu)_+$ and $(\rho_\mu)_-$, all for the unperturbed ten-membered perimeter.

TABLE XX

PROTON COUPLING CONSTANTS AND SPIN POPULATIONS OF THE
1,6-METHANO[10]ANNULENE RADICAL ANION

μ	$a_{H\mu}$	$a_H^{CH_2}$	$c_{a+,\mu}^2$	$c_{a-,\mu}^2$	$(\rho_\mu)_+$	$(\rho_\mu)_-$
1,6	—	1.15	0.200	0.000	0.270	−0.070
2,5,7,10	2.70	—	0.019	0.181	−0.037	0.237
3,4,8,9	0.10	—	0.131	0.069	0.152	0.048

There is qualitative agreement between the coupling constants and the theoretical spin populations $c_{a-,\mu}^2$ and $(\rho_\mu)_-$ computed for the singly occupied orbital ψ_{a-}. However, application of the McConnell relation [Eq. (1) in Section I,A] with $|Q| = 23$–30 gauss gives calculated coupling constants for the ring protons which are much larger than the measured values. Since the orbital ψ_{a-} of the ten-membered perimeter is identical with the lowest antibonding orbital of naphthalene, one would expect the coupling constants for the two sets of four equivalent protons in **19**$^{\ominus}$ to be of about the same magnitude as those observed for the corresponding protons of the naphthalene radical anion: that is, $a_{H\mu} \approx 4.9$ and 1.8 gauss for $\mu = 2,5,7,10$ and 3,4,8,9, respectively. The splittings listed in Table XX, however, are much smaller, the deviations from the expected values being far larger than those usually observed (see Fig. 1 in Section I,B). Apparently, the nonplanarity of the ten-membered ring in **19**$^{\ominus}$, although it seems to have little effect on the cyclic conjugation, does strongly affect the coupling constants of the ring protons.

As mentioned in Section I,A and E, the 1s-spin population, which gives rise to hyperfine splittings by ring protons (α-protons) in planar π-radicals, is due to an indirect mechanism, the so-called π–σ-spin polarization. This spin population has therefore a sign opposite to that of the π-spin population ρ_μ at the adjacent carbon center μ. [The opposite sign of the 1s-spin population manifests itself in the negative sign of the Q value of the McConnell relation; see Eq. (1) in Section I,A.] On the other hand, in nonplanar radicals like the anion **19**$^{\ominus}$, there is a lack of complete orthogonality between the π-electron system and the C—H σ-bonds, which permits a direct leakage of spin population from the π-orbital into the 1s-AO's of the ring hydrogens. The 1s-spin population brought about by this direct mechanism has the same sign as the π-spin population ρ_μ at the carbon center μ (like the 1s-spin population at the β-protons of alkyl substituents; see Section I,E). As a result, the 1s-spin population, which is due to spin polarization and has an opposite sign, will be partially cancelled, and the observed absolute value of the proton coupling constant will be reduced. Such a hypothesis has been confirmed in principle by calculations[81] which make use of the actual geometry of **19**$^{\ominus}$.[77] It is also supported by the magnitude of the coupling constant of a ^{13}C nucleus in one of the four equivalent centers 2, 5, 7, and 10. This value, 6.9 gauss,[82] is very close to the corresponding value of 7.1 gauss observed for the naphthalene radical anion.

The experimental data listed in Table XX were obtained for solutions of **19**$^{\ominus}$ in DME with Na$^{\oplus}$ as gegenion at $-80°C$. They are very sensitive to the temperature as well as to the nature of the solvent and the gegenion.[82] It can be shown that the spectrum of **19**$^{\ominus}$ in DME/Na$^{\oplus}$, over the range -120 to $+40°C$, must be ascribed to a radical anion which is not associated with the gegenion, and thus the variations in the coupling constants cannot be a consequence of ion-pairing. Figures 25 and 26 illustrate the temperature-dependence of the coupling constants under the experimental conditions given above.[82]

[81] F. Gerson and R. Gleiter, unpublished results (1969).
[82] F. Gerson, J. Heinzer, and E. Vogel, in preparation.

As can be seen, the magnitude of $a_{H\mu}$ decreases for $\mu = 2,5,7,10$ and increases for $\mu = 3,4,8,9$ with increasing temperature. One is tempted to rationalize these results by assuming that the splitting of the energy levels of ψ_{a+} and ψ_{a-} is small in **19**$^{\ominus}$, and consequently ψ_{a+} can contribute to the spin populations by

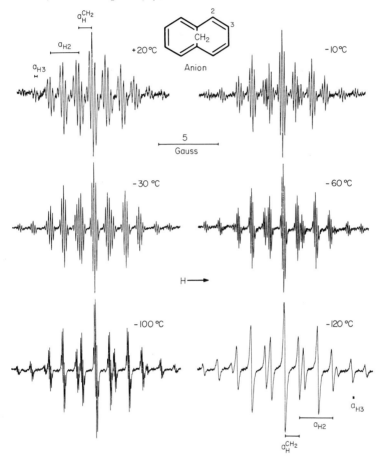

FIG. 25. ESR spectra of the 1,6-methano[10]annulene radical anion taken at various temperatures. Solvent: DME; gegenion: Na$^{\oplus}$.

means of thermal mixing. This contribution should increase with increasing temperature and should lead to changes like those observed experimentally.

Although the absolute values of the coupling constants are slightly different for the system THF/K$^{\oplus}$, in which **19**$^{\ominus}$ is associated with a K$^{\oplus}$ gegenion, a similar temperature-dependence is found. The additional hyperfine splitting by the ^{39}K nucleus in this system is about 0.08–0.09 gauss. The ESR spectra of

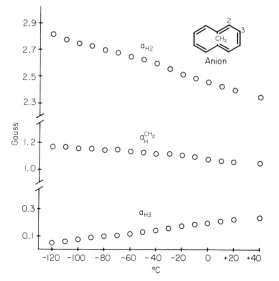

FIG. 26. Proton coupling constants of 1,6-methano[10]annulene vs temperature.

19⊖ in DME/K⊕ and THF/Na⊕ are rather complex, since in the temperature range from −40° to +30°C, a transition occurs from nonassociated to ion-paired radical anions.[82]

B. 1,6-Imino- and 1,6-Oxido[10]annulenes

20   X = NH
21   X = O

The methylene group in **19** can be replaced by a heteroatom bridge without any essential change in the electronic structure of the cyclic π-system.[78,79] Conjugation between this system and the lone pair(s) of the heteroatom in 1,6-imino-[83] and 1,6-oxido[10]annulene[84] (**20** and **21**, respectively) must therefore be markedly inhibited by the special geometrical arrangement of the bridge (Fig. 27). The effect of the bridging NH- and O-groups on the ten-

[83] E. Vogel, W. Pretzer, and W. A. Böll, *Tetrahedron Letters* **40**, 3613 (1965).

[84] E. Vogel, M. Biskup, W. Pretzer, and W. A. Böll, *Angew. Chem.* **76**, 785 (1964); *Angew. Chem. Intern. Ed. Engl.* **3**, 642 (1964); F. Sondheimer and A. Shani, *J. Am. Chem. Soc.* **86**, 3168 (1964).

membered perimeter is thus expected to be essentially inductive, and the ESR spectra of the radical anions **20**$\ominus$ and **21**$\ominus$ should give information on the direction of this effect. As will be shown below, the spin distribution in the two radical anions is very similar to that in the hydrocarbon analog **19**$\ominus$, so that the singly occupied orbital of **20**$\ominus$ and **21**$\ominus$ is also ψ_{a-}, as in the case of **19**$\ominus$. Consequently the overall inductive effect of the heteroatom bridges in **20** and **21** must have the same direction as that of the methylene group in **19**,

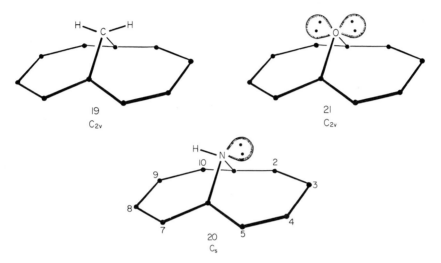

FIG. 27. Molecular models of 1,6-bridged [10]annulenes.

i.e., it must also be electron-repelling. An effect of this kind can come about only if the repulsion between the lone-pair electrons[85,86] and the π-system overrides the electron attraction of the electronegative heteroatoms which is propagated through the σ-bonds.

Table XXI lists the experimental data for **20**$\ominus$[87] and **21**$\ominus$[80,82]; those for **19**$\ominus$ are repeated for comparison. All values refer to DME solutions at $-80°C$ with Na$\oplus$ as gegenion; none of the radical anions **19**$\ominus$, **20**$\ominus$, or **21**$\ominus$ is believed to be associated with the cation at this temperature.

The assignment of the coupling constants of the ring protons in **20**$\ominus$ and **21**$\ominus$ is based on the ESR spectra of the radical anions of the 2,5,7,10-tetra-deuterio derivatives, which show unambiguously that the larger coupling constants for both anions must be assigned to the protons at centers 2,5,7,

[85] D. T. Clark, J. N. Murrell, and J. M. Tedder, *J. Chem. Soc.* p. 1250 (1963); D. T. Clark, *Chem. Commun.* p. 390 (1966).

[86] J. H. Hammons, *J. Org. Chem.* **33**, 1123 (1968).

[87] F. Gerson, J. Heinzer, and E. Vogel, *Helv. Chim. Acta* **53**, 95 (1970).

and 10.[80, 87] However, the imino compound **20** has lower symmetry than the methano and oxido analogs (**19**) and (**21**): C_s instead of C_{2v}, as indicated by Fig. 27. The lower symmetry makes each set of four equivalent protons separate into two pairs; consequently the radical anion **20**$^\ominus$ has four coupling constants for four pairs of equivalent ring protons, instead of two coupling constants for two sets of four equivalent protons. There are two possible assignments for the larger $a_{H\mu}$ values ($\mu = 2,5$ and $7,10$) and, likewise, the two smaller $a_{H\mu}$ values can be assigned in either of two ways ($\mu = 3,4$ and $8,9$). The fact that the splittings of 0.14 and 2.86 gauss lie close to the corresponding values for **19**$^\ominus$, and those of 0.28 and 3.28 are similar to the values for **21**$^\ominus$

TABLE XXI

RING PROTON COUPLING CONSTANTS OF 1,6-BRIDGED
[10]ANNULENES (X = BRIDGING GROUP)

μ	$a_{H\mu}$ (X = CH$_2$)	$a_{H\mu}$ (X = NH)	$a_{H\mu}$ (X = O)
2,5,7,10	2.70	2.86^a; 3.28^a	3.44
3,4,8,9	0.10	0.14^a; 0.28^a	0.42

^{a} Each value for two equivalent protons.

suggests the following assignment (see Table XXI and Fig. 27): $a_{H2} = a_{H5} = 3.28$; $a_{H3} = a_{H4} = 0.28$; $a_{H7} = a_{H10} = 2.86$; and $a_{H8} = a_{H9} = 0.14$ gauss. This assignment is based on the assumption that the chemical environment of the ring protons which are next to the lone pair in **20**$^\ominus$ should resemble the environment of the corresponding protons in **21**$^\ominus$, whereas the ring protons next to the N—H bond should be like those in **19**$^\ominus$.

The coupling constant of the NH proton in **20**$^\ominus$, a_H^{NH}, is 0.56 gauss at $-80°C$ (compared with the coupling constant of the CH$_2$ protons in **19**$^\ominus$, $a_H^{CH_2}$, which is 1.15 gauss). No splitting is observed from the ^{14}N nucleus of the NH bridging group; the coupling constant a_N must thus be smaller than 0.05 gauss.[87]

The increase in magnitude of the coupling constants in the sequence **19**$^\ominus$, **20**$^\ominus$, **21**$^\ominus$ may imply a parallel flattening of the ten-membered perimeter, if the arguments used to rationalize the "reduced" values of the $a_{H\mu}$'s in **19**$^\ominus$ are also valid for the two heteroatom-bridged analogs.

Extensive studies of temperature-dependence have been possible for **20**$^\ominus$, but not for **21**$^\ominus$, which is too unstable at temperatures above $-70°C$.[82] The results obtained in the case of **20**$^\ominus$ are similar to those found for **19**$^\ominus$. An increase in temperature again makes the larger proton coupling constants $a_{H\mu}$ ($\mu = 2,5$ and $7,10$) decrease and the smaller ones ($\mu = 3,4$ and $8,9$) increase; the coupling constant a_H^{NH} shows only a slight change.

With respect to stability, the half-lives of the radical anions of the methano, imino, and oxido compounds are days, hours, and minutes, respectively. The main conversion product from **20**$^\ominus$ and **21**$^\ominus$ is naphthalene, which has been identified by the observation of its radical anion.[80, 87]

C. Cycl[3,2,2]azine

| 22a | 22 | 22b |

Cycl[3,2,2]azine (**22**) is the best known example of a compound consisting of a cyclic π-electron system of carbon centers held together by a central sp^2-hybridized nitrogen atom.[88] This description implies a close relation between cycl[3,2,2]azine and bridged annulenes; i.e., the compound **22** can be treated as an amino-substituted ten-membered perimeter (**22a**). However, the extensive conjugation between the nitrogen lone pair and the cyclic π-electron system suggests that the model of a perturbed perimeter is not entirely adequate. Cycl[3,2,2]azine may instead be considered as isoelectronic with the nonalternant aceindylenyl anion (**22b**). In fact, some of its properties can be well rationalized in terms of **22a**, whereas others are more satisfactorily accounted for by **22b**.[89] Thus **22** is related both to odd nonalternants (Section VI) and to annulenes (Section IV).

Figure 28 shows the ESR spectra of the radical anions of cycl[3,2,2]azine (**22**) and its 1,4-dideuterio derivative, produced by the reduction of the parent compounds with Li in DME.[90] Identical spectra are obtained when Na is used as the reducing agent, while the use of K yields slightly different spectra.[15]

If one tentatively applies the perimeter model (**22a**) to cycl[3,2,2]azine, the two lowest antibonding orbitals,

$$\psi_{a+} = 0.362(\phi_1 + \phi_4 - \phi_8 - \phi_{10}) + 0.138(\phi_2 + \phi_3 - \phi_5 - \phi_7)$$
$$+ 0.447(\phi_6 - \phi_9)$$

[88] R. J. Windgassen, Jr., W. H. Saunders, Jr., and V. Boekelheide, *J. Am. Chem. Soc.* **81**, 1459 (1959).

[89] F. Gerson, E. Heilbronner, N. Joop, and H. Zimmermann, *Helv. Chim. Acta* **46**, 1940 (1963); V. Boekelheide, F. Gerson, E. Heilbronner, and D. Meuche, *ibid.* p. 1951.

[90] F. Gerson and J. D. W. van Voorst, *Helv. Chim. Acta* **46**, 2257 (1963).

and

$$\psi_{a-} = 0.263(\phi_1 - \phi_4 + \phi_8 - \phi_{10}) - 0.425(\phi_2 - \phi_3 - \phi_5 + \phi_7)$$

of the ten-membered perimeter must again be considered (Figure 29). It is obvious, that, for reasons of symmetry, the orbital of the nitrogen lone pairs, ψ_n, cannot interact with the antisymmetric orbital ψ_{a-}, which has a nodal plane passing through the nitrogen, but can mix with the symmetric one, ψ_{a+}. The mixing between ψ_n and ψ_{a+} will lead to two modified orbitals ψ_n' and ψ_{a+}', the

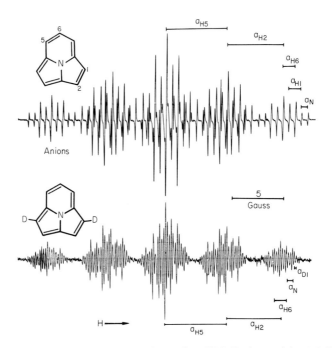

FIG. 28. ESR spectra of the radical anions of cycl[3,2,2]azine and its 1,4-dideuterio derivative. Solvent: DME; gegenion: Li⊕; temperature: −60°C.

energy levels of which are lowered and raised, respectively, with regard to those of ψ_n and ψ_{a+}. The orbital ψ_{a+}' should thus have higher energy than ψ_{a-} which will remain unaffected in this model. As ψ_n' must house the nitrogen lone pair, ψ_{a-} will be the lowest vacant orbital of the neutral compound **22**, and the singly occupied orbital of the radical anion **22**⊖ (see Fig. 29).

The unpaired electron should thus occupy an orbital in which the electron distribution is very similar to that in the antisymmetric orbital ψ_{a-} of a ten-membered perimeter. Comparison of the experimental data for **22**⊖[90] with the values $c_{a-,\mu}^2$ and $(\rho_\mu)_-$ indicates that this model is appropriate (see Table XXII).

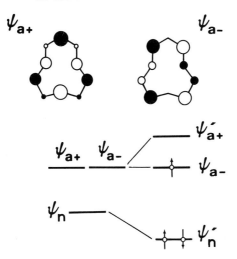

FIG. 29. Degenerate lowest antibonding orbitals of the ten-membered perimeter. Splitting and occupancy in the cycl[3,2,2]azine radical anion. Schematic representation as in Fig. 2.

Since the perimeter orbital ψ_{a-} has a nodal plane through the center 6 and the nitrogen, it is identical with the lowest antibonding orbital of aceindylenyl anion. Thus in this special case, HMO calculations yield the same spin populations with either the perimeter (22a) or the aceindylenyl (22b) model. The McLachlan method, however, gives slightly different spin populations, as can be seen from Table XXII. In particular, the ρ_μ values for $\mu = 5,7$ become somewhat larger than those for $\mu = 2,3$, so that the assignment of the largest coupling constant (6.02 gauss) to the pair of protons at the former centers, and the next

TABLE XXII

PROTON AND ^{14}N COUPLING CONSTANTS, AND SPIN POPULATIONS OF CYCL[3,2,2]AZINE RADICAL ANION

μ	$a_{H\mu}$	a_N	$c_{a-,\mu}^2$	$(\rho_\mu)-$	$\rho_\mu{}^a$
1,4	1.13	—	0.069	0.048	0.045
2,3	5.34	—	0.181	0.237	0.233
5,7	6.02	—	0.181	0.237	0.239
6	1.20	—	0.000	−0.070	−0.072
N	—	0.60	—	—	−0.017

^{a} Calculated with nitrogen as a part of the π-electron system (aceindylenyl anion model): $\alpha_N = \alpha + \beta$.

largest one (5.34 gauss) to the two protons at the latter centers, seems preferable
to the reverse assignment.

D. syn-1,6:8,13-Bisoxido[14]annulene

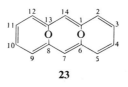

23

[14]Annulene is the third representative of the $(4r + 2)$-membered perimeters
$(r = 3)$. Like [10]annulene, a [14]annulene of symmetry D_{2h}, i.e., with C—C—C
angles of 120 degrees, should be subject to considerable strain resulting from
van der Waals interference of the inner hydrogens (see Forms A and B in
Fig. 30). The successful synthesis of [14]annulene by Sondheimer and Gaoni[91]

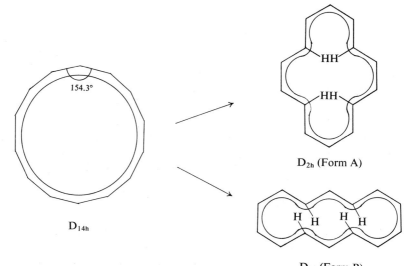

FIG. 30. Fourteen-membered perimeters of symmetry D_{14h} and D_{2h}.

indicates that the steric strain in Form A of [14]annulene is less severe than in
[10]annulene. However, this steric strain is sufficient to cause the perimeter
to be far from planar.[92] The instability and other properties of [14]annulene
point to olefinic character,[91, 92] although recent proton resonance studies at

[91] F. Sondheimer and Y. Gaoni, *J. Am. Chem. Soc.* **82**, 5765 (1960).

[92] L. M. Jackman, F. Sondheimer, Y. Amiel, D. A. Ben-Efraim, Y. Gaoni, R. Wolovsky,
and A. A. Bothner-By, *J. Am. Chem. Soc.* **84**, 4307 (1962).

low temperatures indicate the presence of a diamagnetic ring current.[93] To date no ESR spectrum of a [14]annulene radical ion has been reported.

The van der Waals interference is eliminated in pyrene, in which the four inner hydrogens are removed, but this hydrocarbon can no longer be considered as a perimeter because of the extensive conjugation of the inner double bond with the outer π-system. This conjugation is not present in 15,16-dihydropyrene, the dimethyl derivative, which will be discussed in Section IV,E.

The second possible arrangement of the fourteen-membered perimeter, Form B, is subject to a still greater amount of steric interference of the inner

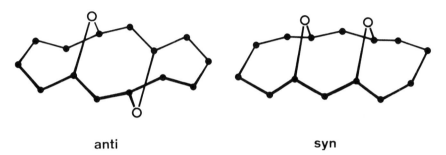

anti   syn

FIG. 31. Molecular models of *anti-* and *syn*-1,6:8,13-bisoxido[14]annulenes.

hydrogens, this interference being comparable to that in [10]annulene; consequently, Form B should be very difficult to synthesize. Again, replacement of the inner hydrogens by direct crosslinks leads not to a true perimeter, but to anthracene. However, introduction of two bridges connecting center 1 with 6 and center 8 with 13 results in a doubly bridged perimeter which is the next higher homolog of the singly bridged [10]annulenes discussed before. The bridging groups can be the same or similar as in the case of [10]annulenes and can be situated on the same side (*syn*; symmetry C_{2v}) or on opposite sides (*anti*; symmetry C_{2h}) of the plane approximately defined by the π-centers. Molecular models clearly show that the fourteen-membered ring in the *anti*-isomers is far from planar, whereas in the *syn*-isomers this ring can be more nearly planar than the perimeter of 1,6-bridged [10]annulenes (see Fig. 31).[94] The physical properties and chemical behavior of the isomers are consistent with the difference in coplanarity of the fourteen-membered rings; only the *syn*-isomers have the properties characteristic of aromatic compounds.[79, 95]

[93] Y. Gaoni, A. Melera, F. Sondheimer, and R. Wolovsky, *Proc. Chem. Soc.* p. 397 (1964).

[94] P. Ganis and J. D. Dunitz, *Helv. Chim. Acta* **50**, 2369 (1967).

[95] E. Vogel, M. Biskup, A. Vogel, and H. Günther, *Angew. Chem.* **78**, 755 (1966); *Angew. Chem. Intern. Ed. Engl.* **5**, 734 (1966).

The radical anion of *syn*-1,6:8,13-bisoxido[14]annulene (**23**),[95] which can readily be generated electrolytically or by alkali metal reduction, has been investigated by ESR spectroscopy.[96] The distance between the two bridging oxygens (2.5 Å)[94] is well suited for formation of a complex of **23**⊖ with the gegenion (Fig. 31). Accordingly, the ESR spectra of **23**⊖ in DME solution

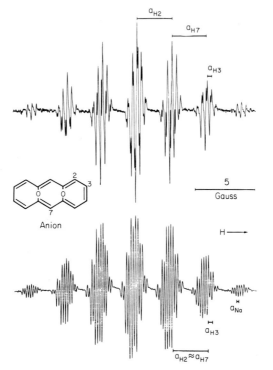

FIG. 32. ESR spectra of *syn*-1,6:8,13-bisoxido[14]annulene radical anion. Upper spectrum: nonassociated radical anion. Solvent: DMF; gegenion: $(C_2H_5)_4N^{\oplus}$; room temperature. Lower spectrum; associated radical anion. Solvent: DME; gegenion: Na⊕; temperature: −80°C.

with K⊕ or Na⊕ as gegenion plainly demonstrate the existence of ion-pairs. On the other hand, if **23**⊖ is generated electrolytically in DMF with $(C_2H_5)_4N^{\oplus}$ as gegenion, the hyperfine structure is due to a nonassociated radical ion. Figure 32 shows the spectra of **23**⊖ as a nonassociated and an ion-paired species.[96]

In Table XXIII the proton coupling constants are given for the electrolytically generated, nonassociated radical anion **23**⊖. They differ only slightly from those observed for the associated species in DME/Na⊕ or DME/K⊕. The

[96] F. Gerson, J. Heinzer, and E. Vogel, *Helv. Chim. Acta* **53**, 103 (1970).

additional splittings due to the ^{23}Na and ^{39}K nuclei amount to 0.20 and 0.04 gauss, respectively. The assignment for the two sets of four equivalent protons ($\mu = 2,5,9,12$ and $3,4,10,11$) is based on the ESR spectra of 2,5,9,12-derivatives of **23** $^\ominus$. Table XXIII also contains the values $c_{a+,\mu}{}^2$, $c_{a-,\mu}{}^2$, $(\rho_\mu)_+$, and $(\rho_\mu)_-$,

TABLE XXIII

PROTON AND ^{13}C COUPLING CONSTANTS, AND SPIN POPULATIONS OF *syn*-1,6:8,13-BISOXIDO[14]ANNULENE RADICAL ANION

μ	$a_{H\mu}$	$c_{a+,\mu}{}^2$	$c_{a-,\mu}{}^2$	$(\rho_\mu)_+$	$(\rho_\mu)_-$	$a_{C\mu}$
2,5,9,12	2.95	0.116	0.027	0.152	−0.009	5.7^a
3,4,10,11	0.37	0.056	0.088	0.043	0.100	^{b}
7,14	2.88	0.143	0.000	0.201	−0.058	7.3
1,6,8,13	—	0.007	0.135	−0.045	0.188	4.9^a

^{a} Assignment based on the relationship between $a_{C\mu}$ and calculated spin populations $(\rho_{\mu+})$.
^{b} Corresponding ^{13}C satellites not detected.

calculated for the two degenerate lowest antibonding orbitals of the fourteen-membered perimeter:

$$\psi_{a+} = -0.084(\phi_1 + \phi_6 + \phi_8 + \phi_{13}) - 0.341(\phi_2 + \phi_5 + \phi_9 + \phi_{12})$$
$$+ 0.236(\phi_3 + \phi_4 + \phi_{10} + \phi_{11}) + 0.378(\phi_7 + \phi_{14})$$

and

$$\psi_{a-} = 0.368(\phi_1 - \phi_6 + \phi_8 - \phi_{13}) - 0.164(\phi_2 - \phi_5 + \phi_9 - \phi_{12})$$
$$- 0.296(\phi_3 - \phi_4 + \phi_{10} - \phi_{11})$$

The two orbitals are shown schematically in Fig. 33.

Comparison of the experimental data with the calculated values clearly indicates that the symmetric orbital, ψ_{a+}, and not the antisymmetric one, ψ_{a-}, is occupied by the odd electron in **23** $^\ominus$. The orbital ψ_{a+} must therefore be of lower energy than ψ_{a-}. This result is predicted theoretically if arguments analogous to those employed for the radical anion of 1,6-oxido[10]annulene (**21** $^\ominus$) (Section IV,B), are applied to the higher homolog (**23** $^\ominus$). Again, the electronic effect of the bridging is considered. If the two oxygen atoms have an overall electron-repelling effect, like the bridging oxygen in **21** $^\ominus$, ψ_{a+} will be destabilized less than ψ_{a-}: $c_{a+,\bar\mu}{}^2 = 0.007$ and $c_{a-,\bar\mu}{}^2 = 0.135$, where $\bar\mu = 1,6,8,13$. The unpaired electron in **23** $^\ominus$ should thus occupy ψ_{a+}, this conclusion being in agreement with experiment.[96]

The coupling constants a_{H_μ} of **23**$^\ominus$ are smaller than those expected by the use of the McConnell relation [Eq. (1)]. However, the apparent reduction of the $|Q|$ value is much less pronounced than in the case of **19**$^\ominus$, **20**$^\ominus$, and **21**$^\ominus$. This fact can be related to the degree of nonplanarity by the application of the arguments of Section IV,A. Accordingly to these arguments, the doubly bridged [14]annulene **23** should be more nearly planar than the singly bridged [10]annulenes **19, 20,** and **21,** a prediction which has essentially been confirmed by X-ray analysis.[77, 94]

Coupling constants of ^{13}C nuclei in all but one of the sets of equivalent carbon centers were also measured for the radical anion **23**$^\ominus$.[96] They are listed in the last column of Table XXIII; their assignment is partly based on the relative intensities of corresponding ^{13}C satellites and partly on calculations.

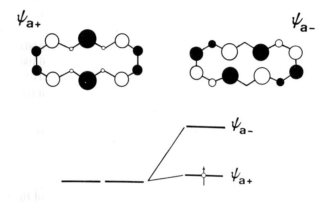

FIG. 33. Degenerate lowest antibonding orbitals of the fourteen-membered perimeter. Splitting and occupancy in *syn*-1,6:8,13-bisoxido[14]annulene radical anion. Schematic representation as in Fig. 2.

As has been observed for the radical anions of 1,6-bridged [10]annulenes **19**$^\ominus$ and **20**$^\ominus$, the large coupling constants a_{H_μ} ($\mu = 2,5,9,12$ and $7,14$) decrease and the smaller one ($\mu = 3,4,10,11$) increases with increasing temperature. In contrast to **19**$^\ominus$ and **20**$^\ominus$,[82] the temperature dependence is, however, very weak; for instance, a rise in temperature from -80 to $+25°$C causes changes of from only 0.02 to 0.04 gauss in the coupling constants.[96] This observation would suggest less mixing of ψ_{a+} and ψ_{a-} in **23**$^\ominus$ than of the corresponding lowest antibonding orbitals in **19**$^\ominus$ and **20**$^\ominus$; the implication is that the splitting of the energy levels of ψ_{a+} and ψ_{a-} should be larger for **23**$^\ominus$ than for **19**$^\ominus$ and **20**$^\ominus$.

E. *trans*-15,16-DIMETHYL-15,16-DIHYDROPYRENE

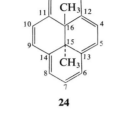

24

As previously mentioned, the dimethyl derivative **24** of 15,16-dihydro-pyrene[97] is an example of a bridged fourteen-membered perimeter of Form *A* (Fig. 30). The π-electron centers in **24** lie within 0.05 Å of coplanarity,[98] so that cyclic conjugation is expected to be practically unimpaired. (See Fig. 34,

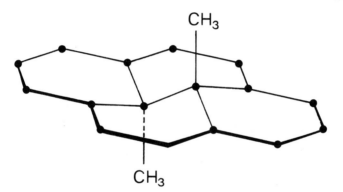

FIG. 34. Molecular model of *trans*-15,16-dimethyl-15,16-dihydropyrene.

which shows a molecular model of the compound.) In fact, the physical and chemical properties of **24** (electronic and proton resonance spectra, ease of substitution) point to a high degree of aromatic character.[99, 100] Not only the radical anion **24**⊖, but also the radical cation **24**⊕ is readily prepared by

[97] For the synthesis of *trans*-15,16-dimethyl-15,16-dihydropyrene, see V. Boekelheide and J. B. Phillips, *J. Am. Chem. Soc.* **89**, 1695 (1967).

[98] A. W. Hanson, *Acta Cryst.* **18**, 599 (1965).

[99] H.-R. Blattmann, V. Boekelheide, E. Heilbronner, and J.-P. Weber, *Helv. Chim. Acta* **50**, 68 (1967).

[100] J. B. Phillips, R. J. Molyneux, E. Sturm, and V. Boekelheide, *J. Am. Chem. Soc.* **89**, 1704 (1967).

standard chemical methods, the former by alkali metal reduction in DME, and the latter by oxidation in concentrated sulfuric acid.[101]

The availability of both radical ions gives information about the effect of the bridging on two sets of doubly degenerate perimeter orbitals: the lowest antibonding set, ψ_{a+} and ψ_{a-}, and the highest bonding set, ψ_{b+} and ψ_{b-}. These

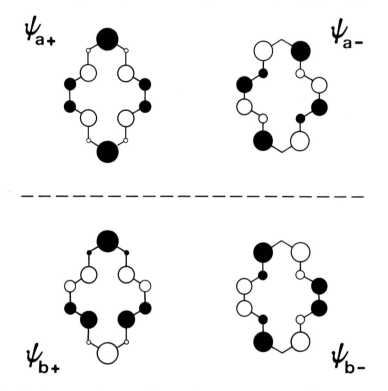

FIG. 35. Degenerate lowest antibonding (ψ_{a+} and ψ_{a-}) and degenerate highest bonding (ψ_{b+} and ψ_{b-}) orbitals of the fourteen-membered perimeter. Schematic representation as in Fig. 2.

orbitals, which are schematically depicted in Fig. 35, are represented by the following functions:

$$\psi_{a+} = -0.084(\phi_1 + \phi_3 + \phi_6 + \phi_8) + 0.378(\phi_2 + \phi_7)$$
$$+ 0.236(\phi_4 + \phi_5 + \phi_9 + \phi_{10}) - 0.341(\phi_{11} + \phi_{12} + \phi_{13} + \phi_{14})$$

$$\psi_{a-} = -0.368(\phi_1 - \phi_3 + \phi_6 - \phi_8) - 0.296(\phi_4 - \phi_5 + \phi_9 - \phi_{10})$$
$$+ 0.164(\phi_{11} - \phi_{12} + \phi_{13} - \phi_{14})$$

[101] F. Gerson, E. Heilbronner, and V. Boekelheide, *Helv. Chim. Acta* **47**, 1123 (1964).

$$\psi_{b+} = 0.084(\phi_1 + \phi_3 - \phi_6 - \phi_8) + 0.378(\phi_2 - \phi_7)$$
$$- 0.236(\phi_4 - \phi_5 - \phi_9 + \phi_{10}) - 0.341(\phi_{11} + \phi_{12} - \phi_{13} - \phi_{14})$$

and

$$\psi_{b-} = 0.368(\phi_1 - \phi_3 - \phi_6 + \phi_8) + 0.296(\phi_4 + \phi_5 - \phi_9 - \phi_{10})$$
$$+ 0.164(\phi_{11} - \phi_{12} - \phi_{13} + \phi_{14})$$

It can be seen that

$$c_{a+,\,\mu}{}^2 = c_{b+,\,\mu}{}^2 \quad \text{and} \quad c_{a-,\,\mu}{}^2 = c_{b-,\,\mu}{}^2$$

these equalities being specific examples of the "pairing" properties of the orbitals in all alternant systems.

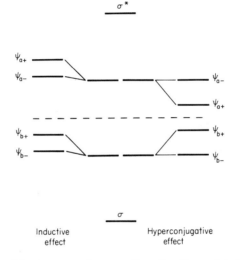

FIG. 36. Splitting of the degenerate lowest antibonding (ψ_{a+} and ψ_{a-}) and the degenerate highest bonding (ψ_{b+} and ψ_{b-}) perimeter orbitals by the inductive and the hyperconjugative effects in *trans*-15,16-dimethyl-15,16-dihydropyrene.

To a first-order approximation, the effect of the perturbation on the relative energies of the degenerate orbitals depends on the squares of the coefficients at the bridged centers $\bar{\mu} = 11,12,13,14$: $c_{a+,\bar{\mu}}{}^2 = c_{b+,\bar{\mu}}{}^2 = 0.116$, and $c_{a-,\bar{\mu}}{}^2 = c_{b-,\bar{\mu}}{}^2 = 0.027$. Since the central alkyl bridge in **24** would be expected to be electron-repelling, the symmetric orbitals ψ_{a+} and ψ_{b+} should be more strongly destabilized than the antisymmetric ones ψ_{a-} and ψ_{b-} (Fig. 36, left). On the basis of this assumption, the unpaired electron in the radical anion **24**$^\ominus$ should occupy the more stable of the two antibonding orbitals, ψ_{a-}. The radical cation **24**$^\oplus$, on the other hand, should be formed by removal of an electron from the less stable of the two bonding orbitals, ψ_{b+}, so that this orbital should house the unpaired electron.

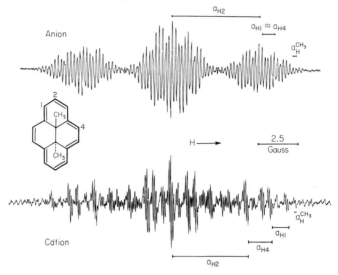

FIG. 37. ESR spectra of *trans*-15,16-dimethyl-15,16-dihydropyrene radical ions. Anion, solvent: DME; gegenion: Na⊕; temperature: −60°C. Cation, solvent: concentrated H_2SO_4; room temperature.

The ESR spectra of the two radical ions are shown in Fig. 37. The coupling constants $a_{H\mu}$ of the ring protons in both radical ions[101] are listed in Table XXIV, together with the HMO and McLachlan values for the two pairs of degenerate orbitals. Hyperfine splittings by the γ-protons in the two methyl groups of the central bridges, $a_H^{CH_3}$ could also be observed; they amount to 0.20 and 0.09 gauss for **24**⊖ and **24**⊕, respectively.[101]

The calculated spin populations for the two equivalent centers 2 and 7 are very large for the symmetric orbitals ψ_{a+} and ψ_{b+}, and quite small for ψ_{a-} and

TABLE XXIV
RING PROTON COUPLING CONSTANTS AND SPIN POPULATIONS[a] OF
trans-15,16-DIMETHYL-15,16-DIHYDROPYRENE RADICAL IONS

μ	$a_{H\mu}^\ominus$	$a_{H\mu}^\oplus$	$c_{a+,\mu}^2$ $= c_{b+,\mu}^2$	$c_{a-,\mu}^2$ $= c_{b-,\mu}^2$	$(\rho_\mu^\ominus)_+$ $= (\rho_\mu^\oplus)_+$	$(\rho_\mu^\ominus)_-$ $= (\rho_\mu^\oplus)_-$
1,3,6,8	0.78	1.03[b]	0.007	0.135	−0.045	0.188
2,7	5.46	4.78	0.143	0.000	0.201	−0.058
4,5,9,10	0.78	1.50[b]	0.056	0.088	0.043	0.100

[a] Note that according to the notation in this chapter (see Section I,F) the superscript symbols ⊕ and ⊖ stand for the charge of the radical ion, whereas the subscript symbols + and − refer to orbital symmetry.
[b] Assignment uncertain.

ψ_{b-} (a nodal plane passes through the two centers in both antisymmetric orbitals). As the coupling constants of the protons at these two centers, unlike those of the other ring protons, can be assigned with certainty, these values are the most informative about the symmetry of the singly occupied orbital. From Figs. 35 and 37, and from Table XXIV, it is evident that the dominant splitting in the ESR spectra of **24**⊖ and **24**⊕ arises from two equivalent protons, i.e., those in positions 2 and 7. This result is particularly obvious in the spectrum of **24**⊖, which separates into three groups with relative intensities 1:2:1. Clearly the unpaired electron occupies symmetric orbitals in both radical ions: ψ_{a+} in the anion and ψ_{b+} in the cation.

Whereas the experimental results confirm the theoretical prediction for the radical cation **24**⊕, they are in direct disagreement with theory for the anion **24**⊖. Several arguments, all of them based on the special geometry of the molecule (Fig. 34) have been advanced to account for the failure of the model in this case.[101] The C—C—C angles at the bridged (11, 12, 13, and 14) and bridging (15 and 16) carbon atoms[98] indicate that the former possess less s-character than pure sp^2-hybridized π-centers, and that the latter have more s-character than sp^3 hybrids. The inductive effect of the alkyl bridge should therefore be weak, and its direction may depend on the electron demand of the perimeter; i.e., it may be electron-repelling in the radical cation and electron-attracting in the anion. A second possible explanation of the discrepancy between theory and experiment involves hyperconjugation. The σ-bonds between the carbons of the two methyl groups and carbons 15 and 16 are almost perpendicular to the plane approximately defined by the centers along the perimeter (Fig. 34), so that hyperconjugation between these bonds and the π-electron system can play an important part. For reasons of symmetry, hyperconjugation of this kind will affect only the symmetric orbitals, ψ_{a+} and ψ_{b+}, and may lead to a sequence of energies for the four orbitals which is in agreement with experiment (Fig. 36, right).

F. 1,8-Bisdehydro[14]annulene

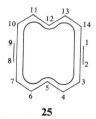

25

The introduction of two triple bonds into the fourteen-membered perimeter to yield 1,8-bisdehydro[14]annulene, **25**,[102] separates the two remaining inner

[102] F. Sondheimer, Y. Gaoni, L. M. Jackman, N. A. Bailey, and R. Mason, *J. Am. Chem. Soc.* **84**. 4595 (1962).

TABLE XXV

PROTON COUPLING CONSTANTS AND SPIN POPULATIONS OF
1,8-BISDEHYDRO[14]ANNULENE RADICAL ANION

μ	$a_{H\mu}$	$c_{a+,\mu}{}^2$	$c_{a-,\mu}{}^2$	$(\rho_\mu)_+$	$(\rho_\mu)_-$
3, 7, 10, 14	4.54^a	0.116	0.027	0.152	−0.009
4, 6, 11, 13	1.15^a	0.007	0.136	−0.045	0.188
5, 12	5.15	0.143	0.000	0.201	−0.058

^{a} Assignment based on comparison with $(\rho_\mu)_+$ values.

hydrogens sufficiently to remove strain due to van der Waals interference. The hydrocarbon is thus planar and has a high degree of aromaticity, as evidenced by proton resonance spectra[102] and X-ray structure determina-

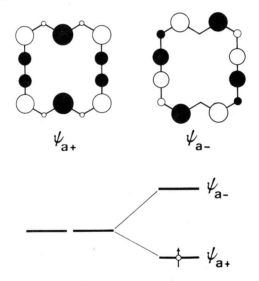

FIG. 38. Degenerate lowest antibonding orbitals of the fourteen-membered perimeter. Splitting and occupancy in the 1,8-bisdehydro[14]annulene radical anion. Schematic representation as in Fig. 2.

tion.[103] It is easily reduced to the radical anion $25^\ominus$ by alkali metals in DME or THF, or electrolytically in DMF with $(t\text{-}C_4H_9)_4N^\oplus I^\ominus$ as supporting salt.[104] The coupling constants of the ring protons, given in Table XXV for the system DME/K$^\oplus$, are practically independent of temperature, solvent, and gegenion.

[103] N. A. Bailey and R. Mason, *Proc. Roy. Soc.* **A290**, 94 (1966).
[104] N. M. Atherton, R. Mason, and R. J. Wratten, *Mol. Phys.* **11**, 525 (1966).

The two degenerate antibonding orbitals ψ_{a+} and ψ_{a-} of the fourteen-membered perimeter are again represented schematically in Fig. 38, the perimeter being drawn this time in the form appropriate for **25**. These orbitals have the following expressions:

$$\psi_{a+} = 0.236(\phi_1 + \phi_2 + \phi_8 + \phi_9) - 0.341(\phi_3 + \phi_7 + \phi_{10} + \phi_{14})$$
$$- 0.084(\phi_4 + \phi_6 + \phi_{11} + \phi_{13}) + 0.378(\phi_5 + \phi_{12})$$

and

$$\psi_{a-} = -0.296(\phi_1 - \phi_2 + \phi_8 - \phi_9) + 0.164(\phi_3 - \phi_7 + \phi_{10} - \phi_{14})$$
$$- 0.368(\phi_4 - \phi_6 + \phi_{11} - \phi_{13})$$

If the triple bonds are treated as perturbations which increase the bond integrals $\beta_{1,2}$ and $\beta_{8,9}$, stabilization is expected for ψ_{a+} ($c_{a+,1} \times c_{a+,2} = c_{a+,8} \times c_{a+,9} = +0.056$) and destabilization for ψ_{a-} ($c_{a-,1} \times c_{a-,2} = c_{a-,8} \times c_{a-,9} = -0.088$). The unpaired electron in the radical anion **25**$^\ominus$ should therefore go into ψ_{a+}, and the experimental results are in excellent agreement with this prediction.[104] The largest coupling constant $a_{H\mu}$ is that of two equivalent protons, and must accordingly be assigned to the pair at centers 5 and 12, while each of the two remaining $a_{H\mu}$ values belongs to a set of four equivalent protons. Consequently, the singly occupied orbital must be ψ_{a+}, and not ψ_{a-}, the latter having a nodal plane through C-5 and C-12. On the basis of this conclusion, the coupling constant of 4.54 gauss should be assigned to the protons at $\mu = 3$, 7,10, and 14, and the smallest one, 1.15 gauss, to those at $\mu = 4,6,11$, and 13.

G. [18]ANNULENE-1,4:7,10:13,16-TRISULFIDE

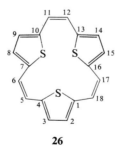

26

[18]Annulene is the fourth member of the homologous series of $(4r + 2)$-membered perimeters ($r = 4$), The hydrocarbon of symmetry D_{6h} synthesized by Sondheimer is planar and exhibits unhindered cyclic conjugation of the π-electrons.[92, 105] To date no ESR spectrum of either of the [18]annulene radical ions has been reported.

[105] F. Sondheimer and R. Wolovsky, *Tetrahedron Letters* NT. 3, p. 3 (1959).

[18]Annulenes containing three heteroatom (oxygen or sulfur) bridges in the positions 1,4:7,10; and 13,16 have also been prepared.[106] These compounds, however, are derivatives of [18]annulene only in a formal sense, since there is strong conjugation across the bridges. A model of three furan or thiophene fragments coupled with three ethylenes should be more suitable for some of the bridged [18]annulenes of this structure. In particular, such a model is appropriate for the trisulfide **26**, which cannot be planar because of the steric interaction of the three bulky sulfur atoms inside the ring.

The radical anions of three bridged [18]annulenes, namely those containing three oxygen, two oxygen and one sulfur, and three sulfur bridges, have been

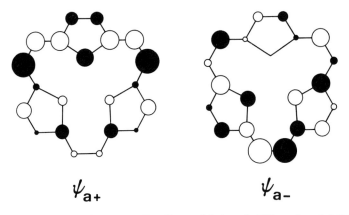

FIG. 39. Degenerate lowest antibonding orbitals of [18]annulene-1,4:7,10:13,16-trisulfide.

investigated by ESR spectroscopy. All three compounds readily yield radical anions on alkali metal reduction in DME or THF. In the first two cases, however the resolution of the ESR spectra was insufficient for a reliable analysis.[107] Only the radical anion of the third compound, [18]annulene-1,4:7,10:13,16-trisulfide, **26**, gave rise to a well-resolved spectrum, which was analyzed in terms of six coupling constants, each for two equivalent protons.[108] Because of the lack of experimental evidence, it was necessary to base the assignment of all of these coupling constants on MO calculations.

A simple HMO model was adopted in which the Coulomb integral of the sulfur atoms was set equal to $\alpha + \beta$, and the bond integrals of the C—S linkages and those connecting the thiophene and ethylene fragments were

[106] G. M. Badger, J. A. Elix, G. E. Lewis, U. P. Singh, and T. M. Spotswood, *Chem. Commun.* p. 269 (1965); G. M. Badger, G. E. Lewis, U. P. Singh and T. M. Spotswood, *ibid.* p. 492; G. M. Badger, J. A. Elix, and G. E. Lewis, *Australian J. Chem.* 18, 70 (1965).

[107] F. Gerson and J. H. Hammons, unpublished results (1968).

[108] F. Gerson and J. Heinzer, *Helv. Chim. Acta* 51, 366 (1968).

taken as 0.7β. The reduction in the strength of the bonds between the fragments corresponds to a twist of 45 degrees, and is consistent with values estimated from Stuart-Briegleb molecular models. The HMO calculation gives twelve bonding and nine antibonding orbitals, the former being occupied in the neutral compound **26** by 24 π-electrons. In radical anion **26**$^\ominus$ the unpaired electron will therefore go into one of the two lowest antibonding orbitals, ψ_{a+} or ψ_{a-}, which are degenerate in this model because of the threefold symmetry axis of the π-electron system. The two orbitals are depicted in Fig. 39.

TABLE XXVI

PROTON COUPLING CONSTANTS AND SPIN POPULATIONS[a] OF
[18]ANNULENE-1,4:7,10:13,16-TRISULFIDE RADICAL ANION

μ	$a_{H\mu}{}^{b}$	$c_{a+,\mu}{}^{2}$	$c_{a-,\mu}{}^{2}$	$(\rho_\mu)_+$	$(\rho_\mu)_-$
2,3	0.86	0.035	0.023	0.015	0.028
5,18	2.96	0.080	0.071	0.059	0.102
6,17	0.43	0.138	0.012	0.177	-0.016
8,15	0.61	0.050	0.008	0.067	-0.024
9,14	1.92	0.001	0.056	-0.017	0.060
11,12	4.40	0.008	0.142	0.005	0.156

[a] Calculated with $\alpha_S = \alpha + \beta$, and $\beta_{CS} = \beta' = 0.7\,\beta'$ (β' is the integral for the bonds between the thiophene and ethylene fragments).
[b] Assignment based on correlation with $(\rho_\mu)_-$ values.

Correlation of the coupling constants with the spin populations $(\rho_\mu)_+$ and $(\rho_\mu)_-$ derived from the corresponding HMO values, $c_{a+,\mu}{}^2$ and $c_{a-,\mu}{}^2$, provides evidence as to which of the two orbitals is occupied by the unpaired electron. The relevant experimental and theoretical values[108] are given in Table XXVI. Since the correlation of $a_{H\mu}$ with $(\rho_\mu)_-$ is significantly better than that of $a_{H\mu}$ with $(\rho_\mu)_+$, the unpaired electron is tentatively assumed to occupy ψ_{a-}. If this assumption is valid, the coupling constants should be assigned to the pairs of equivalent protons on the basis of the spin populations $(\rho_\mu)_-$, as indicated in Table XXVI. An experimental confirmation of this assignment, for instance by the study of deuterated derivatives, would be desirable.

An important result of the ESR study of **26**$^\ominus$ is the pairwise equivalency of the ring protons, i.e., the lack of a threefold symmetry axis. To the best of our knowledge, no X-ray analysis of **26** has been done, but molecular models suggest several possible nonplanar conformations (see Fig. 40). Only one conformation, in which all three sulfur atoms are situated on the same side of the molecule, has C_{3v} symmetry. In the other three conformations, which

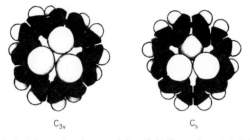

C_{3v} C_s

FIG. 40. Stuart–Briegleb molecular models of [18]annulene-1,4:7,10:13,16-trisulfide. The photographs are reproduced from a paper by G. M. Badger, J. A. Elix, and G. E. Lewis, *Australian J. Chem.* **18**, 70 (1965).

are equivalent, there are only two sulfur atoms on one side of the molecule, so that the symmetry is reduced to C_s.

The ESR spectrum of **26**$^\ominus$ can be interpreted in terms of a conformation of either symmetry C_s or C_{3v}. The C_s conformation can readily account for the observed hyperfine structure if the life-time of one of the three equivalent conformations having this symmetry is longer than ca. 10^{-7} sec (the inverse of the differences in the coupling constants, expressed in sec^{-1}). On the other hand, a distorted C_{3v} conformation cannot be excluded. It is possible that the life-time of a Jahn–Teller distortion might be greater than the critical range of 10^{-7} sec because of association of the gegenion with a particular thiophene fragment.

H. METAHEXAPHENYLENE

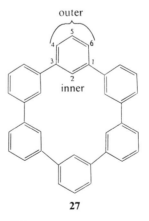

27

Metahexaphenylene **(27)**[109] has been included in Section IV, since it can formally be considered as a derivative of [18]annulene. In fact, the compound

[109] H. A. Staab and F. Binnig, *Tetrahedron Letters* **7**, 319 (1964).

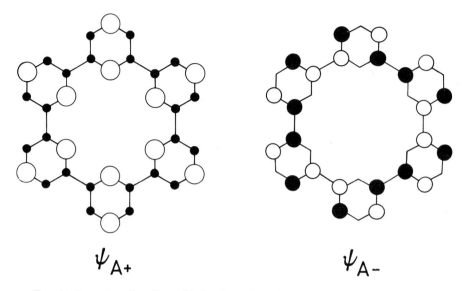

$$\psi_{A+} \qquad\qquad \psi_{A-}$$

FIG. 41. Lowest antibonding orbitals of metahexaphenylene as linear combinations of corresponding benzene orbitals. Schematic representation as in Fig. 2.

is nonplanar and the bonds between the six benzene rings are rather weak[109]; so that a model of weakly coupled π-systems, similar to that applied to ortho-tetraphenylene (**5**) (Section II,D), is probably more appropriate. Arguments analogous to those used for **5** suggest that a linear combination ψ_{A-} of the six

TABLE XXVII

PROTON COUPLING CONSTANTS AND SPIN POPULATIONS OF
METAHEXAPHENYLENE RADICAL ANION

μ	$a_{H\mu}$	$c_{A+,\,\mu}^{2}$	$c_{A-,\,\mu}^{2}$
Inner	0.15^a	0.056	0.000
Outer (like 4,6)	1.69^b	0.014	0.042
Outer (like 5)	0.30^a	0.056	0.000

^{a} Each value for six equivalent protons; assignment uncertain.
^{b} Value for twelve equivalent protons.

antisymmetric benzene orbitals ψ_{a-} should be more stable than the corresponding combination ψ_{A+} of the six symmetric orbitals ψ_{a+} (see Fig. 41). Here too, the first-order stabilization energies relative to the degenerate benzene orbitals amount to $+0.167\beta'$ for ψ_{A+} and $+0.500\beta'$ for ψ_{A-}, where β' is the integral for the weak bond joining the fragments.

The experimental data for the radical anion $27^{\ominus}$ prepared with K in DME[110] are compared in Table XXVII with the HMO values calculated for ψ_{A+} and ψ_{A-}. Clearly, the coupling constants of the ring protons correlate better with $c_{A-,\mu}^2$ than with $c_{A+,\mu}^2$, and indicate that the unpaired electron occupies the orbital ψ_{A-}.

V. Radialenes

[n]Radialenes are hydrocarbons consisting of a ring of n sp^2-hybridized carbon atoms and n cross-conjugated, radially disposed double bonds. HMO calculations yield low bond orders for the endocyclic bonds and indicate only weak cyclic conjugation.[111] Unsubstituted or alkyl-substituted [n]radialenes are known with $n = 3$, 4, and 6.[112-115] Their physical and chemical properties agree with the prediction that the degree of aromaticity of [n]radialenes should be rather small. Radical anions produced from hexamethyl[3]radialene and two symmetric hexaalkyl derivatives of [6]radialene have been studied by ESR spectroscopy.

A. HEXAMETHYL[3]RADIALENE

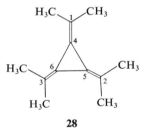

28

The compound 28^{115} is a stable derivative of [3]radialene.[112a] The six methyl groups of **28** should not prevent the carbon atoms of the molecule from

[110] P. H. H. Fischer and K. H. Hauser, Z. Naturforsch. **19a**, 816 (1964).

[111] E. Weltin, F. Gerson, J. N. Murrell, and E. Heilbronner, Helv. Chim. Acta **44**, 1400 (1961); E. Heilbronner, Theoret. Chim. Acta **4**, 64 (1966).

[112] G. W. Griffin and L. I. Peterson, J. Am. Chem. Soc. **84**, 3398 (1962).

[112a] P. A. Waitkus, E. B. Sanders, L. I. Peterson, and G. W. Griffin, J. Am. Chem. Soc. **89**, 6318 (1967).

[113] H. Hopff and A. K. Wick, Helv. Chim. Acta **44**, 19 (1961).

[114] H. Hopff and A. Gati, Helv. Chim. Acta **48**, 1289 (1965).

[115] G. Köbrich and H. Heinemann, Angew. Chem. **77**, 590 (1965); Angew. Chem. Intern. Ed. Engl. **4**, 594 (1965).

lying in a plane. The HMO energy diagram of the nonalternant [3]radialene system is shown in Fig. 42, left. The lowest antibonding orbital, which takes up the unpaired electron in the radical anion, is nondegenerate:

$$\psi_a = 0.533(\phi_1 + \phi_2 + \phi_3) - 0.221(\phi_4 + \phi_5 + \phi_6)$$

Its energy amounts to $E_a = \alpha - 0.414\beta$.

The ESR spectrum of the radical anion $\mathbf{28}^{\ominus}$, produced by reaction of $\mathbf{28}$ with K in DME, should consist of 19 equidistant lines, of which 13 were observed.[116] The lines are spaced 7.57 gauss apart, and the intensity distribution is characteristic of 18 equivalent protons. Clearly, these lines arise from hyperfine interaction of the unpaired electron with the protons of the six equivalent methyl groups. The coupling constant $a_{\mathrm{H}}^{\mathrm{CH_3}}$ is consistent with the HMO value $c_{a,1}^2 = c_{a,2}^2 = c_{a,3}^2 = 0.285$ for the spin population at the substituted exocyclic center, the proportionality factor of 26.6 gauss being in the range observed for β-alkyl protons in other radical anions. Splitting due to a $^{13}\mathrm{C}$ nucleus in one of the six methyl groups is readily observable. The $^{13}\mathrm{C}$ coupling constant of 4.65 gauss is also compatible with the spin population of 0.285 on the adjacent carbon center.

B. 1,2,3,4,5,6-HEXAALKYL[6]RADIALENES

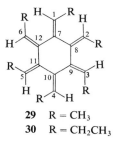

29   R = CH$_3$
30   R = CH$_2$CH$_3$

In contrast to hexamethyl[3]radialene, $\mathbf{28}$, the 1,2,3,4,5,6-hexamethyl-[113] and hexaethyl derivatives[114] of [6]radialene, $\mathbf{29}$ and $\mathbf{30}$, cannot assume a planar arrangement of the ring carbon atoms for steric reasons. The six-membered ring in the two compounds probably resembles the chair form of cyclohexane rather than the regular hexagonal form of benzene. Cyclic conjugation in $\mathbf{29}$ and $\mathbf{30}$ should therefore be even weaker than in $\mathbf{28}$. The

[116] F. Gerson, E. Heilbronner, and G. Köbrich, *Helv. Chim. Acta* **48**, 1525 (1965).

HMO energy diagram of the alternant [6]radialene system is shown in Fig. 42, right. The lowest antibonding orbital,

$$\psi_a = 0.377(\phi_1 + \phi_2 + \phi_3 + \phi_4 + \phi_5 + \phi_6) - 0.156(\phi_7 + \phi_8 + \phi_9 + \phi_{10} + \phi_{11} + \phi_{12})$$

is nondegenerate and has the same energy as the corresponding orbital of [3]radialene. Thus, the electron affinity of the alkyl derivatives **29** and **30** of [6]radialene should be comparable to that of **28**.

The radical anions **29**$^\ominus$ and **30**$^\ominus$, prepared in the same way as **28**$^\ominus$, give rise to ESR spectra[117] which are consistent with the assumption that the unpaired electron occupies ψ_a. The spectrum of the radical anion of the hexamethyl derivative, **29**$^\ominus$, is similar in its simplicity to that of **28**$^\ominus$. With extreme amplification one can observe 21 equidistant lines which on the basis of the intensity distribution are due to 24 equivalent protons. Within the limits of the

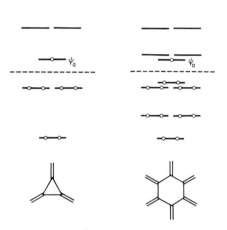

FIG. 42. HMO energy diagrams of [3]- and [6]radialenes. Occupancy in the radical anions.

experimental resolution, the absolute values of a_{H_μ} and $a_H^{CH_3}$ for the six α-protons and the 18 β-protons of the six methyl groups, respectively, must therefore be equal. This result is not surprising, since, as mentioned in Section I,E, β-protons of freely rotating methyl groups and α-protons should have coupling constants of similar magnitude if they are bound to centers having equal spin populations (the same centers, in the case of **29**$^\ominus$). The value of the coupling constants, $a_{H_\mu} \approx a_H^{CH_3} = 3.82$ gauss, is almost exactly half of the splitting found for the methyl protons in **28**$^\ominus$. The squares of the coefficients at the exocyclic

[117] F. Gerson, *Helv. Chim. Acta* **47**, 1941 (1964).

centers for the lowest antibonding orbitals of the [6]- and [3]radialene systems, 0.142 and 0.285, respectively, are also in a 1:2 ratio. This remarkable agreement between the experimental data and the values calculated on the basis of the simplest possible model is not merely a consequence of the symmetries, since the two radical ions differ greatly in steric conformation, degree of cyclic delocalization, charge distribution, and number of alkyl substituents per center. It must be assumed that in this case, as in many others, the agreement is largely fortuitous, and is due to cancelation of the various factors not considered in the simple model.

The coupling constant of a ^{13}C nucleus in one of the six methyl groups of the **29**$^{\ominus}$, which amounts to 2.00 gauss,[117] is again fairly close to half of the corresponding value (4.65 gauss) measured for **28**$^{\ominus}$.

The ESR hyperfine structure of the radical anion of 1,2,3,4,5,6-hexaethyl-[6]radialene, **30**$^{\ominus}$, is due to three sets of six equivalent protons having coupling constants of 4.68, 3.64, and 3.12 gauss. While one of the sets consists of the six equivalent α-protons, the two other sets must be made up of the twelve methylene β-protons, which must be divided into two sterically different sets of six. The γ-protons of the six ethyl groups would be expected to give rise to only a very small splitting, which has not been observed in the ESR spectrum. The assignment of the three coupling constants cannot be made with any certainty. However, if one assumes that the values for the α-protons should be similar in **29**$^{\ominus}$ and **30**$^{\ominus}$, the coupling constant of 3.64 gauss would be assigned to these protons. The two remaining values would belong to the two sets of β-protons, the larger one presumably being due to those which—on a time average—are in the more favorable position for hyperconjugation with the π-system (see Section I,E).

VI. Hydrocarbons with an Odd Number of Centers

In Sections II–V radical ions of compounds which have an even number of carbon centers were considered. These compounds are diamagnetic if they are neutral and form radical anions or cations by uptake or removal of one π-electron. Hydrocarbons with an odd number of centers, on the other hand, are paramagnetic in the neutral state, and are referred to as "odd" or neutral radicals. Two-electron reduction of these odd radicals yields radical dianions. Paramagnetic dianions of this type have been prepared from nonalternant systems like tropylium and two of its benzo derivatives, and from fluorenyl and 4,5-methinephenanthrene, which can be regarded as dibenzo and phenanthreno derivatives of cyclopentadienyl.

A. Tropylium and Its Benzo Derivatives

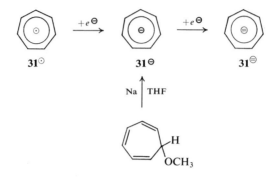

The neutral tropylium radical $31^{\odot}$ [118] and the tropylium dianion $31^{\ominus}$ [119] are both known. The radical dianion $31^{\ominus}$ was apparently produced for the first time by reaction of tropylium bromide with Na–K alloy in DME, but its ESR spectrum was erroneously ascribed to the neutral radical $31^{\odot}$.[120] A more convenient method of preparation of $31^{\ominus}$ is the reaction of tropyl methyl ether with alkali metals in ethereal solvents[119]; the anion $31^{\ominus}$ [121] is thought to be an intermediate. The coupling constant of the seven equivalent protons is 3.52 gauss,[119] a value which is appreciably smaller than the 3.92 gauss observed for the neutral tropylium radical $31^{\odot}$,[118] despite the fact that the spin population in both radicals is $1/7$ at each center.

It is especially important to note that, as in the case of benzene and cyclo-octatetraene radical anions, unequivocal values of $|Q|$ can be obtained for $31^{\odot}$ and $31^{\ominus}$, since both radicals carry a proton at every center, and their symmetries preclude negative spin densities. The value found for $31^{\ominus}$, $|Q| = 7 \times 3.52 = 24.64$ gauss, is in the range characteristic of radical anions. In contrast, the value for $31^{\odot}$, $|Q| = 7 \times 3.92 = 27.44$ gauss, is significantly larger, and typical of neutral radicals.[122] This result is consistent with the common assumption that the $|Q|$ value of the McConnell relation depends on the π-electron charge, decreasing in the sequence radical cations > neutral radicals > radical anions > radical dianions.[3]

The dianion $31^{\ominus}$ is associated in THF or DME with two $Na^{\oplus}$ gegenions, as is shown by the additional ^{23}Na splitting of each of the eight hyperfine lines

[118] G. Vincow, M. L. Morrell, W. V. Volland, H. J. Dauben, Jr., and F. R. Hunter, *J. Am. Chem. Soc.* **87**, 3527 (1965).

[119] N. L. Bauld and M. S. Brown, *J. Am. Chem. Soc.* **87**, 4390 (1965).

[120] J. dos Santos-Veiga, *Mol. Phys.* **5**, 639 (1962).

[121] H. J. Dauben, Jr., and M. R. Rifi, *J. Am. Chem. Soc.* **85**, 3041 (1963).

[122] J. R. Bolton, *in* "Radical Ions" (E. T. Kaiser and L. Kevan, eds.), pp. 9–12. Wiley (Interscience), New York, 1968.

into seven components (a_{Na} = 1.76 gauss).[119, 123] No such splitting was observed for K$^{\oplus}$, either because of rapid intermolecular exchange of the gegenion or because the ^{39}K splitting is too small to have been resolved at the line width attained.

$32^{\ominus}$ $33^{\ominus}$

The radical dianions, $32^{\ominus}$ and $33^{\ominus}$, of benzotropylium and 1,2:4,5-dibenzo-tropylium have also been produced by alkali metal reduction of the corre-

TABLE XXVIII

PROTON COUPLING CONSTANTS AND SPIN
POPULATIONS OF BENZOTROPYLIUM RADICAL
DIANION

μ	$a_{H\mu}$	$c_{a*,\mu}^{2}$	ρ_μ
1,5	2.65^a	0.071	0.034
2,4	6.28	0.232	0.325
3	1.29	0.000	−0.061
6,9	0.28^a	0.014	−0.029
7,8	2.65	0.071	0.067

^{a} Assignment uncertain.

TABLE XXIX

PROTON COUPLING CONSTANTS AND SPIN
POPULATIONS OF DIBENZOTROPYLIUM RADICAL
DIANION

μ	$a_{H\mu}$	$c_{a*,\mu}^{2}$	ρ_μ
1,10	0.70	0.006	−0.029
2,9	3.06	0.098	0.132
3,8	0.70	0.006	−0.025
4,7	2.12	0.075	0.089
5,6	5.34	0.192	0.211
11	0.92	0.000	−0.018

sponding methyl ethers.[123] Like the parent dianion $31^{\ominus}$, each of these dianions appears to be associated with two Na$^{\oplus}$ gegenions. The coupling constants of the protons are given in Tables XXVIII and XXIX, along with the calculated

[123] N. L. Bauld and M. S. Brown, *J. Am. Chem. Soc.* 89, 5417 (1967).

spin populations; the assignment of the coupling constants is based on the computed values. It should be noted that in both radical dianions, $32^{\ominus}$ and $33^{\ominus}$, the singly occupied orbital is the second lowest antibonding HMO of the benzotropylium and the 1,2:4,5-dibenzotropylium system, respectively. This orbital is therefore denoted in Tables XXVIII and XXIX as ψ_{a*} (and not as ψ_a).

The coupling constant of the ^{23}Na nuclei of the gegenions, a_{Na}, amounts to 1.04 gauss for $32^{\ominus}$ and to 0.70 gauss for $33^{\ominus}$.[123]

B. FLUORENYL AND 4,5-METHINEPHENANTHRENE

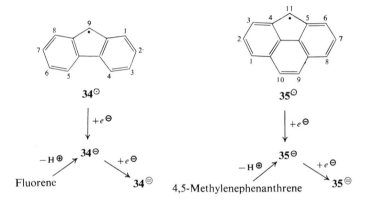

It is known that fluorene reacts with K in THF to yield the diamagnetic fluorenyl anion $34^{\ominus}$,[124] which on further reaction is converted to the radical dianion $34^{\ominus}$.[125] An analogous reaction takes place with 4,5-methylenephenanthrene,[125,126] which yields first the diamagnetic 4,5-methinephenanthrene anion $35^{\ominus}$ and subsequently the radical dianion $35^{\ominus}$.

The coupling constants of the protons in the two radical dianions $34^{\ominus}$ and $35^{\ominus}$ were assigned on the basis of the spin populations calculated for the lowest antibonding HMO's ψ_a of the fluorenyl and 4,5-methinephenanthrene systems. These HMO's are the singly occupied orbitals in $34^{\ominus}$ and $35^{\ominus}$ (see Fig. 43 in the case of $34^{\ominus}$). For the protons at centers $\mu = 1,8$ and 2,7 of fluorenyl, the assignment was verified by comparison of the ESR data of $34^{\ominus}$ with those of the radical dianions of 1- and 2-methyl derivatives.[125] The experimental and computed values are given in Tables XXX and XXXI.[125] The coupling constants of the protons in $35^{\ominus}$ (Table XXXI) are very similar to those of the corresponding protons in the phenanthrene radical anion.

[124] See, e.g., G. W. H. Scherf and R. K. Brown, *Can. J. Chem.* **38**, 697 (1960).

[125] E. G. Janzen, J. G. Pacifici, and J. L. Gerlock, *J. Phys. Chem.* **70**, 3021 (1966).

[126] E. G. Janzen and J. G. Pacifici, *J. Am. Chem. Soc.* **87**, 5504 (1965).

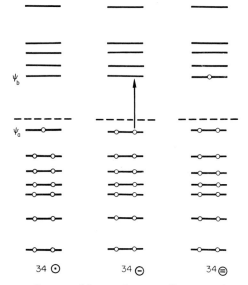

FIG. 43. HMO energy diagram of fluorenyl system. Occupancy in neutral radical (34⊙), diamagnetic anion (34⊖), and radical dianion (34⊜). The arrow marks the lowest energy electronic transition of the anion.

TABLE XXX

PROTON COUPLING CONSTANTS AND SPIN
POPULATIONS OF FLUORENYL RADICAL DIANION

μ	$a_{H\mu}$	$c_{a,\mu}^2$	ρ_μ
1,8	3.05	0.126	0.137
2,7	3.05	0.121	0.143
3,6	0.35	0.005	−0.048
4,5	4.53	0.166	0.222
9	0.53	0.021	0.016

Radical dianions produced in an analogous way from carbazole (*N*-analog of fluorene), 9-phenylfluorene, and 4,5-iminophenanthrene were also investigated.[125] An ESR spectrum was observed as well on two-electron reduction of the Koelsch radical [9,9′-bis(fluorenyl)phenylmethyl].[126] This spectrum, which differs from that of the neutral radical, can be ascribed to the corresponding radical dianion. Its hyperfine structure points to localization of the odd electron on one fluorenyl moiety of the molecule.

TABLE XXXI

PROTON COUPLING CONSTANTS AND SPIN
POPULATIONS OF 4,5-METHINEPHENANTHRENE
RADICAL DIANION

μ	$a_{H\mu}$	$c_{a,\mu}{}^2$	ρ_μ
1,8	3.02	0.099	0.130
2,7	0.53	0.002	−0.042
3,6	3.02	0.116	0.157
9,10	4.96	0.172	0.196
11	0.36	0.000	−0.026

C. COMPARISON OF SOME ESR AND UV SPECTRAL DATA FOR THE FLUORENYL SYSTEM

In the case of the fluorenyl system, proton coupling constants have also been reported for the neutral radical $34^{\odot}$ [127]; the assignments were based on the spin populations calculated for the highest bonding HMO ψ_b of the fluorenyl system, this HMO being the singly occupied orbital in $34^{\odot}$ (see Fig. 43). The results obtained for the neutral radical $34^{\odot}$, together with those given in Table XXX for the radical dianion $34^{\ominus}$, permit one to make an interesting comparison with UV spectral data of methyl-substituted fluorenyl anions.[86] In these derivatives of the diamagnetic monoanion $34^{\ominus}$, the electronic transition of lowest energy corresponds to promotion of an electron from the highest bonding HMO ψ_b to the lowest antibonding HMO ψ_a of the fluorenyl system (see again Fig. 43). According to first-order perturbation theory, the electron-repelling inductive effect of a methyl group at center μ in fluorenyl anion, $34^{\ominus}$, will destabilize ψ_a and ψ_b by amounts proportional to $c_{a,\mu}{}^2$ and $c_{b,\mu}{}^2$, respectively. Since the transition energy equals the difference in energy between ψ_a and ψ_b, a methyl group in $34^{\ominus}$ should produce a UV spectral shift which is proportional to the difference between the squares of the coefficients ($\tilde{\nu}$ = wave numbers in cm^{-1}):

$$\Delta\tilde{\nu}_\mu \propto c_{a,\mu}{}^2 - c_{b,\mu}{}^2$$

This model correctly predicts the direction of the spectral shift $\Delta\tilde{\nu}_\mu$ for all five isomeric methylfluorenyl anions.[86]

The comparison of the UV shifts for the derivatives of $34^{\ominus}$ with the ESR data for $34^{\ominus}$ and $34^{\odot}$ can now be made by means of the following arguments:

[127] D. R. Dalton and S. A. Liebman, *J. Am. Chem. Soc.* **91**, 1194 (1969).

The two orbitals ψ_a and ψ_b involved in the relevant electronic transition of **34**$^\ominus$ are just those which are singly occupied in **34**$^\ominus$ and **34**$^\odot$, respectively. The proton coupling constants $a_{H\mu}$ are thus proportional to $c_{a,\mu}^2$ for **34**$^\ominus$ and to $c_{b,\mu}^2$ for **34**$^\odot$, as stated before. Although the Q values for neutral radicals and radical dianions are expected to be slightly different (see Section VI,A), the difference should be small enough that a proportionality relation should also be valid for the corresponding differences in the experimental and theoretical values. That is:

$$\Delta a_{H\mu} = a_{H\mu}{}^\ominus - a_{H\mu}{}^\odot \propto c_{a,\mu}{}^2 - c_{b,\mu}{}^2$$

where $a_{H\mu}{}^\ominus$ and $a_{H\mu}{}^\odot$ stand for the coupling constants of the protons in **34**$^\ominus$ and **34**$^\odot$, respectively.

Consequently, there should be a correlation between the UV shifts for the methyl derivatives of **34**$^\ominus$ and the differences in the coupling constants for **34**$^\ominus$ and **34**$^\odot$:

$$\Delta\tilde{\nu}_\mu \propto \Delta a_{H\mu}$$

As Table XXXII shows, there is in fact a rough proportionality, although the correlation is certainly not quantitative. The approximate nature of the relation

TABLE XXXII

FLUORENYL SYSTEM: PROTON COUPLING CONSTANTS OF RADICAL DIANION AND NEUTRAL RADICAL, ALONG WITH THE SHIFTS EFFECTED BY METHYL SUBSTITUTION ON THE LONGEST WAVELENGTH UV TRANSITION OF THE ANION

μ	$a_{H\mu}{}^\ominus$	$a_{H\mu}{}^\odot$	$\Delta a_{H\mu}$	$\tilde{\nu}_\mu{}^a$	$\Delta\tilde{\nu}_\mu{}^a$
1,8	3.05	3.75[b] (1.89)	−0.70 (+1.16)	19,760	+330
2,7	3.05	0.49	+2.56	19,600	+170
3,6	0.35	1.89[b] (3.75)	−1.54 (−3.40)	19,070	−360
4,5	4.53	0.49	+4.04	20,050	+620
9	0.53	7.0	−6.5	18,060	−1370

[a] All numbers in cm^{-1}. The value for unsubstituted fluorenyl anion is 19,430 cm^{-1}.

[b] Assignment uncertain. The alternative assignment (values in parentheses) is supported by the comparison with the UV shifts.

is hardly surprising, in view of the neglect of the hyperconjugative effect of the methyl substituent and the numerous other oversimplifications involved in the theoretical treatment of these systems.

Acknowledgments

Support from the "Schweizerischer Nationalfond" (Project Nr. 4651) during the writing of this chapter is gratefully acknowledged.

Acknowledgment is also made to the donors of the Petroleum Research Fund, administered by the American Chemical Society, for partial support of this work.

Thanks are expressed to Dr. G. Moshuk for reading the manuscript and to Mrs. H. Leu for typing it. The authors would also like to thank the editors of *Helv. Chim. Acta, Mol. Phys., J. Chem. Phys.,* and *Australian J. Chem.* for permission to reproduce figures.

3

Diamagnetic Susceptibility Exaltation as a Criterion of Aromaticity

HYP J. DAUBEN, JR.,[1] JAMES D. WILSON, AND JOHN L. LAITY

I. Introduction

The first systematic, extensive investigation into the magnetic properties of organic compounds was made by Pascal and his many collaborators during the first quarter of this century,[1a] and it was these investigators who discovered

[1] Deceased.
[1a] For a compilation of references to Pascal's work, see Selwood,[1b] p. 91.
[1b] P. W. Selwood, "Magnetochemistry," 2nd ed. Wiley (Interscience) New York, 1956.

the anomalous magnetic properties of benzene and its derivatives. Pascal found that the magnetic susceptibilities[2] of organic compounds are approximately additive functions of their constituent groups, and he constructed a system for estimating susceptibilities based on this fact.[3] However, he found that the "double bonds" of benzene had to be handled differently from those of ordinary olefins: the susceptibility found for benzene was nearly as large as that predicted for a *saturated* compound composed of six carbon and six hydrogen atoms, whereas the susceptibilities of nonaromatic unsaturated compounds were seemingly diminished by a substantial increment for each double bond present. Thus benzene possessed a rather larger magnetic susceptibility than was expected from a comparison with the values of alkenes.

After the developments in the theory of structure and bonding in covalent compounds which took place in the 1920's and 1930's, it was recognized that this magnetic susceptibility "exaltation" of benzene was associated with the special properties of the aromatic nucleus. Pacault, in his restructuring of the Pascal System,[4] formalized this by introducing a special constant, Λ, for the susceptibility contribution of the aromatic system. The expression for the susceptibility of a benzene derivative then became[4]

$$\chi_M = \chi_{\text{additive}} + \Lambda$$

where χ_{additive} was the "Pascal susceptibility" of the basic cyclopolyene, and Λ the contribution due to the presence of aromaticity in the ring-nucleus.

Very soon thereafter Pink and Ubbelohde used this concept to test for aromatic character in cyclooctatetraene.[5] They measured the susceptibility exhibited by that compound, and found it was accounted for within experimental uncertainty by the Pascal System *without* the inclusion of a contribution from exaltation. They concluded, therefore, that cyclooctatetraene was not aromatic. The wider significance of their work appears to have been generally overlooked, except by Craig who extended it to a few additional compounds.[6, 6a] Instead interest became focused on the magnetic *anisotropy* of benzenoid aromatics, which is also partially a product of the special electronic properties

[2] Note that, except where indicated, throughout this essay values of the molar diamagnetic susceptibility χ_M will be given in units of $-\chi_M \times 10^{-6}$ cm^3 mole^{-1}.

[3] P. Pascal, *Ann. Chim. Phys.* [8] **19**, 5 (1910); see also C. K. Ingold, "Structure and Mechanism in Organic Chemistry," pp. 187–189. Cornell Univ. Press, Ithaca, New York, 1953; also Selwood[1b].

[4] A. Pacault, *Ann. Chim. (Paris)* [12] **1**, 567 (1946); *Rev. Sci.* **86**, 38 (1948); A. Pacault, J. Hoarau, and A. Marchand, *Advan. Chem. Phys.* **3**, 171–238 (1961).

[5] R. C. Pink and A. R. Ubbelohde, *Trans. Faraday Soc.* **44**, 708 (1948).

[6] D. Craig, *in* "Non-Benzenoid Aromatic Compounds" (D. Ginsburg, ed.), pp. 38–42. Wiley (Interscience), New York, 1959.

[6a] D. Craig, *Theoret. Org. Chem., Papers Kekule Symp., London,* 1958 p. 20. Academic Press, New York, 1959.

of these systems. However, anisotropy is not unique to aromatic compounds[7] and is in any case hard to measure[8]; subsequently it has not turned out to be a useful indicator for aromatic character.

Recently we resurrected the concept of magnetic susceptibility exaltation as a criterion of aromaticity and did the work necessary to establish its validity.[9] We have adopted Pacault's terminology and symbolism,[4] and define exaltation, Λ, as the difference between the susceptibility (χ_M) found for a compound and that estimated (χ_M') for a *cyclopolyene* of the same structure:

$$\Lambda = \chi_M - \chi_M' \tag{1}$$

The estimation of χ_M' can be made using any of several empirical systems currently available; we have used that of Haberditzl.[10] Aromatic compounds, invariably, are found to exhibit large (compared with the uncertainty in Λ) exaltations and, with the exception of some small ring compounds, non-aromatic systems have been found to exhibit zero exaltation. Thus we have concluded that exaltation is truly a manifestation of the presence of fully delocalized electrons in a molecule and a reliable criterion of aromaticity.

In this review we shall consider the theoretical and empirical bases of exaltation, give the values which have been found for a number of interesting compounds, and finally present interpretations of some of the data.

II. Theory

Magnetic susceptibility is a measure of the extent to which a material is either attracted to or repelled by a magnetic field and is commonly determined by measuring the force exerted on a sample by a field of known strength.[11] A material repelled by a magnetic field is said to be diamagnetic; one attracted is paramagnetic. The susceptibility of a compound can be expressed as one of three quantities: the susceptibility per unit volume, κ, which is dimensionless; the susceptibility per unit mass, $\chi = \kappa/D$, which has the units $cm^3 \ gm^{-1}$; and the susceptibility per mole, $\chi_M = M\chi$, which has the units $cm^3 \ mole^{-1}$. Various techniques can give one or the other of the first two quantities directly; the last is always derived. The molar susceptibility, χ_M, is the quantity of primary interest to us.

[7] A. A. Bothner-By and J. A. Pople, *Ann. Rev. Phys. Chem.* **16**, 43 (1965).

[8] See, for example, Selwood,[1b] Chapters II and VII.

[9] H. J. Dauben, Jr., J. D. Wilson, and J. L. Laity, *J. Am. Chem. Soc.* **90**, 811 (1968); **91**, 1991 (1969).

[10] W. Haberditzl, *Sitzber. Deut. Akad. Wiss. Berlin, Kl. Chem., Geol., Biol.* No. 2 (1964); *Angew. Chem. Intern. Ed. Engl.* **5**, 288 (1966).

[11] Thorough discussions of this topic can be found in Selwood,[1b] Chapters I–IV; also L. N. Mulay, "Magnetic Susceptibility." Wiley (Interscience), New York, 1963.

Diamagnetism arises in the action of magnetic fields on the electrons of a molecule or atom.[11] Classically, this was considered to be analogous to the effect of a magnetic field on a circular conductor: a current induced in the conductor gives rise to a magnetic field in opposition to the applied field; similarly, the action of a field on the electrons in their orbits was to induce a net current, such that its associated field opposed the applied field. The quantum mechanical expression for susceptibility[12] [Eq. (2)],

$$\chi_M = \frac{Ne^2}{6mc} \sum \bar{r}^2 + \tfrac{2}{3}N \sum \frac{m^0(n';n)^2}{h\nu(n';n)} \qquad n' \neq n \qquad (2)$$

derived by Van Vleck, includes this classical, or "Langevin" diamagnetism, plus a term that essentially corrects for the departure from spherical or cylindrical symmetry following the formation of molecules from atoms.[13] In this expression $m^0(n';n)$ is an off-diagonal element of the angular momentum matrix, $\nu(n';n)$ is the frequency of the $n' \to n$ transition, N is Avogadro's number, m and e are the mass and charge, respectively, of an electron, and the other symbols have their usual meaning. Thus the value of χ_M exhibited by a system depends on three quantities: the number of electrons in the system and the volume they effectively occupy; the deviation of the system from spherical or cylindrical symmetry (the numerator of the second term approaches zero as the shape of the system approaches a perfect sphere or cylinder); and the inverse of the energy of electronic transitions within the system. The second term is sometimes called the Van Vleck paramagnetism; it makes a relatively small ($\lesssim 20\%$ of χ_M) contribution to the susceptibilities of compounds of light elements,[13] but a progressively larger contribution to the susceptibilities of compounds of heavier elements.[14]

Magnetic susceptibility exaltation arises as a consequence of the special properties of fully delocalized electrons.[15] External magnetic fields act on these electrons—the existence of which characterizes aromatic compounds— to induce an opposing field much larger than that induced in the electrons of ordinary covalent bonds. This was first rationalized by Pauling in terms of a classical circulation of free electrons around the benzene ring,[16] and from this source sprang the much-abused name "ring current."[17]

[12] J. H. Van Vleck, "The Theory of Electric and Magnetic Susceptibilities." Oxford Univ. Press, London and New York, 1932.

[13] Y. G. Dorfman, "Diamagnetism and the Chemical Bond." Elsevier, Amsterdam, 1965; "Diamagnetismus und chemische Struktur." Teubner, Leipzig, 1964.

[14] See Selwood,[1b] p. 83.

[15] L. Salem, "Molecular Orbital Theory of Conjugated Systems," Chapter 4. Benjamin, New York, 1966.

[16] L. Pauling, *J. Chem. Phys.* **4,** 673 (1936).

[17] We recognize that the concept of a physical "ring current' is a simplistic and inaccurate description of the phenomena involved. The phenomena themselves are quite real, however, and the term remains wholly useful.

London developed a simple wave-mechanical description of this effect.[18] The contribution of the ring current to the magnetic susceptibility is called the London diamagnetism, K_L. It is strongly anisotropic, being large in the direction normal to the ring-plane and small or zero in the plane.[15] It is responsible for a large fraction (about one-half in benzene)[19] of the large susceptibility anisotropies observed for benzenoid aromatic compounds, and is the source of the diamagnetic exaltation.

The molar susceptibility can be written as an average of three orthogonal contributions[20]:

$$\chi_M = \tfrac{1}{3}(\chi_x + \chi_y + \chi_z) \tag{3}$$

If we assume that possession of London diamagnetism is the only way in which an aromatic differs magnetically from a nonaromatic molecule, and thus that the susceptibility of the aromatic differs only by the contribution of K_L, then (taking z normal to the molecular plane) χ_z(aromatic) $\simeq \chi_z$(nonaromatic) $+ K_L$ and Eq. (4) follows.

$$\chi_M(\text{aromatic}) \simeq \chi_M(\text{nonaromatic}) + \tfrac{1}{3}K_L \tag{4}$$

There is evidence to suggest that this assumption is nearly, but not exactly, correct. However, the only requisite is a systematic difference between the diamagnetic susceptibilities of aromatic and nonaromatic compounds, and this has been experimentally verified.[9] Further discussion of this point follows.

Thus theory implies that aromatic compounds should be distinguishable from similar nonaromatic compounds by their numerically larger susceptibilities. In practice this distinction is not easily made by direct comparisons,[21] but it can be readily accomplished by comparing measured susceptibilities with those obtained through an accurate susceptibility-estimation method, assuming the compound to be nonaromatic. Because of their London diamagnetism the susceptibilities found for aromatics will be larger than those estimated[2] ($\chi_M - \chi_M' = \Lambda > 0$). That is, aromatic compounds will exhibit the phenomenon of diamagnetic susceptibility *exaltation*; by definition, nonaromatic compounds will not. This is the basis for using diamagnetic susceptibility exaltation as a criterion of aromatic character.

Obviously, the utility of this method of identifying aromatic compounds depends on the accuracy with which the method adopted estimates the susceptibilities of nonaromatic compounds. We showed[9] that the method of

[18] F. London, *J. Phys. Radium* **8**, 397 (1937).
[19] See discussion of this point below.
[20] See Selwood,[1b] p. 32; also Bothner-By and Pople.[7]
[21] It can be done, however, but not satisfactorily. Compare the susceptibilities of benzene, $\chi_M = 54.8$, and cyclooctatetraene, 53.1. Having two additional methine groups cyclooctatetraene should exhibit a larger susceptibility than benzene, but the London diamagnetism of benzene makes up the difference.

Haberditzl[10] estimates the susceptibilities of nonaromatic *hydrocarbons*, at least, very well indeed (the uncertainties in χ_M and χ_M' are comparable). Some of the data used to support this conclusion appear in Table IV. For heterocycles the situation is less clear. As will be discussed below, the uncertainty in χ_M' is larger than might be desirable, although not too large for our present purposes.

III. Method

To apply the criterion of magnetic susceptibility exaltation to a compound, two kinds of data are needed: the actual and estimated molar magnetic susceptibilities. It is pertinent to discuss how these data can be obtained.

A. The Measurement of Magnetic Susceptibility

The text of Selwood[1b] and the several reviews by Mulay and Mulay[22] provide a thorough compilation of the numerous techniques available for measuring magnetic susceptibilities, and a detailed, critical evaluation of several NMR methods is available.[23] This section will therefore consist only of brief statements of the strengths and limitations of several of the most significant methods.

1. *Gouy Balance*

There are numerous modifications of the basic Gouy balance technique, all of which determine the volume magnetic susceptibility (κ) of a sample in a uniform magnetic field by measurement of force. The equipment needed is relatively simple, but rather large (several grams) samples are usually required. This method can be used over a wide range of temperatures for measurements on solids, liquids, or gases, and the required weighings have been made easier by the advent of semiautomatic balances. Accuracy of $\pm 1\%$ or better is readily obtained.

2. *Faraday Balance*

The many variations of the Faraday balance method all measure the weight of the sample both in and out of a *nonuniform* magnetic field with a constant field gradient. The great advantage of this technique is that the mass susceptibility (χ) is obtained directly, so that a density determination is not required. Only a small sample (a few milligrams) is needed, and measurements are

[22] L. N. Mulay, *Anal. Chem.* **34**, 343R (1962); L. N. Mulay and I. L. Mulay, *ibid.* **36**, 404R (1964); **38**, 501R (1966); **40**, 440R (1968).

[23] J. L. Laity, Ph.D. Thesis, Univ. of Washington, 1968.

possible over a wide temperature range. This method is ordinarily limited to powdered solids, but high accuracy ($\pm 0.1\%$) can be obtained. This is probably the best technique commonly available for measuring the susceptibilities of solids.

3. Induction Methods

One example of several electrical induction methods has been given by Broersma,[24] who described an alternating current induction apparatus for the rapid and highly precise ($< \pm 0.1\%$) measurement of the magnetic suscepti- bilities of organic compounds. While Broersma's results (obtained by a combination of techniques) are generally considered among the most accurate ever obtained, the induction apparatus is relatively complex (although no magnetic field is used), and the measurements are restricted to liquids over a small temperature range.

4. The Sphere-Cylinder NMR Method

Since high-resolution NMR spectrometers are now found in most labora- tories, NMR methods for the determination of susceptibilities should prove increasingly popular. We have found two methods most satisfactory.

The NMR technique that gives the greatest accuracy, precision, and ease of measurement for diamagnetic (and weakly paramagnetic) liquids is that of Frei and Bernstein,[25] as modified by Mulay and Haverbusch.[26] Besides an NMR spectrometer, the only equipment required is a specially constructed glass cell[26] that fits within a regular NMR tube. The chemical shift separation of the two absorbances of the liquid within spherical and cylindrical portions of the reference cell is directly proportional to the volume magnetic suscepti- bility of the substance in the outer NMR tube. With careful and quite tedious calibration,[23, 25-27] it is possible to measure volume diamagnetic susceptibilities to an accuracy of about $\pm 0.2\%$. The choice of reference liquid is most im- portant,[23] with tetramethylsilane giving the best results for diamagnetic organic compounds.

5. The Concentric Cylinder NMR Method

A second useful technique for measuring volume magnetic susceptibilities by NMR was developed by Reilly et al.[28] and Douglass and Fratiello.[29]

[24] S. Broersma, J. Chem. Phys. 17, 873 (1949); Rev. Sci. Instr. 20, 660 (1949).
[25] K. Frei and H. J. Bernstein, J. Chem. Phys. 37, 1891 (1962).
[26] L. N. Mulay and M. Haverbusch, Rev. Sci. Instr. 35, 756 (1964).
[27] D. J. Frost and G. E. Hall, Mol. Phys. 10, 191 (1966).
[28] C. A. Reilly, H. H. McConnell, and R. G. Meisenheimer, Phys. Rev. 98, 264A (1955).
[29] D. C. Douglass and A. Fratiello, J. Chem. Phys. 39, 3161 (1963).

A glass cell consisting of two concentric cylinders (available from Wilmad Glass Co.) is employed. With the cell *not* being spun, the width of the signal derived from the reference substance (toluene in our work), which is held in the outer tube, is directly proportional to the susceptibility of the sample within the inner concentric cylinder. Tedious calibration with samples of known susceptibility is again required.[28-30]

This method is somewhat less precise and more time-consuming than the sphere-cylinder NMR technique, but has the advantage of being applicable to solids, liquids, and gases. The sample tubes can be calibrated for direct determination of density, a necessity for determinations of the mass susceptibilities of powdered solids.

B. The Estimation of Magnetic Susceptibility

Only two methods of estimating magnetic susceptibilities of organic compounds are widely used: the Pascal System of atomic constants,[3] and the Haberditzl "Semi-Empirical Increment System."[10] Both of these methods derive from the observation by Pascal that the magnetic susceptibility of an organic compound can be estimated as the sum of contributions from its parts. This is expressed by Eq. (5)

$$\chi_M' = \sum_i f_i \chi_i \qquad (5)$$

where f_i, for example, is the number of times any particular structural element (bond, atom, electron pair, or other special feature) of susceptibility χ_i is repeated in the molecule, and there are a number i such increments.

In the Pascal System, each type of atom has been assigned a fundamental value; the susceptibility of a molecule is the sum of the values of all its atoms, plus the sum of whatever "corrections" due to special structural features are needed. The Haberditzl System assigns a susceptibility increment to each type of bond and electron grouping, with the susceptibility being the sum of the contributions of all the bonds (lone pairs, etc.). Since magnetic susceptibility arises in the electrons, and not the nuclei,[31] there is more theoretical justification for susceptibility being an additive function of the bonds in a molecule rather than the atoms, but the two approaches are fundamentally the same. The Haberditzl system uses more parameters and is therefore somewhat more accurate. We have adopted the Haberditzl System, and values of the bond increments used are given in Table I.

[30] J. R. Zimmerman and M. R. Foster, *J. Phys. Chem.* **61**, 282 (1958).
[31] H. F. Hameka, *J. Chem. Phys.* **37**, 3008 (1963); **34**, 1896 (1961).

TABLE I

VALUES OF STRUCTURAL ELEMENT INCREMENTS OF THE HABERDITZL SYSTEM

Structural element	χ_M $(-10^{-6}$ cm^3 mole$^{-1})$	Structural element	χ_M $(-10^{-6}$ cm^3 mole$^{-1})$
		Bonds[a]	
C*—C*	2.4[b]	C*πC*	2.2[b]
C*—N*	3.2[c]	C*πN*	2.2[c]
C*—O*	2.8[c]	C*πO*	2.2[c]
C*—S*	4.1[c]	N*πN*	2.2[c]
N*—N*	3.2[c]	N*πO*	2.2[c]
N*—O*	3.6[c]	N*—H	3.6[c]
C*—H	3.2[b]	N$^+$—H	2.0[c]
C*—H (terminal)	3.6[b]	C*—C	2.6[b]
		"Core" electrons[d,e]	
C	0.15	O	0.08
N	0.10	S	0.90
		"Lone-pairs"[a,e]	
N*	2.30	O$^+$	1.40[c]
O*	1.75	S	5.15
		Groups[c]	
CH$_3$ on N$^+$	9.1	"Van Vleck" corrections	
CH$_3$ on C*	14.5	C≡N	−6.0
"Phenylation"[f]	35.0	C≡O	−6.0
"Annelation"[f]	20.3	"6π" cation	−8.0
		Benzene-substitution increments[g]	
CH$_3$	11.3	F	3.6
C$_2$H$_5$	22.4	Cl	15.2
n-C$_3$H$_7$	34.1	Br	24.1
i-C$_3$H$_7$	34.5	I	37.2
t-C$_4$H$_9$	46.5	OH	5.4
CH$_2$=CH	13.4	OCH$_3$	15.7
C$_6$H$_5$	48.5	NH$_2$	8.1
t-C$_4$H$_9$C≡C	60	N(CH$_3$)$_2$	35
C$_6$H$_5$CH$_2$	60	p-CH$_3$C$_6$H$_4$	61

[a] The hybridization and charge of atoms connected by the bonds in question are indicated by superscripts. Unadorned atomic symbols denote sp^3 hybridized atoms; an asterisk denotes sp^2 hybridization; and plus and minus signs denote the appropriate charge.

[b] Value of Haberditzl.[10]

[c] Value estimated by present authors. See text and Dauben et al.,[9] Laity,[23] and Wilson.[32]

[d] Electrons in orbitals not available for bonding (e.g., the $1s^2$ electrons of C, N, and O).

[e] Values of Baudet. See Baudet et al.[33]

[f] See text. Normal benzene exaltation excluded.

[g] These values indicate the increase in the susceptibility of benzene effected by substitution of each atom or group; i.e., this value equals $\chi_M(C_6H_5X) - \chi_M(C_6H_6)$. The increments are approximately additive and can be combined to obtain the susceptibilities of complex substituted aromatic compounds. See Laity.[23]

1. *Extension of the Haberditzl System*

As originally developed, the Haberditzl System is not suitable for use with heterocyclic unsaturated compounds, since no increments have been derived for the various C—X and X—Y bonds between sp^2-hybridized atoms. We extended the system[32] to cover these types of structural elements, essentially in the same way as the system originally developed. Susceptibility data were obtained for any available compounds that included the bond or other element of interest. The susceptibility of as much of the compound as possible was estimated using the original Haberditzl increments; the remainder was then divided among the unaccounted-for structural elements in a self-consistent fashion. The increment values obtained this way are rather crude, since they are based on quite a small number of data, and should be regarded as provisional. They are also given in Table I.

By necessity, aromatic amines and ethers had to be used as model compounds for most of these bonds. It is not obvious that they are in fact adequate models, for, in general, the heteroatoms in these compounds are not hybridized sp^2. The effect of this difference is not easily assessed, and the data were used as they stand. At this time, the results should be considered approximate.

Note that in Table I increments are given separately for "lone-pair" and "core" (completed-shell) electrons; the values given are from Baudet et al.[33] Haberditzl combined these numbers, but the system just given is more flexible.

2. *The Effects of Van Vleck Paramagnetism*

The above increment system as presently constructed suffers from a limitation, inherent in the choice of model compounds, that stems from the shape-dependence of the Van Vleck paramagnetism. Since all the model compounds (open-chain and cyclic alkanes and alkenes with more than four carbon atoms) usually take the shape of a short, irregular cylinder, the Haberditzl system contains a built-in contribution from the Van Vleck paramagnetism appropriate to molecules of this shape. Thus the system can be expected to predict with accuracy the susceptibilities of molecules which are irregularly cylindrical, but should be less accurate (due to the variations in Van Vleck paramagnetism) for molecules differently shaped (e.g., more spherical, accurately cylindrical, or long and flat).[34] We are particularly concerned with the increase in the Van Vleck paramagnetism observed in long, flat molecules,[35] e.g., polyphenyls

[32] J. D. Wilson, Ph.D. Thesis, Univ. of Washington, 1966.

[33] J. Baudet, J. Tillieu, and J. Guy, *Compt. Rend.* **244**, 1756 (1957).

[34] Thus Haberditzl had to derive a set of increment-values for spherical molecules of the adamantane type different from those used for linear and cycloalkanes.[10]

[35] See Dorfman,[13] Chapter VI.

and linear and bent "acenes." In an attempt to correct for this we have introduced the "annelation" and "phenylation" increments, the first (A) to be applied when a benzene ring is fused across *two* atoms of an aromatic system, as in deriving naphthalene from benzene, and the second (P) when a ring-hydrogen is replaced by a phenyl group. Thus the value of χ_M' for anthracene is given by χ_M' (anthracene) $= \chi_M'$ (naphthalene) $+ A = 61.4 + 20.3 = 81.7$,[2] and for biphenyl, $\chi_M' = \chi_M'$ (benzene) $+ P = 41.1 + 35.0 = 76.1$. These values were derived[32] from magnetic anisotropy data, and are approximate.

Another effect of the Van Vleck paramagnetism which must be considered is that associated with the unshared electron pairs of carbonyl[36] and azine[37] groups. In both of these, the low-energy $n \rightarrow \pi^*$ transition gives rise to a local paramagnetism, the average size of which was estimated from a combination of theoretical estimates,[36-38] magnetic anisotropy data for anthracene, acridine and phenazine,[7] and the Haberditzl ketone increment.[10]

Dorfman[13] has discussed the effect of the distortion of an electronic system by the presence of a charge on it. In cations, the diamagnetic susceptibility is reduced. We have made an estimate of the magnitude of this effect in tropenylium ions by a roundabout route,[39] and this rather uncertain value has been adopted for all cations with 6 π-electrons.

Also in Table I are given values for certain substituent groups on an aromatic nucleus. These represent the increase in susceptibility on substitution of each group for hydrogen and were obtained by taking the difference in the susceptibilities of the appropriately substituted benzene and benzene itself.[23] Thus, for toluene, $\chi_M' = \chi_M'$ (benzene) $+ \chi_M'$ (methyl) $= 54.8 + 11.3 = 66.1$.

3. *Small-Ring Compounds*

Compounds which contain small rings present serious problems, in that none of the methods now in use predict their susceptibilities at all well.[23] This is illustrated by the data in Table II, in which it can be seen that the diamagnetic susceptibilities of cyclopropanes and cyclopentanes are underestimated and those of cyclobutanes overestimated. (This rather peculiar alternation was first observed by Barter *et al.*[40]; Burke and Lauterbur found a similar sequence in the ^{13}C and ^{1}H NMR chemical shifts of these compounds, with the protons and carbon atoms being shielded in C_3H_6 and deshielded in C_4H_8.[41]) No satisfactory explanation for these observations has been presented, although the rather high diamagnetism of the cyclopropyl system

[36] J. A. Pople, *Discussions Faraday Soc.* **34**, 7 (1962).
[37] J. D. Baldeschwieler and E. W. Randall, *Proc. Chem. Soc.* p. 303 (1961).
[38] V. M. S. Gil and J. N. Murrell, *Trans. Faraday Soc.* **60**, 248 (1964).
[39] H. J. Dauben, Jr., J. L. Laity, and J. D. Wilson, to be published; see also Laity.[23]
[40] C. Barter, R. G. Meisenheimer, and D. P. Stevenson, *J. Phys. Chem.* **64**, 1312 (1960).
[41] J. J. Burke and D. C. Lauterbur, *J. Am. Chem. Soc.* **86**, 1870 (1964).

has been attributed[42] to the existence of a ring current in the strongly distorted bonds of this ring.

An alternative explanation is that the variations in susceptibility are related to changes in the Van Vleck paramagnetism for these systems. Since cyclopropane is flat, and cyclopentane nearly so, the magnitude of this paramagnetic term should be smaller in these compounds than in the larger cycloalkanes; the folded conformation of cyclobutane may induce a larger contribution from this

TABLE II

MAGNETIC SUSCEPTIBILITIES OF SMALL RING COMPOUNDS

Compound	χ_M (−10⁻⁶ cm³ mole⁻¹)	$\chi_M{}'$	$\varLambda$
Cyclopropane	39.2[a]	34.1	5.1
1,2-Bis(cyclopropyl)ethane	95.5 ± 0.5[b]	85.0	10.5
1,3,3-Trimethylcyclopropene	62.8 ± 0.3[b]	61.1	1.7
Cyclobutane	40.5[a]	45.4	−3.9
Cyclobutylcyclobutane	82.6 ± 0.3[b]	85.2	−2.6
Bicyclo[4.2.0]octa-2,4-diene	65.0 ± 1[b]	66.7	−1.7
Hexamethylbicyclo[2.2.0]hexa-1,5-diene	117.3 ± 0.4	117.5	−0.2
Cyclopentane	59.2[a]	56.8	2.4
Cyclopentene	49.5[c]	47.0	2.5

[a] C. Barter, R. G. Meisenheimer, and D. P. Stevenson, *J. Phys. Chem.* **64**, 1312 (1960).

[b] H. J. Dauben, Jr. and J. L. Laity, unpublished work (1968); J. L. Laity, Ph.D. Thesis, Univ. of Washington, 1968.

[c] H. J. Dauben, Jr., J. D. Wilson, and J. L. Laity, *J. Am. Chem. Soc.* **91**, 1991 (1969).

effect. Note that our data neither support nor deny the possibility of a diamagnetic ring current in cyclopropane; its exaltation can be rationalized on several grounds, none of which is strongly supported.

For our present purposes this failure of the Haberditzl System when applied to small-ring compounds is of concern only to the extent that it introduces a small uncertainty in $\chi_M{}'$ of five-membered ring compounds. It cannot be ascertained from the data available whether or not introducing unsaturation into small rings removes or reduces the discrepancies between χ_M and $\chi_M{}'$ observed for the saturated compounds, but it appears that in five-membered compounds the discrepancy may be reduced. In any event, we have chosen not to correct values of $\chi_M{}'$ for this, and the following data should be considered in this light.

[42] D. J. Patel, M. E. H. Howden, and J. D. Roberts, *J. Am. Chem. Soc.* **85**, 3218 (1963).

4. *The Least-Squares Increment System*

During the course of an investigation of the susceptibilities of alkyl benzenes, Laity found that both the Haberditzl and improved Pascal systems predicted the susceptibilities of these compounds significantly more poorly than might

TABLE III

A Least-Squares Semiempirical Increment System[a]

Group	χ $(-10^{-6} \text{ cm}^3 \text{ mole}^{-1})$	Group	χ $(-10^{-6} \text{ cm}^3 \text{ mole}^{-1})$
C_1—H	4.34	C_1*—H	4.18
C_2—H	4.07	C_2*—H	3.96
C_3—H	3.87	$C_1 \pi C_2$	2.87
C_1—C_1,	3.07	$C_2 \pi C_2$	2.85
C_1—C_2,		C—C*	1.98
C_1—C_3,		C*—C*	1.24
or C_2—C_2			
C_1—C_4,	2.88		
or C_2—C_3			
C_3—C_3,	3.37		
C_2—C_4,			
C_3—C_4,			
or C_4—C_4			
O_1—H	3.35	C—C(O_1)	3.37
C—O_1	3.82	C(O)—C(O)	4.41
C—O_2	2.28	C*—N_1	3.26
N_1—H	2.98	C*—N_2	2.00
N_2—H	1.99	C*—N_3	3.40
C_2—N_1	3.26	C_1—N_1	5.34
C_2—N_2	2.28	C_1—N_2	5.43
C_2—N_3	1.76	C_1—N_3	3.08
Benzene ring exaltation			14.54

[a] The numerical subscript below an atom gives the number of carbon atoms attached to that atom, e.g., two C_2—H bonds are present in the grouping

Starred symbols refer to sp^2 hybridized atoms; unstarred are in an sp^3 state of hybridization. The lines connecting atoms represent σ-bonds; π-bonds are denoted by π symbol placed between the atoms. Atomic symbols followed by another symbol in parentheses, as in C—C(O_1), refer to atoms with an atom of the kind parenthesized attached to them. Thus the above example refers to the carbon–carbon σ-bond in the grouping C—C—OH.

be expected in view of the large quantity of good susceptibility data available. He utilized these values and data for aliphatic hydrocarbons, alcohols, ethers, and amines to set up a new increment system, similar to that of Haberditzl but more refined.

As in the other systems used for estimating susceptibilities, it is assumed that Eq. (5) holds. Thus the susceptibility of a compound is represented as the sum of the susceptibility increments (of number n) associated with the different kinds of structural elements which comprise the molecule; in setting up a susceptibility-estimation system these increments are treated as unknowns. With the observed molar diamagnetic susceptibilities of many compounds (of number q), one obtains a system of q equations in n unknowns ($q > n$). These equations can be treated by a least-squares method, and are easily solved by matrix methods with the aid of a computer (one example of a computer program for the least-squares solution of an over-determined system of simultaneous equations is given by Wiberg).[43]

When applied to the data assembled by Laity this empirical method led to excellent agreement between observed and calculated molar susceptibilities,[44] but the increment values obtained tended to be large (in absolute value), physically meaningless, and strongly dependent on the particular set of data used in the treatment. To circumvent these difficulties Baudet's theoretical estimates of the susceptibilities of several bonds were included in the treatment as experimental data. This tactic was successful, and a very good set of increments was obtained. A fuller description of this work and a demonstration of its success in predicting the susceptibilities of the several classes of compounds to which it can be applied can be found in Laity's thesis.[23] The increment values obtained are given in Table III.

IV. Exaltation Data and Their Interpretation

In the preceding pages the discovery and theoretical foundation of magnetic susceptibility exaltation have been described. We shall present in this section the evidence that was amassed to show exaltation to be a valid and useful criterion of aromaticity, demonstrate its application to several problem compounds, and then discuss some further applications to which it can be extended.

A. EXALTATION AS A CRITERION FOR AROMATICITY

The validity of the criterion was first tested experimentally by computing the exaltations of an extensive series of cycloalkanes, cycloalkenes, cyclopolyenes,

[43] K. B. Wiberg, "Computer Programming for Chemists," p. 47. Benjamin, New York, 1965.

[44] Agreement between the calculated and observed susceptibilities of aromatic compounds is only found when the calculated value includes the exaltation, i.e., $\chi_M = \chi_M' + \Lambda$.

and a large number of benzenoid aromatic hydrocarbons.[9] The results of these and other computations are given in Table IV; they will be summarized and briefly highlighted here.

Among the monocyclic neutral hydrocarbons studied, only cyclopentadiene, benzene, cycloheptatriene, and their derivatives exhibit significant exaltations. Benzene is of course aromatic; if it did not have an exaltation, nothing would be expected to. Cycloheptatriene is also aromatic; this conclusion is supported by spectral[45] and thermochemical[46] evidence as well as our present results. We hypothesize that conjugation is effected by overlap of the indented π-orbitals of the 1- and 6-carbon atoms in the slightly buckled ring,[45] and the magnitude of the exaltations of the methyl-substituted cycloheptatrienes supports this hypothesis. However, we cannot satisfactorily rationalize the exaltation of cyclopentadiene. We cannot decide between the possibility of hyperconjugative π-electron delocalization[47] and the chance that the Haberditzl method has failed here. The data are inconclusive. In any event, this single ambiguity does not invalidate the conclusion, which is supported by an enormous weight of other data, that exaltation is a phenomenon peculiar to aromatic systems.

The benzenoid hydrocarbons all exhibit satisfactorily large values of Λ. Note that, in agreement with theory,[48] the magnitude of Λ appears to be a function of the size of the aromatic system.

Similarly, the nonbenzenoid aromatics[49] (azulene, azupyrene, acepleiadylene, the bridged [10]annulenes, and the dihydropyrenes) exhibit large exaltations, whereas the pseudoaromatic compounds[49, 50] (cyclooctatetraene, [16]annulene, the fulvenes, heptalene, heptafulvalene, and the dibenzopentalenes) all exhibit essentially zero exaltation (Fig. 1). These results show that large diamagnetic susceptibility exaltation is a property common to both benzenoid and nonbenzenoid aromatic hydrocarbons, but one that is absent from cycloalkenes and pseudoaromatic compounds.

[45] R. E. Davis and A. Tulinsky, *Tetrahedron Letters*, p. 839 (1962); M. Tratteborg, *J. Am. Chem. Soc.* **86**, 4265 (1964).

[46] R. B. Turner, *Theoret. Org. Chem., Papers Kekule Symp.*, *London*, 1958, p. 67. Academic Press, New York, 1959.

[47] R. S. Mulliken, *J. Chem. Phys.* **7**, 339 (1939); C. A. Coulson, "Valence," 2nd ed., p. 312. Oxford Univ. Press, London and New York, 1961; G. W. Wheland, "Resonance in Organic Chemistry," pp. 672–675. Wiley, New York, 1955.

[48] B. Pullman and A. Pullman, "Les théories électroniques de la chimie organique," Chapter IX. Masson, Paris, 1952.

[49] See Fig. 1 for the structures of some of these compounds.

[50] We consider "pseudoaromatic"[6a] cyclic compounds that possess fully conjugated peripheries but not the ground state stabilization and other properties which are the results of full delocalization. However, their properties are sufficiently distinctive (primarily, because they can be easily transformed into derivatives of aromatic systems by simple redox or semiaddition reactions[32]) such that "polyenic" does not adequately describe them.

TABLE IV

Diamagnetic Exaltation Data

Compound	χ_M	$\chi_M{}'$	Λ
		$(-10^{-6} \text{ cm}^3 \text{ mole}^{-1})$	
Monocyclic hydrocarbons			
Cyclopentane	59.2	56.8	2.4
Cyclopentene	49.5 ± 0.5^b	47.0	2.5
Cyclopentadiene	$44.5^{a,\,b}$	38.0	6.5
5,5-Dimethylcyclopentadiene	67.5 ± 0.6^b	62.7	4.8
Cyclohexane	68.1^a	68.1	0.0
Cyclohexene	57.5^a	58.3	−0.8
1,3-Cyclohexadiene	48.6^a	49.3	−0.7
1,4-Cyclodexadiene	48.7^a	48.5	0.2
Benzene	54.8^a	41.1	13.7
Cycloheptane	78.9 ± 0.7^b	79.5	−0.6
Cycloheptene	69.3 ± 0.6^b	69.7	−0.4
1,3-Cycloheptadiene	61.0 ± 0.6^b	60.7	0.3
1,4-Cycloheptadiene	61.0 ± 0.4^b	59.9	1.1
1,3,5-Cycloheptatriene	59.8 ± 1.0^b	51.7	8.1
1,6-Dimethyl-1,3,5-cycloheptatriene	84.3 ± 0.3^b	76.0	8.3
3,7,7-Trimethyl-1,3,5-cycloheptatriene	95.6 ± 0.4^b	88.5	7.1
7,7′-Bis(cycloheptatrienyl)	$119 \ \ \pm 3^b$	100	19
Cyclooctane	91.4^a	90.8	0.6
Cyclooctene	80.5 ± 0.6^b	81.0	−0.5
1,3-Cyclooctadiene	72.8 ± 0.8^b	72.0	0.8
1,5-Cyclooctadiene	71.5 ± 0.7^b	71.2	0.3
1,3,5-Cyclooctatriene	65.1 ± 0.8^b	64.0	1.1
Cyclooctatetraene	$53.9^{c,\,b}$	54.8	−0.9
1,4,7-Cyclononatriene	72.5 ± 1^b	72.8	0.0
Cyclododecene	$127 \ \ \pm 2^b$	126.4	0.6
[16]Annulene	$105 \ \ \pm 2^b$	110	−5
Benzenoid aromatic compounds			
Benzene	54.8^a	41.1	13.7
Toluene	66.1^a	53.3	12.8
Styrene	68.2^a	55.6	12.6
Indene	$80.5^{a,\,b}$	61.4	19.1
Fluorene	110.5^a	84.8	25.7
Triphenylmethane	166^a	125	41
Stilbene	120^a	82	28
1,4-Diphenylbutadiene	130^a	106	24
Biphenyl	103.3^a	77.1	26.2
p-Diphenylbenzene	152^a	113	39
4,4′-Diphenylbiphenyl	201^a	149	52
Biphenylene (**1**)	88 ± 3^b	74	14

TABLE IV (*continued*)

Compound	χ_M	$\chi_M{}'$ $(-10^{-6} \text{ cm}^3 \text{ mole}^{-1})$	Λ
Benzenoid aromatic compounds			
Naphthalene	91.9[a]	61.4	30.5
Anthracene	130.3[a]	81.7	48.6
Phenanthrene	127.9[a]	81.7	46.2
Tetracene	168[a]	102	66
Chrysene	167[a]	102	65
Pentacene	205[a]	122	83
Dibenz[*a,h*]anthracene	193[d]	122	71
Fluoranthene	138[a]	96	42
Pyrene (**2**)	155[a]	98	57
Triphenylene	157[a]	107	50
Perylene	171[e]	121	50
Benzo[*a*]pyrene	194[a]	119	75
Coronene	243[a]	140	103
Ovalene	354[a]	181	173
Nonbenzenoid aromatic and pseudoaromatic compounds			
Azulene	91.0[f]	61.4	29.6
Azupyrene (**3**)	151 ± 4[b]	98	53
Acepleiadylene (**4**)	155 ± 5[b]	98	57
1,6-Methano[10]annulene (**5**)	111.9 ± 0.4[b]	75.1	36.8
1,6-Oxido[10]annulene (**6**)	108.0 ± 0.5[b]	69.1	38.9
trans-15,16-Dimethyl-15,16-dihydropyrene	210 ± 15[b]	129	81
1,3,6,8,15,16-Hexamethyl-15,16-dihydropyrene	250 ± 20[b]	178	72
Acenaphthylene	111.6[a]	72.3	39.3
Acenaphthene	109.3[a]	82.4	26.9
Acepleiadiene	135 ± 3[b]	106	29
2-Phenyl-5,7-dimethylpleiapentalene	179 ± 4[b]	149[a]	30[a]
3,5-Dimethylaceheptalene	112 ± 3[b]	112	0
Pentafulvene (**7**)	43.0[g]	41.9	1.1
Cyclooctatetraene	53.9[c, b]	54.8	−0.9
[16]Annulene	105 ± 2[b]	110	−5
Heptafulvalene (**8**)	94 ± 3[b]	92.0	2
Heptalene (**9**)	72 ± 7[b]	78.2	−6
9-10-Dimethyldibenzopentalene (**10**)	132[h]	146[a]	−14[a]
7,7-Dimethylbenzofulvene	105[h]	103[a]	2[a]
7-Phenylbenzofulvene	131[h]	130[a]	1[a]
Heterocyclic compounds			
Pyrrole	47.6[a]	37.4	10.2
2,4-Dimethylpyrrole	69.9[a]	60.0	9.6

TABLE IV (*continued*)

Compound	χ_M	$\chi_M{}'$	Λ
	$(-10^{-6}$ cm^3 mole$^{-1})$		
Heterocyclic compounds			
2,3,5-Trimethylpyrrole	82.3[a]	71.3	11.0
Furan	43.1[a]	34.2	8.9
Thiophene	57.4[a]	44.4	13.0
Pyrazole	42.6[b]	36.0	6.6
3,5-Dimethyl-1,2-oxazole	59.7[a]	51.5	8.2
1,3-Thiazole	50.6[a]	38.3	12.3
3,4-Dimethyl-1,2,5-oxadiazole	57.2[a]	46.2	11.0
1,3,4-Thiadiazole	37.3[a]	32.2	5.1
N-Phenylsydnone	88.1[i]	77.1[q]	11.0[q]
N-p-Tolylsydnone	98.0[i]	88.4[q]	9.6[q]
Pyridine	49.2[a]	35.8	13.4
Pyrazine	37.6[a]	30.5	7.1
Borazine	49.6[j]	41.9[r]	7.7
Hexamethylborazine	119.0[j]	109.6[r]	9.4
Indole	85.0[a]	57.7	27.3
Quinoline	86.0[a]	56.1	29.9
Isoquinoline	83.9[a]	56.1	27.8
Carbazole	117.4[a]	78.0	39.4
Acridine	123.3[a]	76.4	46.9
Phenothiazine	114.8[a]	82.2	32.6
Mesoporphyrin dimethyl ester	585[k]	352	233
Protoporphyrin dimethyl ester	595[k]	370	225
Ketoaromatics			
N-Ethylpyridone	74 ± 1[l]	67.3	7
Cytosine (4-amino-2-pyrimidol)	55.8[a]	51.4	4.4
Thymine (6-methyl-2,4-pyrimidiol)	57.1[a]	59.9	−2.8
Barbituric acid	53.8[a]	55.4	−1.4
Uric acid	66.2[a]	62.9	3.3
Tropone	50.0[m]	46.2	7.8
Tropolone	61.0[a]	51.6	9.4
Coumarin	82.5[a]	65.3	17.2
Xanthone	108.1[a]	85.6	22.5
N,N'-Dimethyl aminotroponimine	96 ± 2[n]	89.7	6.3
Aromatic cations[s]			
Tropenylium	56 ± 1[n]	39.0	17
Methyltropenylium	66 ± 2[n]	50.3	16
Phenyltropenylium	104 ± 1[n]	87.5[q]	16.5[q]
Hydroxytropenylium	60 ± 2[n]	44.5	15.5
Triphenylcyclopropenium	172 ± 6[n]	132[t]	40
Triphenylcarbenium	156 ± 5[n]	122[t]	34

TABLE IV (continued)

Compound	χ_M	χ_M'	Λ
		$(-10^{-6} \text{ cm}^3 \text{ mole}^{-1})$	
Aromatic cations[s]			
Pyridinium	46 ± 2[n]	41.4	5
1,3,5-Trimethylpyrylium	77 ± 3[n,p]	68.0	9
N-Methylpyridinium	52 ± 2[n]	47.0	5
N-Methylthiazolium	53 ± 2[n]	48.0	5
Quinolinium	85 ± 2[n]	61.3	24

[a] G. W. Smith, "A Compilation of Diamagnetic Susceptibilities," Gen. Motors Corp. Res. Rept. GMR-317. Gen. Motors Corp., Detroit, Michigan, 1960; "Supplement to GMR-317," GMR-396, 1963.

[b] H. J. Dauben, Jr., J. D. Wilson, and J. L. Laity, J. Am. Chem. Soc. 90, 811 (1968); 91, 1991 (1969); also see Wilson[32] and Laity.[23]

[c] S. Shida and S. Fujii, Bull. Chem. Soc. Japan 24, 173 (1951).

[d] K. Lonsdale and K. S. Krishnan, Proc. Roy. Soc. A156, 597 (1936).

[e] H. Shiba and G. Hazato, Bull. Chem. Soc. Japan 22, 92 (1949).

[f] W. Klemm, Ber. 90, 1051 (1957).

[g] J. Thiec and J. Weimann, Bull. Soc. Chim. France p. 177 (1956).

[h] E. D. Bergmann, J. Hoarau, A. Pacault, B. Pullman, and A. Pullman, J. Chim. Phys. 49, 472 (1952).

[i] Y. Matsunaga, Bull. Chem. Soc. Japan 30, 177 (1957).

[j] H. Watanabe, K. Ito, and M. Kubo, J. Am. Chem. Soc. 82, 3294 (1960).

[k] R. Havemann, W. Haberditzl, and P. Grzegorzewski, Z. Physik. Chem. (Leipzig) 217, 91 (1961).

[l] J. D. Wilson, unpublished work (1967).

[m] T. Nozoe, Proc. Japan Acad. 28, 477 (1952); Chem. Abstr. 48, 2678c (1952).

[n] H. J. Dauben, Jr., J. L. Laity, and J. D. Wilson, to be published; see also Laity[23] and Wilson.[32]

[p] G. Havemann, W. Haberditzl, and V. Koeppel, Z. Physik. Chem. (Leipzig) 218, 288 (1961).

[q] Normal phenyl exaltation included in χ_M' and not Λ.

[r] Estimated assuming the increments $B^*-N^* \approx C^*-C^*$, $B\pi N \approx C\pi C$, and $B^*-H \approx C^*-H \approx N^*-H$. As a result the uncertainty in Λ is rather high.

[s] The susceptibilities of the cations are obtained by subtracting the anion susceptibilities from the measured salt susceptibilities. See Laity[23] for more detail.

[t] No Van Vleck correction for charge included.

We discussed above that the uncertainty in values of χ_M' for heterocycles is somewhat larger than it is for hydrocarbons. Despite this, the aromatic heterocycles all exhibit substantial exaltations. It is particularly noteworthy that two unusual ring systems, the sydnones and borazines, both show exaltations, and that values of exaltation are smaller among the "ketoaromatics" (tropone, pyridone, cytosine, etc.), reflecting the low efficiency of cross-conjugation. Note also the large exaltation of the porphyrin ring system.

The cationic aromatic systems all show exaltations, although those of the aromatic "onium" ions are small enough to cause some doubts. These doubts arise because only a very rough estimate of the effect of charge on the Van

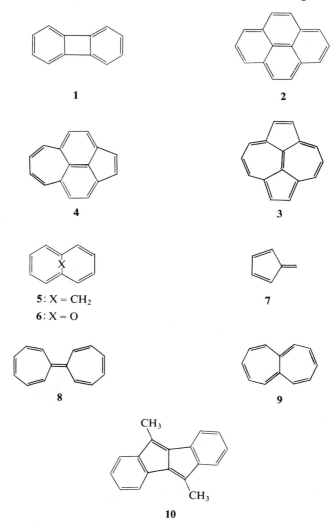

FIG. 1. The structures of certain aromatic and pseudoaromatic compounds.

Vleck paramagnetism (see Section III) can be made for these systems, and this appears to be comparable with Λ. Theory predicts[38] that the exaltation of pyridine and similar systems should be reduced on protonation; this appears to be borne out.[51]

[51] For pyridine the decrease in Λ on protonation amounts to 8 ± 2, while for quinoline it is 6 ± 2, the same within experimental error.

Such results demonstrate that there exists a strong correlation between the exhibition of exaltation and the possession of aromatic character. For six-membered and larger ring compounds, there are no exceptions to this rule: the susceptibilities of all the obviously nonaromatic cyclopolyenes are estimated within experimental error by the Haberditzl method; the susceptibilities of all the verifiably aromatic compounds are significantly underestimated by that method. All aromatics—neutral hydrocarbons, heterocycles, hydrocarbon ions, heterocyclic ions—exhibit exaltation, and no other kinds of compounds do so within the limits of the method used to get χ_M'. These results totally justify our conclusion that exaltation is a valid criterion of aromaticity. The method provides a clear demarcation between aromatic and nonaromatic compounds.

The exaltation criterion requires no data that are very difficult to obtain, as bond-length criteria do. It is easy to apply and needs no sensitive evaluation of the magnitude of accompanying effects, as do the NMR criteria. Because of these advantages, and because it is now amply verified and documented, magnetic susceptibility exaltation stands as the most satisfactory experimental test for aromatic character yet devised.

B. THE DETERMINATION OF AROMATIC CHARACTER IN CERTAIN UNUSUAL SYSTEMS

The utility of any criterion for aromaticity must lie in its ability to classify nontrivial systems as either aromatic or nonaromatic. A particularly important distinction is that between aromatic and pseudoaromatic nonbenzenoid hydrocarbons. Disagreements among the various, early, approximate theories when they were applied to these kinds of compounds[52] initially gave impetus to this field of research; only recently has any consensus been reached on their classification. We observed that the nonbenzenoid hydrocarbons generally conceded to be aromatic showed exaltation, whereas those thought not to be aromatic (the *pseudo*aromatic compounds) did not. While these results only confirm previous conclusions, the demonstration of a clear demarcation between the two classes of compounds is particularly welcome.

Some further conclusions will be drawn from these exaltation data, however it is worth noting here that the values of exaltation of azulene and naphthalene are the same within experimental uncertainty, as are the exaltations of pyrene, azupyrene, and acepleiadylene. We infer from this that, as far as the π-electrons are concerned, frameworks made up of equal numbers of five- and seven-membered rings are equivalent to those made up entirely of six-membered rings. This result is predicted by relatively simple molecular-orbital calculations,[15] and being correct it confirms the general soundness of their application to π-electron molecules.

[52] For example, see Craig's article.[6]

We might also note that several pseudoaromatic compounds ([16]annulene, **9**,[49] and **10**) have values of Λ which are significantly negative, i.e., $\chi_M < \chi_M'$.[2] This phenomenon arises because of first-order π-orbital paramagnetism[15, 53] (or "paramagnetic ring current").[54]

We observed above that the borazine and sydnone heteroaromatic systems exhibit exaltations typical of aromatic compounds. These data confirm the conclusions of Watanabe *et al.*[55] and Matsunaga,[56] respectively, concerning these systems, and further demonstrate the broad utility of exaltation as a test for aromaticity.

1. *Acenaphthylene and Similar Compounds*

Considerable interest attends the class of perifused, tricyclic systems of which acenaphthylene is the best known example. The four simplest, acenaphthylene, pleiadiene, pleiapentalene (perhaps more accurately "aceazulene"),

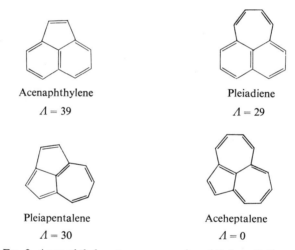

Acenaphthylene
$\Lambda = 39$

Pleiadiene
$\Lambda = 29$

Pleiapentalene
$\Lambda = 30$

Aceheptalene
$\Lambda = 0$

FIG. 2. Acenaphthylene-type compounds and their exaltations.

and aceheptalene (Fig. 2), exhibit chemical properties typical of both "aromatic" (e.g., electrophilic substitutions) and "olefinic" (e.g., electrophilic additions) compounds, and NMR spectra that do not permit a clearer distinction.[32] Yet exaltation data clearly show the first three to be aromatic

[53] G. Wagnière and M. Gouterman, *Mol. Phys.* **5**, 621 (1962); H. C. Longuet-Higgins, *Chem. Soc. (London), Publ.* **21**, 109 (1967); F. Baer, H. Kuhn, and W. Regel, *Z. Naturforsch.* **22a**, 103 (1967).

[54] J. A. Pople and K. G. Untch, *J. Am. Chem. Soc.* **88**, 4811 (1966).

[55] H. Watanabe, K. Ito, and M. Kubo, *J. Am. Chem. Soc.* **82**, 3294 (1960).

[56] Y. Matsunaga, *Bull. Chem. Soc. Japan* **30**, 227 (1957).

(pleiadiene presumably only because of the naphthalene nucleus) and the last as not.

When applied to these compounds, theory has not performed well. Conflicting predictions concerning the aromatic nature of pleiapentalene and aceheptalene have been derived from computations by Asgar-Ali and Coulson[57] and Jung and Hafner,[58] with the predictions of the latter in good agreement with what we have found. Further, the exaltation of acenaphthylene was seriously overestimated by Pullman and Pullman[48] (see Section IV,C). Why these molecules respond so poorly to theoretical treatment is not clear; more work is definitely indicated. In particular, the complete lack of exaltation in aceheptalene is puzzling; one might have expected, perhaps naively so, that the azulene nucleus would have manifested itself.

2. Homoaromaticity

We know exaltation to be a valid indicator of aromaticity in classical systems (i.e., those with fully conjugated peripheries). It should be an equally effective indicator for nonclassical systems, of which Winstein's "homoaromatic" systems[59] are the best known. Several of these compounds have been examined, and the results are discussed here.

Much of the experimental evidence for the presence of aromaticity in homotropenylium (**11**, $X = H$) and substituted homotropenylium cations consists of their NMR spectra.[60-63] The chemical shift positions of the ring

11

hydrogens fall in the aromatic carbonium ion range, and the large chemical shift differences (5.8 ppm for **11**, $X = H$) between the two hydrogens of the

[57] M. Asgar-Ali and C. A. Coulson, *Mol. Phys.* **4**, 65 (1961).

[58] D. Jung and K. Hafner, personal communication to H. J. Dauben, Jr. (1967).

[59] S. Winstein, *J. Am. Chem. Soc.* **81**, 6524 (1959); S. Winstein and J. Sonnenberg, *ibid.* **83**, 3244 (1961); R. J. Piccolini and S. Winstein, *Tetrahedron* **19**, Suppl. 2, 423 (1963).

[60] S. Winstein, *Chem. Soc. (London), Spec. Publ.* **21**, 5-45 (1967).

[61] We use "homotropenylium" here, replacing the more popular term "homotropylium," and in keeping with systematic tropenylium ion nomenclature. See H. J. Dauben, Jr., and D. F. Rhoades, *J. Am. Chem. Soc.* **89**, 6764 (1967).

[62] J. L. von Rosenberg and R. Pettit, *J. Am. Chem. Soc.* **85**, 2531 (1963); C. E. Keller and R. Pettit, *ibid.* **88**, 604 and 606 (1966); W. Merk and R. Pettit, *ibid.* **90**, 814 (1968).

[63] S. Winstein, H. D. Kaez, C. G. Kreiter, and E. C. Friedrich, *J. Am. Chem. Soc.* **87**, 3267 (1965); S. Winstein, C. G. Kreiter, and J. I. Baumann, *ibid.* **88**, 2047 (1966); M. Brookhart, M. Ogliaruso, and S. Winstein, *ibid.* **89**, 1965 (1967); R. F. Childs and S. Winstein, to be published; also Laity.[23]

methylene group are most striking. This methylene group chemical shift inequality indicates the presence of a substantial magnetic anisotropy of the type characteristic of aromatic rings.

The diamagnetic susceptibilities of several monohomotropenylium ions have been determined[23] and are given in Table V. Application of the Haberditzl increment system shows for homotropenylium cation a predicted susceptibility $\chi_M' = 58.9$, thus implying an exaltation for the ion of $72 - 58.9$ or 13, a value

TABLE V

EXALTATIONS OF "HOMOAROMATIC" SYSTEMS

Compound	$\chi_M{}^a$	χ_M' (-10^{-6} cm^3 mole^{-1})	Λ
Homotropenylium cation (**11**, X = H)	72 ± 2	51.0^b	21
11, X = CH$_3$	83 ± 2	62.3^b	21
11, X = C$_6$H$_5$	120 ± 2	99.5^{b,c}	20
11, X = OH	74 ± 5	56.4^b	19
1,3,5-Cycloheptatriene	60 ± 1	52.1	8
cis,cis,cis-1,4,7-Cyclononatriene (**12**)	72.5 ± 1	72.8	0
2,4,6-Cyclooctatrienone (**13**)	58.5 ± 0.5	58.0	0

^{a} Susceptibilities were determined by an NMR method; see Laity[23] for full experimental details. The susceptibilities of the anions (HSO$_4$$^-$ or FSO$_3$$^-$) have been subtracted from the measured values of the salts to give the cation susceptibilities.

^{b} The χ_M' and Λ values include the Van Vleck paramagnetic correction for positive charge of 8×10^{-6} cm^3 mole^{-1}.

^{c} The normal phenyl exaltation is included in χ_M' and not Λ.

nearly as high as that of benzene. As was noted earlier in this chapter, the susceptibilities of carbonium ions contain a weak (6–8×10^{-6} cm^3 mole^{-1}) Van Vleck paramagnetic term not included in the susceptibility increment systems (such as Haberditzl's) that were derived for neutral molecules. Since this paramagnetic term acts to lower the diamagnetic susceptibilities of cations, the exaltation of homotropenylium ion, as shown in Table V, is actually 21.

The relative exaltations of the substituted homotropenylium ions parallel the results obtained for tropenylium ions (see Section B,6). The methyl- and phenyl-substituted ions have the same exaltation and, thereby, the same aromaticity as the unsubstituted ion. 1-Hydroxyhomotropenylium ion also has the same exaltation, within experimental uncertainty (which is somewhat large for hydroxyhomotropenylium cation due to its slow, irreversible rearrangement to protonated acetophenone[63]). These exaltations establish that the aromatic nature of the homotropenylium cation (as with tropenylium cation) is not altered appreciably by the electronic effects of the methyl, phenyl, or hydroxy substituents.

The presence of homoaromatic C-1 to C-7 σ-type overlap of π-electrons in the structures of these homoaromatic ions results, not only in the possibility of π-electron delocalization about the ring, but also leads to a considerable amount of cyclopropane ring character in the molecules. This effect may explain why the exaltation of homotropenylium ion is substantially greater than that of tropenylium ion. If the exaltation $\Lambda = 5.2$ of a cyclopropane ring is subtracted from the exaltation $\Lambda \simeq 21$ of homotropenylium ion, the result is $\Lambda \simeq 16$, slightly less than the $\Lambda \simeq 17$ of tropenylium ion.

Diamagnetic susceptibility exaltation responds to both the presence of aromatic π-electron delocalization and, to a lesser extent, cyclopropane ring formation, thus making this aromaticity criterion a particularly sensitive tool for the study of monomethylene-bridged homoaromatic systems.

Monohomobenzene character accounts for the appreciable exaltation ($\Lambda \simeq 8$) noted above for the 1,3,5-cycloheptatriene ring system. However, 2,4,6-cyclooctatrienone (**12**, a potential homotropone) and *cis,cis,cis*-1,4,7-cyclononatriene (**13**, a trishomobenzene) exhibit $\Lambda \simeq 0$ and thus have no appreciable aromatic character.

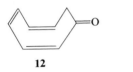

12

13

3. Polymethinium Ions

It has been suggested[64] that formamidinium ion and its vinylogues, the polymethinium ions, might be considered to be "linear" aromatic species, because they are stabilized by resonance and undergo electrophilic substitution reactions.[64,65] We have measured the susceptibilities and calculated the exaltations of a number of these: the results are presented in Table VI. They are found not to exhibit exaltation, and thus cannot be aromatic in the sense we use it.

However, these molecules adopt a uniform *s*-trans configuration in solution,[64,66] and therefore cannot be expected to exhibit exaltation. The ability

[64] H. J. Dauben, Jr., G. Feniak, and R. B. Lund, unpublished work; G. Feniak, Ph.D. Thesis, Univ. of Washington, 1955; R. B. Lund, Ph.D. Thesis, Univ. of Washington, 1960.

[65] J. Kucera and Z. Arnold, *Collection Czech. Chem. Commun.* **32**, 1704 (1967).

[66] S. S. Malhotra and M. C. Whiting, *J. Chem. Soc.* p. 3812 (1960); S. McGlynn and W. T. Simpson, *J. Chem. Phys.* **28**, 297 (1958); G. Scheibe, C. Jutz, W. Seiffert, and D. Grosse, *Angew. Chem.* **76**, 270 (1964); S. Dahne and J. Ranft, *Z. Phys. Chem.* **224**, 65 (1963).

TABLE VI

EXALTATIONS OF SOME POLYMETHINIUM SALTS

$$(CH_3)_2N\!-\!(CH\!=\!CH)_n\!-\!CH\!=\!\overset{+}{N}(CH_3)_3 \; \leftrightarrow \; (CH_3)_2\overset{+}{N}\!=\!CH\!-\!(CH\!=\!CH)_n\!-\!N(CH_3)_2$$

n	Compound (anion)	$\chi_M(salt)^a$	$\chi_M{}'(salt)^b$	Λ
		\(10^{-6} cm^3 mole^{-1}\)		
0	ClO_4^-	101 ± 2	103	-2
1	ClO_4^-	116 ± 2	117	-1
1	I^-	141 ± 3	137	-4
2	ClO_4^-	133 ± 2	131	2

[a] Data from J. L. Laity, Ph.D. Thesis, Univ. of Washington, 1968.
[b] The susceptibility of ClO_4^- was taken to be 33, and I^-, 52. The cation $(CH_3)_2C\!=\!N(CH_3)_2^+$ ($\chi_M = 61 \pm 1$) was used as a model for the susceptibility-increment of $=\!N(CH_3)_2^+$.

to delocalize around a closed path is a prerequisite for the establishment of a ring current, and these compounds obviously lack this ability.

4. Rapid Valence Isomerism vs. Delocalization. Bullvalene

Bullvalene (**14**) is an example of a molecule able to undergo rapid and

14

reversible valence isomerization. Doering and Roth[67] predicted and Schröder et al.[68] confirmed that above about 30°C bullvalene exhibits a single NMR resonance, a consequence of isomerization among 1.2×10^6 identical isomers which is "rapid" (on the NMR time scale) at those temperatures. Bullvalene, however, is not aromatic; it exhibits $\chi_M = 79.2 \pm 0.8$ and $\chi_M{}' = 74.0$, and thus $\Lambda = 5$, a value that can be accounted for either by its inclusion of a cyclopropane ring or by the fact that the compound is nearly spherical, and thus should possess a smaller Van Vleck paramagnetism than the model compounds for the Haberditzl system.[34] The exaltation observed is thus the result of inadequate model compounds and not delocalization.

[67] W. E. Doering and W. R. Roth, *Angew. Chem. Intern. Ed. Engl.* **2**, 115 (1963); *Tetrahedron* **19**, 715 (1963).
[68] G. Schröder, J. F. M. Oth, and R. Merényi, *Angew. Chem. Intern. Ed. Engl.* **4**, 752 (1965); G. Schröder and J. F. M. Oth, *ibid.* **6**, 414 (1967).

In contrast, the results of theoretical calculations by Wulfman[69] suggest that semibullvalene (**15**) has a delocalized ground state and should exhibit exaltation. This prediction awaits experimental testing.

15

5. *The Effects of Severe Ring Strain*

The effects of severe steric strain on aromatic nuclei has been a popular subject for investigation in the past several years. Numerous highly strained benzene derivatives have been synthesized and their properties determined,

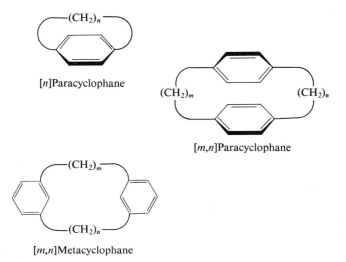

FIG. 3. Cyclophanes.

particularly with a view toward assessing the effect of strain on the π-electronic distribution of aromatic nuclei. Two classes of highly strained aromatics are the cyclophanes (Fig. 3) and benzenes highly substituted with bulky alkyl groups. The exaltations of a number of compounds from both classes were measured in order to determine if exaltation is affected by strain.

It is well known that the cyclophanes include several highly strained benzene derivatives, particularly the [8]-, [9]-, [2.2]-, [2.4]-, and [3.3]paracyclophanes

[69] D. Wulfman, private communication to J. D. Wilson, 1969; see also H. E. Zimmerman, R. W. Binkley, R. S. Givens, G. L. Grunewald, and M. A. Sherwin, *J. Am. Chem. Soc.* **91**, 3316 (1969).

and [2.2]metacyclophane.[70] The exaltation data obtained for these and several other cyclophane systems is given in Table VII. It can be seen that most of the cyclophanes exhibit the exaltation expected for the number of benzene nuclei they include, but that some have significantly reduced exaltation.

TABLE VII

EXALTATIONS OF CYCLOPHANES

Compound[a]	$\chi_M{}^b$	$\chi_M'{}^c$	Λ	$\Lambda/\Lambda_{bz}{}^d$
[12]PCP	184 ± 3	169.8	14	1.0 ± 0.2
[9]PCP	151.0 ± 0.5	135.8	15.2	1.04 ± 0.05
[8]PCP, 3-hydroxy	146 ± 3	130.5	15	1.1 ± 0.2
[6.6]PCP	233 ± 3	203.3	30	2.0 ± 0.2
[4.4]PCP	187 ± 3	157.9	29	2.0 ± 0.2
[1.8]PCP	197 ± 3	169.3	28	1.9 ± 0.2
[3.4]PCP	170 ± 4	146.6	23	1.6 ± 0.3
[2.4]PCP, 2,3-dicarbomethoxy	213 ± 2	188.2	25	1.7 ± 0.2
[3.3]PCP	157 ± 2	135.2	22	1.5 ± 0.15
[2.2]PCP, 4-ethyl	157 ± 2	135.1	22	1.5 ± 0.15
[2.2]metacyclophane	146 ± 2	112.5	33	2.2 ± 0.15

[a] The abbreviation PCP stands for paracyclophane.

[b] H. J. Dauben, Jr., J. L. Laity, and D. J. Cram, unpublished work (1967); J. L. Laity, Ph.D. Thesis, Univ. of Washington, 1968.

[c] The χ_M' values were obtained using the least-squares increment system.

[d] Λ_{bz} represents the exaltation of benzene.

It is most noteworthy that the very strained [8]- and [9]paracyclophanes and [2.2]metacyclophane show no reduction of exaltation; this forces us to conclude that in the cyclophanes strain alone does not distort the π-systems enough to affect the exaltation. The significantly reduced exaltations found in the [2.2]-, [2.4]-, and [3.3]paracyclophanes must therefore stem from some cause other than strain. Other evidence suggests that the π-systems of the two abnormally close benzene rings in these compounds interact rather strongly[71]; this transannular interaction may reduce the exaltation by altering the electronic distribution so that normally the most stable configuration—presumably that which allows the London diamagnetism to be a maximum—is no longer energetically favored.

[70] X-ray crystallographic studies of some of these have been reported: P. K. Gantzel, C. L. Coulter, and K. N. Trueblood, *Angew. Chem.* **72**, 755 (1960); *Acta Cryst.* **13**, 1042 (1960); P. K. Gantzel and K. N. Trueblood, *ibid.* **18**, 958 (1965); C. J. Brown, *J. Chem. Soc.* pp. 3265 and 3278 (1953).

[71] D. J. Cram and R. C. Helgeson, *J. Am. Chem. Soc.* **88**, 3515 (1966), and references therein.

TABLE VIII

SUSCEPTIBILITIES OF ALKYL BENZENES

Compound	χ_M	$\Delta\chi_M bz^c$	n	$\Delta\chi_M/n$
Methyl-substituted benzenes				
Toluene	66.1^a	11.3	1	11.3
o-Xylene	77.8^a	23.0	2	11.5
m-Xylene	76.6^a	21.8	2	10.9
p-Xylene	76.8^a	22.0	2	11.0
Mesitylene	88.3 ± 0.3^b	33.5 ± 0.3	3	11.2 ± 0.1
Durene	100.5 ± 1^b	45.7 ± 1	4	11.4 ± 0.25
Hexamethylbenzene	122.5 ± 1^b	67.7 ± 1	6	11.3 ± 0.2
Ethyl-substituted benzenes				
Ethylbenzene	77.2^a	22.4	1	22.4
1,3-Diethylbenzene	99.6 ± 0.3^b	44.8 ± 0.3	2	22.4 ± 0.15
Hexaethylbenzene	189.0 ± 2^b	134.2 ± 2	6	22.4 ± 0.3
n-Propyl-substituted benzenes				
n-Propylbenzene	89.2^a	34.4	1	34.4
1,4-Di-n-propylbenzene	123.0 ± 1^b	68.2 ± 1	2	34.1 ± 0.5
Hexa-n-propylbenzene	259.0 ± 2^b	204.2 ± 2	6	34.0 ± 0.3
i-Propyl-substituted benzenes				
i-Propylbenzene	89.0^a	34.2	1	34.2
1,4-Di-i-propylbenzene	124.0 ± 1^b	69.2 ± 1	2	34.6 ± 0.5
1,2,4,5-Tetra-i-propylbenzene	192.5 ± 1^b	137.7 ± 1	4	34.4 ± 0.25
Hexa-i-propylbenzene	262.0 ± 3^b	207.2 ± 3	6	34.5 ± 0.3
t-Butyl-substituted benzenes				
t-Butylbenzene	101.5 ± 0.5^b	46.7 ± 0.5	1	46.7 ± 0.5
1,2-Di-t-butylbenzene	146.8 ± 0.5^b	92.0 ± 0.5	2	46.0 ± 0.25
1,4-Di-t-butylbenzene	148.0 ± 1^b	93.2 ± 1	2	46.6 ± 0.5
1,4,4-Tri-t-butylbenzene	194.0 ± 1^b	139.2 ± 1	3	46.4 ± 0.3
1,3,5-Tri-t-butylbenzene	196.0 ± 1^b	141.2 ± 1	3	47.1 ± 0.3
1,2,4,5-Tetra-t-butylbenzene	239.0 ± 5^b	184.2 ± 5	4	46.1 ± 1.3

[a] G. W. Smith, "A Compilation of Diamagnetic Susceptibilities of Organic Compounds," Gen. Motors Corp. Res. Rept. GMR-317. Gen. Motors Corp., Detroit, Michigan, 1960.

[b] H. J. Dauben, Jr., J. L. Laity, and E. M. Arnett, unpublished work (1967); J. L. Laity, Ph.D. Thesis, Univ. of Washington, 1968.

[c] $\Delta\chi_M bz = \chi_M - \chi_M(benzene)$.

Arnett and his co-workers have synthesized several examples of benzenes strained by multiple substitution with large alkyl groups,[72] and the suscepti-

[72] For example, see E. M. Arnett and J. M. Bollinger, *J. Am. Chem. Soc.* **86**, 4729 (1964); E. M. Arnett, J. C. Sanda, J. M. Bollinger, and M. Barber, *ibid.* **89**, 5389 (1967), and references therein.

bilities of some of these were determined to search for reductions of exaltation (Table VIII). The results are most conveniently presented graphically (Fig. 4) by plotting the susceptibility of the compounds against the number of substituents on each ring. From this it is seen that the increase in susceptibility

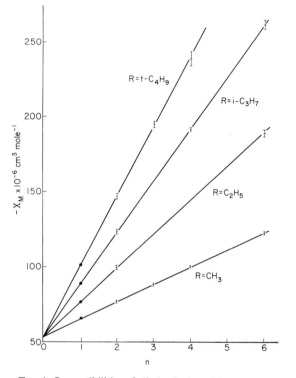

FIG. 4. Susceptibilities of alkyl-substituted benzenes $C_6H_{6-n}R_n$.

with substitution is *linear in each case*, and we infer that, as in the cyclophanes, introduction of distortion into the benzene nucleus by multiple substitution of bulky alkyl groups (as in hexaisopropyl-; 1,2,4-tri-*t*-butyl-; and 1,2,4,5-tetra-*t*-butylbenzene) does not significantly affect the aromaticity of the benzene ring.

The susceptibilities of *t*-butyl-substituted pyrroles confirm that this kind of strain does not affect aromatic character. From the data of Table IX it can be seen that the incremental change in substitution of a *t*-butyl group for a hydrogen atom on the pyrrole ring is the same (46.5), within experimental error, as on benzene, even for the change from unstrained 2,5-di-*t*-butylpyrrole to the strained 2,3,5-tri-*t*-butyl derivative. The exaltation of the pyrrole ring is unaltered by these substitutions.

TABLE IX

EXALTATIONS OF STRAINED BENZENES AND PYRROLES

Compound	χ_M	$\chi_M'{}^c$	Λ
	$(-10^{-6} \text{ cm}^3 \text{ mole}^{-1})$		
Indane	84.8 ± 0.3^b	70.0	14.8
Trindane	$143 \quad \pm 1^b$	128.7	14.3
Tetralin	93.3^a	79.0	14.3
Dodecahydrotriphenylene	$173 \quad \pm 5^b$	155.7	17.0
Pyrrole	47.6^a	37.4	10.2
2,5-Di-t-butylpyrrole	141.0 ± 0.6^b	130.4	10.6
2,3,5-Tri-t-butylpyrrole	186.0 ± 1.5^b	177.0	9.0

[a] G. W. Smith, "A Compilation of Diamagnetic Susceptibilities," Gen. Motors Corp. Res. Rept. GMR-317. Gen. Motors Corp., Detroit, Michigan, 1960.

[b] J. L. Laity, Ph.D. Thesis, Univ. of Washington, 1968.

[c] The χ_M' values were calculated using the least-squares increment system, in which Λ (benzene) = 14.5.

The possibility of bond alternation and diminished aromaticity in indane and related compounds has been a subject of considerable speculation.[73,74] From NMR evidence Meier et al.[73] recently deduced that the ring current is reduced in indane[75] and as-hydrindacene to about 70%, trindane about 40%, and in dodecahydrotriphenylene to about 60% of a full benzene ring current, and have proposed that this phenomenon is general among substituted aromatic compounds (Fig. 5). However, the exaltations of these compounds (Table IX) do not support the above deductions. Clearly, indane, trindane, and dodecahydrotriphenylene, along with the unstrained 1,2,3,4-tetrahydronaphthalene, have—within experimental uncertainty—the same exaltation as benzene. These compounds are therefore normally aromatic. This result is confirmed by an X-ray crystallographic examination of trindane, the most strained of this series of compounds, which has revealed no significant bond alternation or deviation from planarity in the benzene ring.[74]

The results described in the preceding few paragraphs make it abundantly clear that ring strain does not significantly affect the π-electronic distributions of benzene derivatives, since in no case could any change in exaltation be detected where strain alone was involved. This implies, very strongly, that fully delocalized π-systems are not much affected by distortions of their σ-frameworks from normal sp^2 geometry.

[73] H. Meier, E. Muller, and H. Suhr, Tetrahedron 23, 3713 (1967).

[74] E. R. Boyka and P. A. Vaughn, Acta Cryst. 17, 152 (1969).

[75] See Fig. 5.

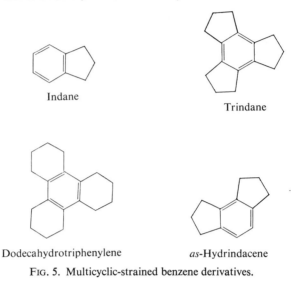

Indane

Trindane

Dodecahydrotriphenylene   *as*-Hydrindacene

FIG. 5. Multicyclic-strained benzene derivatives.

6. *Charge Localization in Substituted Tropenylium Ions*

It is reasonable to suppose that substituents on an aromatic ring might in some cases be able to perturb the π-system sufficiently to alter the exaltation.[76] Laity found that the exaltations of benzene and the homotropenylium ion were unaffected (within the limits of experimental error) by any one simple substituent,[23] but this was not the case for all the tropenylium ions studied. As can be seen from the data of Table X, the exaltations of certain substituted tropenylium ions are markedly smaller than that of the parent system.[39] Furthermore, there is an apparent correlation between the decrease in Λ and the ability of the substituent to stabilize a positive charge by resonance. Strongly electron-donating substituents tend to reduce exaltation. We interpret these data to mean that substantial withdrawal of charge from the ring tends to destabilize the fully delocalized molecular orbitals of the tropenylium ion π-electron system. That is, withdrawal of charge tends to localize the π-electrons, reducing their mobility; this causes the exaltation to be decreased. These results are reflected to a degree in the NMR spectra of the ions, with the chemical shifts of the ring-protons generally moving to higher field in the cations with lower exaltation.[77] It should also be reflected in lower resonance

[76] For a discussion of this point, see S. Sriraman and R. Sabesan, *Trans. Faraday Soc.* **58**, 1080 (1962).

[77] T. J. Pratt, R. B. Medz, W. R. Gresham, H. J. Dauben, Jr., and K. M. Harmon, to be published. See also these Ph.D. Theses from the University of Washington: K. M. Harmon (1958), T. J. Pratt (1964), R. B. Medz (1964), D. F. Rhoades (1969), and W. R. Gresham (1969).

TABLE X

EXALTATIONS OF SUBSTITUTED TROPENYLIUM IONS

Substituent	$\chi_M{}^{a,b}$	χ_M'	Λ
	$(-10^{-6} \text{ cm}^3 \text{ mole}^{-1})$		
Hydrogen	56 $\pm 1^c$	39	17
Methyl	66 ± 2	50	16
t-Butyl	101 ± 1	85.5	15.5
Phenyl	104 ± 1	87.5	16.5^d
Benzyl	114 ± 2	99	15^d
t-Butylethynyl	113 ± 1	99	14
Hydroxy	60 ± 2	44.5	15.5
Methoxy	69.5 ± 0.5	54.5	15.0
1,4-Dimethoxy	83 ± 2	70	13
Chloro	66 ± 1	54	12
Amino	55 ± 1	49	6
Dimethylamino	78 ± 1	74	4
N,N'-Dimethyl-1,2-ethylene-diamino	145 ± 2	139	6

^{a} χ_M(cation) = χ_M(salt) − χ_M(anion). Anion susceptibilities were taken from P. W. Selwood, "Magnetochemistry," 2nd ed., p. 78. Wiley (Interscience), New York, 1956; L. N. Mulay, "Magnetic Susceptibility," p. 1782. Wiley, New York, 1968; see also V. C. G. Trew, S. F. A. Husain, and A. J. Siddiji, *Trans. Faraday Soc.* **61**, 1086 (1965). These values were used: BF_4^-, $\chi_M = 39$; ClO_4^-, 33; SbF_6^-, 77. The standard error is about 1%.

^{b} H. J. Dauben, Jr., J. L. Laity, and J. D. Wilson, to be published; J. L. Laity, Ph.D. Thesis, Univ. of Washington, 1968.

^{c} See also V. I. Belova, M. E. Vol'pin, and Ya. K. Syrkin, *J. Gen. Chem. USSR (English Transl.)* **29**, 688 (1959).

^{d} Normal phenyl exaltation included in χ_M' and not in Λ.

energies and in a significant bond alternation in these species, but no significant amount of data on these properties are yet available.

C. THEORETICAL CALCULATIONS OF EXALTATION

Exaltation is supposed to be a manifestation of the theoretical quantity, the London diamagnetism. Should this truly be the case, and if the two quantities can be simply related, exaltation data can provide powerful tests of the validity of various kinds of quantum–mechanical calculations in this area. One of the more important needs in the theory of aromatic systems has existed for some phenomenon or quantity that is both easily calculable and readily measurable; it now appears that exaltation can fill that need.

We assumed (Section II) that aromatic compounds differ magnetically from nonaromatic compounds only in their possession of the extra diamagnetism due to ring current. Estimates of the value of this quantity for benzene have been made by theoretical methods[15] and from NMR data,[78] and their consensus gives $K_L \simeq 30$. This number can be used to check the accuracy of the

TABLE XI

EXALTATION AND LONDON DIAMAGNETISM VALUES FOR
AROMATIC COMPOUNDS

Compound	Λ/Λ (benzene)[a]	K_L/K_L (benzene)[b]
Benzene	1.0	1.00
Biphenyl	1.9	1.88
Terphenyl	2.9	2.74
Quaterphenyl	3.8	3.62
Biphenylene	1.0	−0.13, 0.72[c]
Naphthalene	2.2	2.22
Azulene	2.2	2.27
Anthracene	3.5	3.53
Phenanthrene	3.4	3.31
Tetracene	4.8	4.88
Chrysene	4.7	4.61
Pentacene	6.1	6.26
Dibenz[a,h]anthracene	5.1	5.58
Acenaphthylene	2.9	3.31
Pyrene	4.2	4.70
Fluoranthene	3.1	3.04
Triphenylene	3.6	4.33
Perylene	3.7	4.17
Coronene	7.5	10.13

[a] Calculated from data of Table IV.

[b] B. Pullman and A. Pullman, "Les théories électroniques de la chimie organique," p. 545. Masson, Paris, 1952 (except as otherwise noted).

[c] Derived from the data of H. P. Figeys, *Chem. Commun.* p. 495 (1967).

above assumption. If it is correct, then from Eq. (4) the observed susceptibility should equal that which a nonaromatic of the same formal structure would have, plus one-third the London diamagnetism. From (4) and (1) it follows that $3\Lambda = K_L$. However, for benzene $3\Lambda = 41$[79] and $K_L = 30$, and hence the

[78] B. P. Dailey, *J. Chem. Phys.* **41**, 2304 (1964).

[79] The value $K_L = -41 \times 10^{-6}$ cm^3 mole^{-1} is the theoretical maximum for the quantity, obtained by the Pauling method[16] for a circle of 1.40 Å radius, and probably is at least 25% too large.[15]

assumption must not be exact. Apparently aromatic compounds possess somewhat larger susceptibilities than nonaromatic compounds, even without the ring current. It seems likely that this is a result of the higher symmetry of the aromatics, since this would tend to decrease the contribution of the Van Vleck paramagnetism and give the result observed.

Even though the assumption we made is not exact, it is likely that $\Lambda \propto K_L$, i.e., that the error in the assumption is systematic. Although K_L data are not available for compounds other than benzene, the ratios K_L/K_L (benzene) have been computed[48] for a large set of condensed aromatic hydrocarbons, and these should fairly well represent the actual ratios.[15] In Table XI they are compared with values of Λ/Λ (benzene) calculated from the data of Table IV; the two sets of data are seen to be in reasonable agreement, and we can conclude that K_L and Λ are approximately proportional.

Theoretical methods of calculation can now be tested very simply by calculating the ratio K_L/K_L (benzene) and comparing this with the ratio of exaltations. Theoretical predictions of that quantity for several compounds have appeared in the literature; a few are presented here to illustrate its use.

1. [10]Annulene

Salem has predicted,[15] for [10]annulene in the shape of naphthalene, K_L/K_L (benzene) = 2.33. The compound itself has been observed only as a transient intermediate,[80] but the well-known[81] 1,6-methano(oxido, amido, etc.)-bridged derivatives are thought to possess π-systems not greatly different from that of their parent. We find for 1,6-methano[10]annulene, Λ/Λ (benzene) = 2.7, and for 1,6-oxido[10]annulene, 2.8. Both numbers are in quite good agreement with the predicted value, especially considering that the two compounds are poorer approximations to flat disks than are most aromatics, and should thus exhibit a slightly different relationship between K_L and Λ.

2. Biphenylene and Pseudoaromatic Compounds

Calculations of K_L/K_L (benzene) have been carried out for different geometries of several important compounds,[48, 82, 83] and the comparison (Table XII) of these with the value of Λ/Λ (benzene) obtained here is particu-

[80] E. E. Van Tamelen and T. L. Burkoth, *J. Am. Chem. Soc.* **89**, 157 (1967); S. Masamune and R. T. Seidner, *Chem. Commun.* p. 542 (1969).

[81] For reviews, see E. Vogel, *Chem. Soc. (London), Spec. Publ.* **21**, 113–147 (1967); F. Sondheimer, I. C. Calder, J. A. Elix, Y. Gaoni, P. J. Garratt, K. Grohman, G. de Maio, J. Mayer, M. V. Sargent, and R. Wolovksy, *ibid.* pp. 75–107.

[82] H. P. Figeys, *Chem. Commun.* p. 495 (1967).

[83] T. Nakajima, in "Molecular Orbitals in Chemistry, Physics, and Biology" (P.-O. Löwdin and B. Pullman, eds.), p. 457. Academic Press, New York, 1964; T. Nakajima and S. Kohda, *Bull. Chem. Soc. Japan* **39**, 805 (1966).

larly illuminating. The values of Pullman and Pullman[48] were obtained assuming equal bond lengths and regular geometry; those of Figeys[82] and Nakajima[83] were made for "minimum energy" bond-alternant models. The close agreement of Nakajima's data with experiment is probably fortuitous, but it is striking that in every case the "irregular" model predicts the ratio better than the "regular" one. These results confirm previous inferences[83, 84] that pseudoaromatic compounds exist as bond-alternant, nonplanar molecules, and demonstrate the rather strong dependence of the calculated value of K_L on the geometry assumed.

TABLE XII

THEORETICAL EXALTATIONS OF PSEUDOAROMATICS

Compound[a]	K_L/K_L (benzene)		Λ/Λ (benzene)
	"Regular"[b]	"Irregular"[c]	
Biphenylene	−0.13	+0.7	+1.0
Dibenzopentalene	−0.75	+0.99	+1.0
Heptalene	−8.34	−0.21	−0.45
Heptafulvalene	—	+0.12	+0.15
Fulvene	+0.08	+0.08	+0.08

[a] See Fig. 1 for structures of these compounds.

[b] B. Pullman and A. Pullman, "Les théories électroniques de la chimie organique," p. 545. Masson, Paris, 1952.

[c] T. Nakajima, in "Molecular Orbitals in Chemistry, Physics, and Biology" (P.-O. Löwdin and B. Pullman, eds.), p. 451. Academic Press, New York, 1964; T. Nakajima and S. Kohda, Bull. Chem. Soc. Japan 39, 804 (1966), except biphenylene, which is from H. P. Figeys, Chem. Commun. p. 495 (1967).

D. APPLICATION TO THE ESTIMATION OF CHEMICAL SHIFT

The ring current model for the magnetic properties of aromatic hydrocarbons has been severely criticized,[85, 86] largely, it appears, because the results of quantitative calculations of the chemical shift due to ring current have been very poor. Nevertheless the concept is justifiably popular, for it provides a very satisfactory qualitative model for the phenomena. We would like to suggest that the quantitative results have been poor at least partly because workers in the field have unjustifiably assumed that the ring current is almost

[84] G. M. Badger, "Aromatic Character and Aromaticity." Cambridge Univ. Press, London and New York, 1969.

[85] J. A. Pople, J. Chem. Phys. 41, 2559 (1964).

[86] J. I. Musher, J. Chem. Phys. 43, 4081 (1965); Advan. Magnetic Resonance 2, 177 (1967).

solely responsible for the generation of the magnetic susceptibility anisotropies observed in aromatic hydrocarbons. (These provide data apparently suitable for semiempirical calculations of $\delta_{\text{ring current}}$.) This assumption has recently been questioned,[15,78] and the very recent results of Hüttner and Flygare[87] make it practically untenable.

However, Dailey has found that the relative ring-proton chemical shifts of benzene, naphthalene, phenanthrene, anthracene, pyrene, perylene, and coronene correlate well with the relative values of the theoretically calculated (HMO) London diamagnetism K_L of these compounds[78,88]; and he has argued that a reasonable value of the actual $\delta_{\text{ring current}}$ for benzene can be obtained by adopting the best theoretical value[15,78] ($K_L = -30 \times 10^{-6}$ cm^3 mole^{-1}) for that quantity. We observed above that $K_L(\text{HMO}) \propto \Lambda$. This suggests that exaltation data could be used to estimate values of the chemical shift due to ring current much more successfully than anisotropy data have. They should be of greatest value in applications on nonalternant systems, where reliable theoretical calculations are more difficult to attain.

V. Suggestions for Future Work

Studies on magnetic susceptibility exaltation came to an end at the University of Washington with the untimely death of Professor Dauben, and neither JDW nor JLL presently plan to continue them elsewhere. Yet we have naturally become aware of some of the unsolved problems and unexplored areas for research in this field which could be of use to others. For this reason we offer the following suggestions for further research.

A. AN ACCURATE PREDICTIVE SYSTEM FOR THE MAGNETIC SUSCEPTIBILITIES OF COMPOUNDS OTHER THAN HYDROCARBONS AND ALCOHOLS

We have stated that present methods for estimating the susceptibilities of complicated heterocompounds are less accurate than is desirable. The main problem is a lack of data, particularly, reliably accurate data for unsaturated but nonaromatic nitrogen, sulfur, and oxygen compounds, and complicated compounds of phosphorus and boron. Someone with access to a variety of such kinds of compounds and to an NMR spectrometer would make a very important contribution measuring and reporting their susceptibilities. Given these data, construction of a system for estimating $\chi_M{}'$ should be a fairly straightforward application of the readily available computer programs for

[87] W. Hüttner and W. H. Flygare, *J. Chem. Phys.* **50**, 2863 (1969).
[88] N. Jonathan, S. Gordon, and B. P. Dailey, *J. Chem. Phys.* **36**, 2443 (1962).

least-squares analysis of numerous simultaneous equations, as Laity has done[23] for hydrocarbons and alcohols.

B. KETO DERIVATIVES OF AROMATICS

The problem with these compounds (e.g., tropone, pyridone, barbituric acid) is twofold: first, the compounds of greatest interest[89] are the biologically active derivatives of pyrimidine and purine, and measurements on these suffer all the uncertainties found for other heterocycles. Second, the contribution of the Van Vleck paramagnetism from this kind of keto group is uncertain, because the energy of the $n \to \pi^*$ transition is strongly modified by being rather more highly polarized C^+—O^- than that of a normal aliphatic ketone (by the ability of the rest of the molecule to stabilize a positive charge). This problem demands both a solution to the problem of estimating χ_M' for heterocycles, discussed above, and a theoretical study of the Van Vleck paramagnetism of dipolar carbonyl groups.

C. INORGANIC SYSTEMS

There has been some discussion of the question of aromaticity among inorganic compounds, and evidence has been obtained that some of them (e.g., $B_{10}H_{10}^{2-}$,[90] S_4N_4,[91] $[C_5H_5]_2Fe$,[92] $[CF_3P]_5$[23]) exhibit exaltation. In principle, exaltation should be fully as capable of identifying delocalization around a cycle (or over the surface of a sphere) in inorganic compounds as in organic compounds, but in practice, it suffers from the same deficiency as when applied to heterocyclic compounds—lack of data for suitable model compounds—only to a more considerable extent. Again, the only way the problem can be solved is·by acquisition of magnetic susceptibility data for all kinds of covalent inorganic compounds.

D. "PARAMAGNETIC RING CURRENTS"

There is theoretical and experimental evidence to suggest that certain pseudo-aromatic compounds can exhibit the phenomenon of π-orbital paramagnetism.[9,54,55] The experimental evidence includes NMR spectral measurements and the "negative exaltations" found for [16]annulene, heptalene, and dibenzopentalene.[9] Those same characteristics that make it the most satisfactory

[89] C. Giessner-Prettre and B. Pullman, *Compt. Rend.* **261**, 2521 (1965); A. Veillard, B. Pullman, and G. Berthier, *ibid.* **252**, 2321 (1961).

[90] W. Lipscomb, "Boron Hydrides," p. 93. Benjamin, New York, 1965.

[91] P. S. Braterman, *J. Chem. Soc.* p. 2297 (1965).

[92] L. N. Mulay and M. E. Fox, *J. Am. Chem. Soc.* **84**, 1308 (1962).

criterion for aromatic character should make exaltation a superior method for examining this interesting phenomenon.

E. THE [18]ANNULENETRI (CHALCOGENIDES)

These four systems, [18]annulenetrioxide, -dioxide sulfide, -oxide disulfide, and -trisulfide, present an interesting series of compounds, in that NMR data would indicate a fundamental difference between the electronic structure of the first two and the last two compounds.[84, 93] The large size of the sulfur atoms is thought to impede coplanarity of the whole system in the last two compounds, and as a result, aromatic character seems to be confined in them to the five-membered heterocyclic nuclei. In contrast, the trioxide and dioxide sulfide appear to be aromatic throughout the macrocycle. If this rationale for the differences in NMR spectra observed for the four is correct, the exaltations of the compounds should be quite different: the compounds in which delocalization takes place over the entire ring should exhibit $\Lambda \gtrsim 100$.

F. PORPHINES, PHTHALOCYANINES, AND OTHER MACROHETEROCYCLES

A few magnetic susceptibility measurements have been reported[94] for these kinds of compounds, enough to show they too exhibit exaltation (Table IV), but these data are yet to be discussed with respect to bonding and delocalization in the systems. It would be instructive to compare the exaltations of a porphine and its tetrahydroderivative, for example. Furthermore, the effect of incorporating (diamagnetic) metal ions into the nucleus has not been studied; this can be expected to alter the exaltation, but in what way is unpredictable.

G. RING CURRENT IN THE FIVE-MEMBERED HETEROCYCLES

Whether pyrrole and furan are "as aromatic as benzene" or not is a long-argued and, as yet, unanswered question. Recently it appeared in discussions concerning the relative magnitude of the ring current (J) in these systems.[95-98]
Theoretical studies by Black, Brown, and Heffernan[98] and by Davies,[95] indicate that values of J are reduced (by about 30%) on going from benzene or pyridine to furan and pyrrole, with the results on thiophene less clear.

[93] G. M. Badger, J. A. Elix, and G. E. Lewis, *Australian J. Chem.* **19**, 1221 (1966); G. M. Badger, G. E. Lewis, and U. P. Singh, *ibid.* p. 257.

[94] R. Havemann, W. Haberditzl, and P. Grzegorzewski, *Z. Physik. Chem. (Leipzig)* **217**, 91 (1961).

[95] D. W. Davies, *Chem. Commun.* p. 258 (1965).

[96] H. P. deJongh and H. Wynberg, *Tetrahedron* **21**, 515 (1965).

[97] R. J. Abraham and W. A. Thomas, *J. Chem. Soc., B* p. 127 (1966).

[98] P. J. Black, R. D. Brown, and M. L. Heffernan, *Australian J. Chem.* **20**, 1305 (1967).

These predictions are supported by the anisotropy data of LeFevre and Murthy[99] and by some NMR data,[96] but the results of their very careful NMR study led Abraham and Thomas[97] to conclude that furan, thiophene, and benzene had very nearly equal J's. This problem can also be approached with exaltation studies, since $J \propto K_L/S \propto \Lambda/S$ (where S is the area covered by the π-system).[15] Because exaltation data are easy to obtain and straightforward to interpret this approach should provide most convincing results. However, available data are not sufficiently accurate (because of the poor models for χ_M') to allow one to draw any meaningful conclusions about relative J values. It appears that among the five-membered compounds J(thiophene) > J(pyrrole) > J(furan), but this conclusion is tentative. Again, solution of the heterocyclic model-compound problem will greatly aid in solving this problem.

ACKNOWLEDGMENTS

As with most research, the work described in this paper could not have been carried out without a great deal of help from a large number of people. We wish to acknowledge their contributions.

Our research was supported by the U.S. Army Research Office (Durham) and the National Science Foundation, who granted a fellowship to one of us (JDW). We thank them for their support.

We received valuable technical advice from Doctors D. C. Douglass and John Letcher, Professors V. Schomaker and H. A. Staab, and Dr. R. F. Zürcher. Professors A. G. Anderson and Y. Pocker and Doctors S. C. Duggan and J. L. Sprung provided critical evaluation of portions of the manuscript. Mr. B. J. Nist was of considerable assistance in susceptibility measurements.

Samples for measurement of susceptibilities of many unusual compounds were lent to us by so many people we cannot list them all here. Many of the most interesting results herein described are the result of these most gracious loans of samples.

Finally we wish to acknowledge the contribution of Mrs. Pat Palazzolo, who typed the manuscript.

[99] R. J. W. LeFevre and D. S. N. Murthy, *Australian J. Chem.* **19**, 1321 (1966).

4

Monocyclic and Polycyclic Aromatic Ions Containing Six or More π-Electrons

P. J. Garratt and M. V. Sargent

I. Introduction

The present chapter is concerned with the cyclic $(4n + 2)$ π-electron systems having six or more π-electrons and bearing one or more formal charges. The related two π-electron systems will be discussed in a future volume. The group of molecules under discussion consists of aromatic ions, and the first member of the group to be recognized was the cyclopentadienyl anion (1), prepared by Thiele in 1901 by treatment of cyclopentadiene with potassium.[1] The same worker also attempted to prepare the corresponding anion 2 from cyclohepta-triene,[1a] and concluded from the failure of this reaction that cycloheptatriene was "homoaromatic."[2] The correct prediction that the third member of the six π-electron series should be the tropylium (cycloheptatrienium) cation (3) was made by Hückel in 1931.[3] The lack of acidity of the methylene protons in cycloheptatriene could now be recognized as due to the lack of stabilization of the cycloheptatrienyl anion (2), which does not have a closed, aromatic electronic configuration. Hückel's ideas were virtually ignored by experimental chemists until, in 1945, Dewar[4] reexamined the experimental data which had been obtained for the mold metabolite stipitatic acid, and concluded that it possessed a seven-membered aromatic ring. This concept initiated a widespread

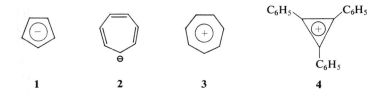

1   2   3   4

activity in the area of tropone chemistry, culminating in the synthesis of the tropylium cation (3) by Doering and Knox in 1954.[5] These authors also reported that this synthesis had previously been carried out by Merling in 1891,[6] but that his preparation had been recognized neither by the author nor by subsequent investigators.

[1] J. Thiele, *Ber.* **34**, 68 (1901).

[1a] J. Thiele, *Ann.* **319**, 226 (1901).

[2] For more recent discussions of the concept of homoaromaticity, see W. von E. Doering, G. Laber, R. Vanderwahl, N. F. Chamberlain, and R. B. Williams, *J. Am. Chem. Soc.* **78**, 5448 (1956); S. Winstein, *Chem. Soc.* (*London*), *Spec. Publ.* **21**, 5 (1967); also Section 7.

[3] E. Hückel, *Z. Physik* **70**, 204 (1941); **72**, 310 (1931); **76**, 628 (1932); **83**, 632 (1933); *Intern. Conf. Phys.*, *London*, 1934, Vol. 2. *Phys. Soc.*, London, 1935.

[4] M. J. S. Dewar, *Nature* **155**, 50, 141, and 479 (1945).

[5] W. von E. Doering and L. J. Knox, *J. Am. Chem. Soc.* **76**, 3203 (1954).

[6] G. Merling, *Ber.* **24**, 3108 (1891).

The fulfillment of Hückel's prediction led some organic chemists to an interest in the molecular orbital theory, and in 1952, Roberts *et al.*[7] published a paper in which the aromaticity of a number of other ions was predicted. The synthesis of the triphenylcyclopropenium cation (**4**) by Breslow,[8] for which a delocalization energy of 2.0 β had been predicted,[7] confirmed that this cation

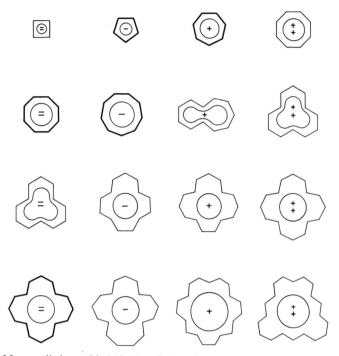

FIG. 1. Monocyclic ions with 6,10,14, and 18 π-electrons. Those in heavy type have been prepared.

was exceptionally stable, as expected for an aromatic system. Investigations leading to the preparation of other aromatic ions are described in the subsequent sections, but these will not include any description of studies on the cyclopentadienyl anion, the tropylium cation, or systems related to these, which have been extensively reviewed elsewhere.[9]

So far only a small number of the possible ions in this series have been prepared (Fig. 1), and although these have been of outstanding theoretical

[7] J. D. Roberts, A. Streitwieser, and C. M. Regan, *J. Am. Chem. Soc.* **74**, 4579 (1952).

[8] R. Breslow, *J. Am. Chem. Soc.* **79**, 5318 (1957); R. Breslow and C. Yuan, *ibid.* **80**, 5991 (1958).

[9] D. Ginsburg, ed., "Nonbenzenoid Aromatic Compounds." Wiley (Interscience), New York, 1959; T. Nozoe, *Progr. Org. Chem.* **5**, 132 (1961).

interest, the syntheses of many more systems may be expected over the next few years. The range of related polycyclic systems is yet more extensive, and even less work has been carried out in this area.

In the following sections the monocyclic systems will be treated first, primarily in order of increasing number of π-electrons (n), and secondarily in order of increasing ring size. The polycyclic systems will then be treated in order of the increasing size of the smallest ring.

II. Monocyclic Systems with Six π-Electrons

The principal members of this series are the cyclopentadienyl anion and the tropylium cation, both of which have an extensive chemistry which will not be discussed here. Attempts have been described to synthesize two other members of this series, the cyclobutadienyl dianion (**5**) and the cyclooctatetraenium dication (**6**).

5   6

A. Cyclobutadienyl Dianion

It was reported by Adam in 1963[10] that treatment of 3,4-diodo-1,2,3,4-tetramethylcyclobutene (**7**) with n-butyllithium in ether at $-70°C$ gave, after treatment with methanol, a mixture of octamethyltricyclo[4.2.0.0²·⁵]octa-3,7-diene (**8**) and 3-methylene-1,2,4-trimethylcyclobutene (**9**) (Fig. 2). Since

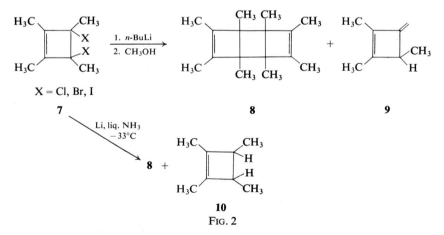

Fig. 2

[10] W. Adam, *Tetrahedron Letters* p. 1387 (1963).

these products were not those expected from the quenching of the tetramethyl-cyclobutadienyl dianion, direct metal halogen interchange was investigated by treatment of the dihalo compound with lithium in liquid ammonia. After aqueous workup **8** was again obtained, but in this case together with 1,2,3,4-tetramethylcyclobutene (**10**). However, quenching the solution with deuterium oxide instead of water gave **10** and not the corresponding dideuterio compound, indicating that the dianion was not present in solution in the liquid ammonia.

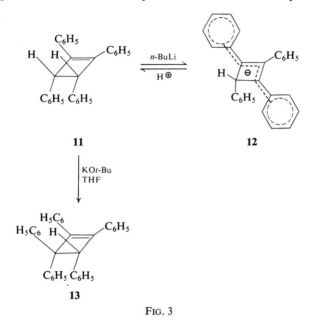

FIG. 3

Treatment of 1,2,3,4-tetramethylcyclobutene with *n*-butyllithium was also unsuccessful, no evidence for the formation of either the anion or dianion being obtained.

Freedman *et al.*,[11] however, found that treatment of *cis*-1,2,3,4-tetraphenyl-cyclobutene (**11**) with *n*-butyllithium in tetrahydrofuran (THF) gave the stable, deep-green tetraphenylcyclobutenyl anion **12**, which incorporated one atom of deuterium on treatment with deuterium oxide (Fig. 3). Protonation of the anion **12** with oxy acids gave the cis-isomer **11**, the product of kinetic control, whereas when the anion is formed as an intermediate in the equilibration of **11** with lithium cyclohexylamide or potassium *t*-butoxide in THF, the product is *trans*-1,2,3,4-tetraphenylcyclobutene (**13**). The formation of the dianion was not observed in these reactions.

[11] H. H. Freedman, G. A. Doorakian, and V. R. Sandel, *J. Am. Chem. Soc.* **87**, 3019 (1965).

B. Cyclooctatetraenium Dication

Bryce-Smith and Perkins[12] reported that treatment of cyclooctatetraene with with stannic bromide and bromine did not give the cyclooctatetraenium dication (6). Similarly, treatment of 7,8-dichlorobicyclo[4.2.0]octa-2,4-diene

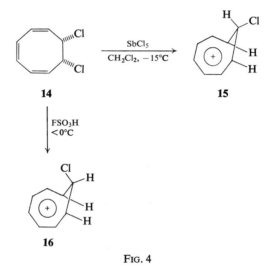

FIG. 4

with stannic chloride, or cycloocta-1,3,5-triene with hydride abstractors, did not give the dication. It now seems clear that the failure to observe 6 in these reactions is probably due to the stability of the homotropylium cation.

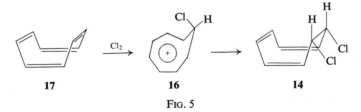

FIG. 5

Huisgen and his co-workers[13, 14] have shown that the 8-chlorohomotropylium cations 15, 16 are formed on treatment of cis-7,8-dichlorocycloocta-1,3,5-triene (14) with antimony pentachloride or fluorosulfonic acid (Fig. 4) and also that the endo-8-chlorohomotropylium cation is an intermediate in the

[12] D. Bryce-Smith and N. A. Perkins, J. Chem. Soc. p. 1339 (1962).
[13] G. Boche, W. Hechtl, H. Huber, and R. Huisgen, J. Am. Chem. Soc. 89, 3344 (1967).
[14] R. Huisgen, G. Boche, and H. Huber, J. Am. Chem. Soc. 89, 3345 (1967).

chlorination of cyclooctatetraene (Fig. 5). Antimony pentachloride gave exclusively the *exo*-8-chlorohomotropylium cation (**15**), whereas fluorosulfonic acid gave exclusively the *endo*-8-chlorohomotropylium cation (**16**). The difference in behavior was considered to be due to exclusive attack at one of the two boat conformations of **14**. Treatment of *trans*-7,8-dichlorocyclooctatriene with either antimony pentafluoride or fluorosulfonic acid gave the *exo*-8-homotropylium cation (**15**). The chlorination of cyclooctatetraene **17** was considered to go exclusively through the *endo*-cation **16**, since the final product is *cis*-7,8-dichlorocycloocta-1,3,5-triene (**14**), whereas addition of chloride ion to the *exo*-cation **15** gives only *trans*-7,8-dichlorocycloocta-1,3,5-triene.

Electrochemical oxidation of cyclooctatetraene[15, 16] in acetic acid containing acetate ion gives a complex mixture of acetates. These products may arise by electrophilic attack of the cyclooctatetraenium dication (**6**) on the acetate anion, but the evidence for **6** as an intermediate is slight.

III. Monocyclic Systems with Ten π-Electrons

The two members of this group which have been prepared are the cyclooctatetraenyl dianion (**18**) and the cyclononatetraenyl anion (**19**). Two related bicyclic ions have also been prepared, in which a one-carbon bridge between the 1,6 atoms minimizes nonbonded interactions, and allows the ions to adopt a conformation of low strain. These ions are the 1,6-methanocyclononatetraenyl anion (**20**) and the 1,7-methanocycloundecapentaenium cation (**21**). The ions **20** and **21** will be treated with the monocyclic systems, the bridge being considered a minor perturbation.

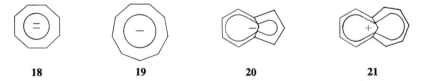

18	**19**	**20**	**21**

A. CYCLOOCTATETRAENYL DIANION AND RELATED SYSTEMS

The facile reaction of cyclooctatetraene (COT, **17**) in ether with alkali metals was reported in the classic paper of Reppe *et al.*[17] These workers established that a dialkali metal salt must be formed, since on carboxylation a

[15] L. Eberson, K. Nyberg, M. Finkelstein, R. C. Petersen, S. D. Ross, and J. J. Uebel, *J. Org. Chem.* **32**, 16 (1967).

[16] M. Finkelstein, R. C. Petersen, and S. D. Ross, *Tetrahedron* **23**, 3875 (1967).

[17] W. Reppe, O. Schlichting, K. Klager, and T. Toepel, *Ann.* **560**, 1 (1948).

dicarboxylic acid was obtained. Treatment of the dilithium derivative with water gave a yellow oil, considered to be 1,3,6-cyclooctatriene (**22**), but later shown by Cope and Hochstein[18] to be a mixture of 1,3,6- and 1,3,5-cyclo-octatrienes, together with minor amounts of other products. These latter workers also demonstrated that the disodium derivative could be prepared by reduction of COT with sodium in liquid ammonia, and that decomposition of this solution with ammonium chloride gave an equimolar mixture of 1,3,6-cyclooctatriene and 1,3,5-cyclooctatriene (**23**) (Fig. 6).

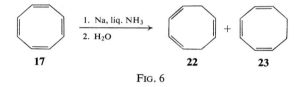

1. Na, liq. NH₃
2. H₂O

17   **22**   **23**

FIG. 6

A number of polarographic studies were carried out with COT,[19-21] and it was shown that the reduction involved a two-electron process that was independent of pH. It was realized that the ease of reduction "required formulation of a structure to explain formation of the stable anion,"[19] but no structure was suggested. However in a discussion of a paper presented to the Chemical Society, London, in 1955, Ubbelohde[22] reported that magnetic susceptibility measurements supported the view that dilithium cyclo-octatetraenide has the structure $Li_2^{\oplus} C_8H_8^{\ominus}$, and suggested that the dianion has aromatic character. No confirmatory report of these experiments appeared, and these findings were not generally known until they appeared in a review in 1960.[23] Similar views were expressed on theoretical grounds by Longuet-Higgins and McEwen.[23a] Independently, Katz had arrived at the same con-clusion; this was reported in 1960 with experimental verification.[24] When a solution of COT in THF is treated with potassium, the nuclear magnetic resonance (NMR) spectrum, taken at intervals, shows a decrease in the signal due to the protons of COT, which remains a sharp singlet, and a broad band appears. The broad band sharpens, until, after 2 moles of potassium have been consumed, it has become a narrow singlet (half width < 1 Hz). The posi-

[18] A. C. Cope and F. A. Hochstein, *J. Am. Chem. Soc.* **72**, 2515 (1950); see also W. R. Roth, *Ann.* **671**, 25 (1964).

[19] R. M. Elofson, *Anal. Chem.* **21**, 917 (1949).

[20] J. H. Glover and H. W. Hodgson, *Analyst* **77**, 473 (1952).

[21] L. E. Craig, R. M. Elofson, and I. J. Ressa, *J. Am. Chem. Soc.* **75**, 480 (1953).

[22] A. R. Ubbelohde, *Chem. & Ind.* (*London*) p. 153 (1956).

[23] M. E. Vol'pin, *Russ. Chem. Rev.* (*English Transl.*) **29**, 129 (1960).

[23a] H. C. Longuet-Higgins and K. L. McEwen, *J. Chem. Phys.* **26**, 719 (1957).

[24] T. J. Katz, *J. Am. Chem. Soc.* **82**, 3784 and 3785 (1960).

tion of the new singlet, which was assigned to the protons of potassium cyclo-octatetraenide (**18**, Fig. 7), was within 2 cycles of the original signal for COT, and indicated a balance between the effects of ring current and the charge on the

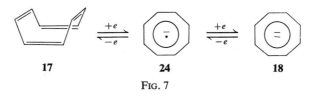

17   **24**   **18**

FIG. 7

carbon atoms. The electron spin resonance (ESR) signal[25] of the cyclo-octatetraenyl anion radical (**24**) supports a planar structure with eight equivalent protons (*vide infra*), and the experimental data indicated that the addition of the first electron was the rate-determining step. This behavior is the exact opposite of that found in the aromatic hydrocarbons, and was interpreted as being due to the barrier which had to be overcome to flatten the ring when the first electron is introduced. The addition of the second electron to the anion is a more facile process as the molecule is already planar. Since the spectrum of the solvent remains well resolved throughout the experiment, it was also concluded that the equilibrium between COT, the dianion **18**, and the anion radical **24** lies on the side of the COT and the dianion (Fig. 8). This behavior, which again is the opposite to that found in aromatic hydrocarbons,

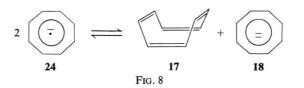

24   **17**   **18**

FIG. 8

was subsequently[26] shown to depend on the presence of alkali metals, since in the absence of alkali metal ions the anion radical does not disproportionate to the dianion and COT. Similar conclusions on the structure of the dianion **18** were subsequently reported by Fritz and Keller.[27]

Dipotassium cyclooctatetraenide exhibits a single NMR proton resonance signal at $\tau 4.3$,[24,27] ($C_8H_8^{\ominus}$ 2 Na$^{\oplus}$ $\tau 4.33$[28]) the infrared (IR) spectrum is simple,[27] showing bands at 2994, 1431, 880, and 684 cm^{-1}, and the ultraviolet (UV) spectrum in THF shows maxima at 236 (2.87), 263 (3.23), and 407 nm

[25] T. J. Katz and H. L. Strauss, *J. Chem. Phys.* **32**, 1873 (1960).
[26] R. D. Allendoerfer and P. H. Rieger, *J. Am. Chem. Soc.* **87**, 2336 (1965).
[27] H. P. Fritz and H. Keller, *Z. Naturforsch.* **16b**, 231 (1961); *Ber.* **95**, 158 (1962).
[28] T. Schaeffer and W. G. Schneider, *Can. J. Chem.* **41**, 966 (1963).

$(\log \epsilon 2.51)$.[27] These observations are consistent with the dianion being a planar, aromatic 10π-electron system of D_{8h} symmetry.

The ESR spectrum[25, 26, 29-31] of the anion radical **24**, produced by either alkali metal or polarographic reduction, shows nine lines in the appropriate ratio for eight equivalent protons, the hyperfine splitting being 3.23 ± 0.03 gauss. The ^{13}C-splitting in the anion radical indicates that the three σ bonds must be in a plane.[32] The position of the equilibrium in Fig. 8 depends on the

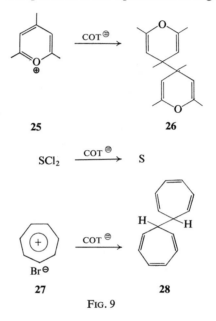

Fig. 9

presence or absence of alkali metal ions, and in N,N-dimethylformamide (DMF) or dimethyl sulfoxide (DMSO) in the absence of such ions dispro-portionation of the radical anion does not occur.[26] The equilibrium constant with potassium ion in THF has a value of about 10^8 toward the right (Fig. 8), whereas in the absence of metal ions it lies toward the left with an equilibrium constant of 10^{-4}. A detailed analysis of the importance of ion pairing in this system has been made.[31]

There appears to be a large free energy of activation in the addition of an electron to COT, and its value is similar to that estimated (21 kcal·mole^{-1}) for flattening the ring. The subsequent addition of an electron to the anion to

[29] H. L. Strauss, T. J. Katz, and G. K. Fraenkel, *J. Am. Chem. Soc.* **85**, 2360 (1963).
[30] A. Carrington and P. F. Todd, *Mol. Phys.* **7**, 1525 (1964).
[31] F. J. Smentowski and G. R. Stevenson, *J. Am. Chem. Soc.* **89**, 5120 (1967).
[32] H. L. Strauss and G. K. Fraenkel, *J. Chem. Phys.* **35**, 1738 (1961).

form the dianion is very rapid, indicating a low free energy of activation for this process. Such views are completely supported by polarographic studies,[26,33] which indicate that a simple reversible addition of two electrons occurs, followed by a rapid, irreversible chemical reaction of the dianion.

The chemistry of the cyclooctatetraenyl dianion has been more extensively investigated than that of any other system described in this chapter. Dipotassium cyclooctatetraenide can transfer electrons to a number of

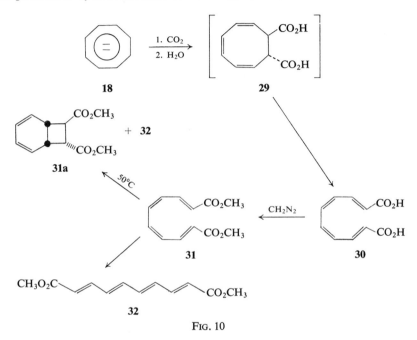

FIG. 10

systems,[34,35] and behaves as a reducing agent in a variety of reactions. Thus the trimethylpyrylium cation (**25**) is reduced by **18** to the dimer, 2,2',4,4',6,6'-hexamethyl-4,4'-bi-4*H*-pyran (**26**),[36] sulfur dichloride is reduced to sulfur,[37] and the tropylium cation (**27**) is reductively dimerized[38] (**28**, Fig. 9).

The dianion reacts with carbon dioxide to give a dicarboxylic acid, which for many years was considered to be 2,4,7-cyclooctatetriene-1,6-dicarboxylic acid, but which has recently been shown to be mainly *trans, cis, cis, trans*-deca-

[33] T. J. Katz, W. H. Reinmuth, and D. E. Smith, *J. Am. Chem. Soc.* **84**, 802 (1962).

[34] D. N. Kursonov and Z. V. Todres, *Dokl. Akad. Nauk. SSSR* **172**, 1086 (1967); *Chem. Abstr.* **67**, 6148j (1967).

[35] T. S. Cantrell and H. Shechter, *J. Am. Chem. Soc.* **89**, 5877 (1967).

[36] K. Conrow and P. C. Radlick, *J. Org. Chem.* **26**, 2260 (1961).

[37] T. J. Katz and P. J. Garratt, unpublished results; see Katz *et al.*[45]

[38] R. W. Murray and M. L. Kaplan, *J. Org. Chem.* **31**, 962 (1966).

2,4,6,8-tetraen-1,10-dioic acid (**30**) (Fig. 10).[39] Esterification with diazomethane gave the corresponding diester **31**, which was rearranged on warming
at ca. 50°C to a 9:1 mixture of the bicyclo derivative **31a** and the all-trans
diester **32**, whereas when **31** was heated above the melting point only **32** was
obtained.

The dianion is alkylated by treatment with methyl iodide to give a mixture
of 5,8-dimethyl-1,3,6-cyclooctatriene (**33**) and 7,8-dimethyl-1,3,5-cycloocta-
triene (**34**).[40] Both of these compounds are thermally unstable. A 1,5-hydrogen
shift occurs with the compound **33** to give the isomeric 3,8-dimethyl-1,3,5-

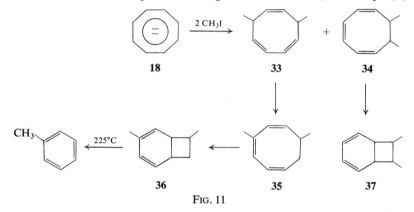

18   **33**   **34**

36   **35**   **37**

FIG. 11

cyclooctatriene (**35**), which at higher temperatures undergoes a Cope
rearrangement to 3,8-bicyclo[4.2.0]octa-2,4-diene (**36**). The latter compound
at 225°C decomposes to give toluene (Fig. 11). The isomer **34** is thermally
extremely labile and could not be isolated pure due to the ease of dimerization.
Cope rearrangement to 7,8-bicyclo[4.2.0]octa-2,4-diene (**37**) also occurs.

The dianion reacts with geminal dichlorides,[41-46] with acyl halides,[47-51]

[39] T. S. Cantrell, *Tetrahedron Letters* p. 5635 (1968).
[40] D. A. Bak and K. Conrow, *J. Org. Chem.* **31**, 3958 (1966).
[41] T. J. Katz and P. J. Garratt, *J. Am. Chem. Soc.* **85**, 2852 (1963).
[42] T. J. Katz and P. J. Garratt, *J. Am. Chem. Soc.* **86**, 4876 (1964).
[43] T. J. Katz and P. J. Garratt, *J. Am. Chem. Soc.* **86**, 5194 (1964).
[44] E. A. LaLancette and R. E. Benson, *J. Am. Chem. Soc.* **87**, 1941 (1965).
[45] T. J. Katz, C. R. Nicholson, and C. A. Reilly, *J. Am. Chem. Soc.* **88**, 3832 (1966).
[46] S. W. Staley and T. J. Henry, *J. Am. Chem. Soc.* **91**, 1239 (1969).
[47] V. D. Azatyan, *Dokl. Akad. Nauk SSSR* **98**, 403 (1954); *Chem. Abstr.* **49**, 12318i (1955);
Arm. Khim. Zh. **20**, 248 (1967); *Chem. Abstr.* **67**, 99706c (1967).
[48] V. D. Azatyan and R. S. Gyuli-Kevkhyan, *Dokl. Akad. Nauk Arm. SSR* **20**, 81 (1955);
Chem. Abstr. **50**, 4051a (1956).
[49] V. D. Azatyan and R. S. Gyuli-Kevkhyan, *Izv. Akad. Nauk Arm. SSR, Khim. Nauk* **14**,
451 (1961); *Chem. Abstr.* **58**, 3327a (1963).
[50] T. S. Cantrell and H. Shechter, *J. Am. Chem. Soc.* **85**, 3300 (1963).
[51] T. S. Cantrell and H. Shechter, *J. Am. Chem. Soc.* **89**, 5868 (1967).

and with aldehydes and ketones[35, 47–49, 52, 53] to give a variety of interesting products.

Reaction of methylene chloride,[43] chloroform,[41, 43, 44] carbon tetrachloride,[43] α,α-dichloromethylmethyl ether,[41, 43] 1,1-dichloroethane,[42] and 1,1,1-trichloroethane[42] with solutions of the dianion in THF led to the corresponding *cis*-bicyclo[6.1.0]nonatrienes (Fig. 12). A recent modification of this

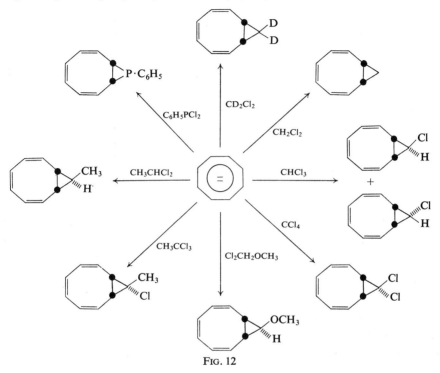

FIG. 12

procedure has been described in which the dianion is generated with sodium in liquid ammonia.[46]

The mechanism of these reactions is not clear. The order of reactivity $CCl_4 > CHCl_3 > CH_2Cl_2$ indicates that the reaction is not a simple nucleophilic substitution, and the predominance of one isomer in these reactions,[43] although originally overestimated in the case of 9-chloro[6.1.0]bicyclononatriene,[44] also supports this view. No products of 1,4-addition were observed, although such products were obtained when the dianion was treated with acyl halides (*vide infra*). The known reducing properties of the dianion lend credence to the concept that the dihalide is converted to a carbenoid

[52] G. Wittig and D. Wittenberg, *Ann.* **606**, 1 (1957).

[53] T. S. Cantrell and H. Shechter, *J. Am. Chem. Soc.* **87**, 136 (1965).

species, which then adds to the COT formed concurrently by oxidation of the dianion (Fig. 13).

The *cis*-bicyclo[6.1.0]nonatrienes are thermally labile, and in general valence tautomerize to the corresponding *cis*-8,9-dihydroindenes. Such a rearrangement is a nonallowed concerted process from orbital symmetry considerations,[54] but the facility with which the reaction occurs suggests that it must proceed by some pathway with a low free energy of activation. Both *syn*- (**38a**) and *anti*-9-chlorobicyclo[6.1.0]nonatriene (**39b**) thermally isomerize

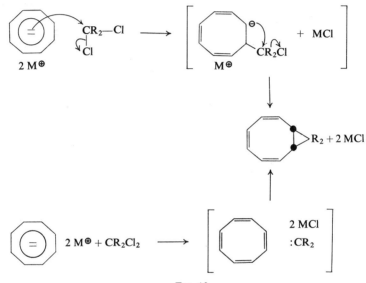

FIG. 13

to the same 7-chloro-*cis*-8,9-dihydroindene (**39**),[43, 44] and *syn*- (**40a**) and *anti*-9-methylbicyclo[6.1.0]nonatriene (**40b**) thermally isomerize to the same mixture of 7-methyl-*cis*-8,9-dihydroindenes (**41**).[55] However, 9,9-dimethyl-bicyclo[6.1.0]nonatriene (**42a**) thermally rearranges at 151°C to 7,7-dimethyl-*trans*-8,9-dihydroindene (**43a**)[46] (Fig. 14). Likewise the 9,9-methylethyl isomers of **42b,c** are converted thermally with loss of stereochemistry at C-9 to the same mixture of *trans*-dihydroindenes **43b,c**.[55a] The difference in behavior between the dialkyl derivatives and the monosubstituted derivative may be due to the preponderance of one particular conformation in the case of the monosub-stituted derivative (Fig 15.).

[54] R. Hoffmann and R. B. Woodward, *Accounts Chem. Res.* **1**, 17 (1968).
[55] P. C. Radlick and W. Fenical, *J. Am. Chem. Soc.* **91**, 1560 (1969).
[55a] S. W. Staley and T. J. Henry, *J. Am. Chem. Soc.* **91**, 7787 (1969).

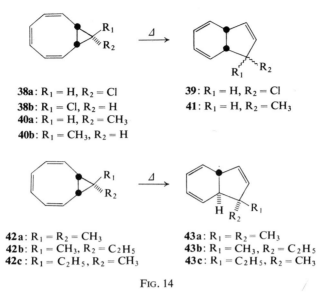

38a: R_1 = H, R_2 = Cl
38b: R_1 = Cl, R_2 = H
40a: R_1 = H, R_2 = CH$_3$
40b: R_1 = CH$_3$, R_2 = H

39: R_1 = H, R_2 = Cl
41: R_1 = H, R_2 = CH$_3$

42a: R_1 = R_2 = CH$_3$
42b: R_1 = CH$_3$, R_2 = C$_2$H$_5$
42c: R_1 = C$_2$H$_5$, R_2 = CH$_3$

43a: R_1 = R_2 = CH$_3$
43b: R_1 = CH$_3$, R_2 = C$_2$H$_5$
43c: R_1 = C$_2$H$_5$, R_2 = CH$_3$

Fig. 14

9-Phenyl-9-phosphobicyclo[6.1.0]nonatriene (**44**) prepared from the dianion and dichlorophenylphosphine[45] does not rearrange to the dihydroindene-type derivative, but gives 9-phenyl-9-phosphobicyclo[4.2.1]nonatriene (**45**). The

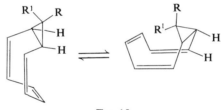

Fig. 15

reaction is stereochemically controlled, and probably proceeds by an orbital symmetry allowed 1,5-bond shift. The compound **45** on treatment with HCl in CHCl$_3$ at 100°C inverts to the more stable epimer **46** (Fig. 16).

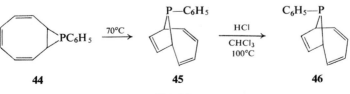

44

45

46

Fig. 16

Azatyan and his co-workers[47-49] reported that the reaction of the dianion **18** with alkyl or acyl halides, and with aldehydes and ketones, gave high boiling products, which were considered to be 7,8-disubstituted cyclooctatrienes. The diols obtained by treatment of **18** with aldehydes or ketones, or as secondary products from the Grignard reaction on the ketones formed by acylation of **18**, could be dehydrated to THF derivatives. Wittig and Wittenberg[52] also assigned the 7,8-bis(α-hydroxybenzylhydryl)-1,3,5-cyclooctatriene structure **(47)** to the compound obtained from the reaction of the dianion with benzophenone. Because of the limited proof for many of these structures, these reactions were reexamined in detail by Cantrell and Shechter.[35, 50, 51, 53] Addition of dilithium cyclooctatetraenide to excess acetyl chloride gives a

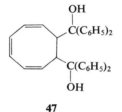

47

complex mixture of products consisting of *trans, cis, cis, trans*-3,5,7,9-dodecatetraene-2,11-dione **(48)**, *syn*-9-hydroxy-9-methylbicyclo[4.2.1]nona-2,4,7-triene **(49)**, *syn*-9-acetoxy-9-methylbicyclo[4.2.1]nona-2,4,7-triene **(50)**, and *syn*-9-acetoxy-9-methylbicyclo[6.1.0]nona-2,4,6-triene **(51)** (Fig. 17). Similar reactions occur with benzoyl chloride and phthaloyl chloride (Fig. 17). The dianion **18** thus appears to react by both 1,2- and 1,4-addition. The initial product of 1,2-addition is a 7,8-disubstituted cyclooctatriene, which subsequently rearranges to the corresponding acyclic tetraene.

With aldehydes and ketones, complex mixtures of products are again formed which appear to result from 1,2- and 1,4-addition. In these cases however the 7,8-disubstituted 1,3,5-cyclooctatrienes generally rearrange to the corresponding 7,8-disubstituted bicyclo[4.2.0]octa-2,4-dienes (Fig. 18).

In all of these reactions the products are those expected from a nucleophilic attack of the dianion on the carbonyl compound, and contrast with the exclusive occurrence of bicyclo[6.1.0]nonatrienes found in the reaction of *gem*-dihalides with the dianion (*vide supra*).

The dianion **18** is reported to react with trimethylsilyl chloride to give 7,8-bis(trimethylsilyl)cycloocta-1,3,5-triene.[56]

Breil and Wilke[57] have reported that sandwich complexes of the cyclo-

[56] V. D. Azatyan, *Izv. Akad. Nauk Arm. SSSR, Khim. Nauk* **17**, 706 (1964); *Chem. Abstr.* **63**, 4323f (1965).

[57] H. Breil and G. Wilke, *Angew. Chem.* **78**, 942 (1966); *Angew. Chem. Intern. Ed. Engl.* **5**, 898 (1966).

octatetraenyl dianion (**18**) with a number of transition metals can be formed. Tri(cyclooctatetraene)dititanium was shown by X-ray crystallographic analysis to have a layered structure of alternating COT molecules and titanium atoms, the two outer COT rings being planar.[57a]

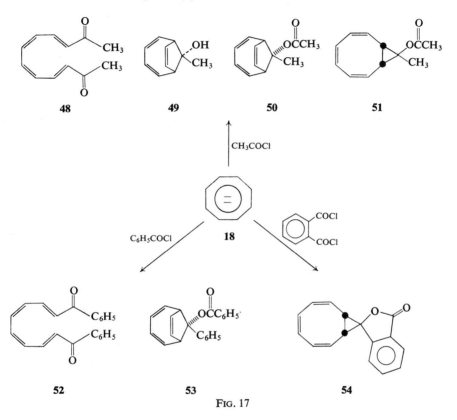

FIG. 17

Recently potassium cyclooctatetraenide has been reported to form a sandwich compound with uranium tetrachloride, which was formulated as bis(cyclooctatetranyl)uranium (**55**) with a D_{8d} or D_{8h} arrangement.[57b] It has been proposed that the *f*-orbitals of uranium participate in the bonding of this compound.[57b] An X-ray crystallographic analysis has confirmed the sandwich structure and has shown that the molecule has D_{8h} symmetry with planar, eclipsed C_8H_8 rings.[57c]

[57a] H. Dietrich and H. Dierles, *Angew. Chem.* **78**, 943 (1966); *Angew. Chem. Intern. Ed. Engl.* **5**, 899 (1966).

[57b] A. Streitwieser and U. Müller-Westerhoff, *J. Am. Chem. Soc.* **90**, 7364 (1968).

[57c] A. Zalkin and K. N. Raymond, *J. Am. Chem. Soc.* **91**, 5667 (1969).

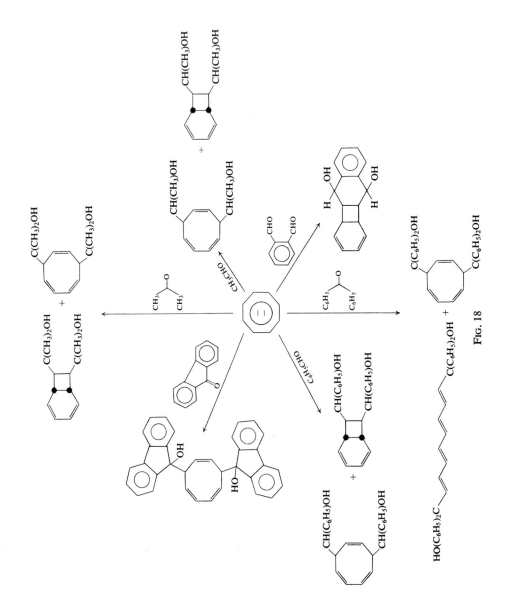

Fig. 18

Irradiation with UV light has been shown to increase the basicity of the dianion.[58] The irradiated dianion reacts with 1-hexyne-1-d_1 to give cyclooctatrienes with extensive incorporation of deuterium, whereas no reaction occurs between 1-hexyne and the dianion in the dark.

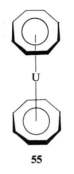

55

Oth[58a] has prepared the dianion (**55b**) of cyclooctatetraenyl *t*-butyl ether (**55a**) (Fig. 18a). The NMR spectrum of the dianion **55b** at $-50°C$ shows a multiplet at $\tau 4.5$, a quintet at $\tau 5.8$, and a doublet at $\tau 8.85$. On cooling to

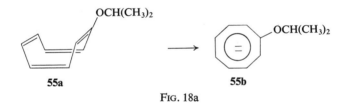

55a   **55b**

FIG. 18a

$-100°C$ the resonance signals broaden and the quintet loses its fine structure and appears as a broad band.

The *sym*-dibenzocyclooctatetraenyl anion (**56**),[59,60] the *sym*-dibenzocyclooctatetraenyl dianion (**57**),[60] and the tetraphenylene anion (**58**),[59] have all been prepared.

The ESR spectrum of the anion **56**, prepared by reduction of *sym*-dibenzocyclooctatetraene with alkali metals or polarographically, shows a hyperfine pattern of three binomial, overlapping quintets, expected for three sets of four equivalent protons.[59,60] The splitting constants are 2.6,[59] (2.54[60]), 1.80,[59,60] and 0.02 gauss,[59,60] and were assigned to the sets of protons at positions 5,2,

[58] J. I. Brauman, J. Schwartz, and E. E. van Tamelen, *J. Am. Chem. Soc.* **90**, 5328 (1968).
[58a] J. F. M. Oth, private communication.
[59] A. Carrington, H. C. Longuet-Higgins, and P. F. Todd, *Mol. Phys.* **8**, 45 (1964).
[60] T. J. Katz, M. Yoshida, and L. C. Siew, *J. Am. Chem. Soc.* **87**, 4516 (1965).

and 1 respectively. From McConnell's relation, $a_H = Q\rho_c$ where ρ_c is the spin density at the neighboring carbon atom and Q is a constant, taken as -23 gauss, the value of the spin densities at carbon atoms 5, 2, and 1 were 0.110, 0.078, and 0.0087, respectively. The extent of interaction of the ethylene bonds with the benzene rings was suggested by Carrington *et al.* to be weak,[59] but Katz and his co-workers[60] concluded that although the calculated data was in somewhat better agreement if this interaction was weak, it did not preclude the possibility that the interaction was strong. In fact the half-wave potential obtained from polarographic measurements indicate that this is a strong interaction.

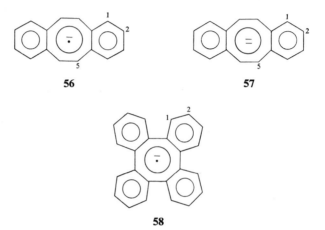

56   57

58

Further reduction of the anion **56** leads to the dianion **57**. The NMR spectrum of *sym*-dibenzocyclooctatetraene (singlets at $\tau 2.95$ and 3.29) disappears, and a new spectrum of the dianion **57** appears, consisting of two A_2X_2 multiplets at $\tau 2.06$ and 3.70, and a singlet at $\tau 2.92$.[60] The downfield shift of the protons in the new spectrum, despite the introduction of two negative charges on the molecule, indicate a larger degree of cyclic delocalization in the dianion **57** than in *sym*-dibenzocyclooctatetraene. Using a modification of the procedure of Schaeffer and Schneider, Katz *et al.*[60] calculated the excess electron densities on carbon atoms 5, 2, and 1 as 0.19, 0.16, and 0.06, respectively.

The dianion **57** reacts with water to give 5,6-dihydro-*sym*-dibenzocyclooctatetraene.

The ease of reduction of *sym*-dibenzocyclooctatetraene and the apparent large diamagnetic ring current effect in the dianion **57** supports the view that the degree of electron delocalization in the central ring is greater in the dianion, and probably the anion, than it is in the hydrocarbon, and suggests that **56** and **57** are planar, aromatic systems.

The tetraphenylene anion (58) was synthesized by reduction of tetraphenylene with potassium in dimethoxyethane at $-90°C$.[59] The hyperfine pattern of nine main lines of binomial intensity, each split into a partially resolved multiplet, was consistent with a system of two sets of eight equivalent protons. The splitting constants of 1.4 and 0.2 gauss were in good agreement with those obtained from the McConnell relation for the lowest antibonding orbital (1.41 and 0.0 gauss), and can be assigned to the two sets of protons at positions 2 and 1, respectively. The degree of interaction between the benzene rings in the tetraphenylene anion was assumed to be weak, but this agreement may not be well founded (*vide supra*).

Benzocyclooctatetraene is also reduced by alkali metals to the benzo-cyclooctatetraenyl dianion.[60a]

B. CYCLONONATETRAENYL ANION

The cyclononatetraenyl anion (19), the next higher homolog of the cyclo-pentadienyl anion (1), was independently synthesized by two groups in 1963.[41,61] Treatment of 9-chlorobicyclo[6.1.0]nonatriene (59) with lithium in THF gave lithium cyclononatetraenide (19a),[41,43,44,61] and treatment of *anti*-9-methoxybicyclo[6.1.0]nonatriene (60) with potassium in THF gave-potassium cyclononatetraenide (19b).[41,43] Sodium cyclononatetraenide (19c) was prepared in lower yield by proton abstraction from bicyclo[6.1.0]-nonatriene (61a) with the methyl sulfinyl carbanion in DMSO,[44] and less readily with other bases (Fig. 19).

These salts were all obtained in solution, but a crystalline tetraethylam-monium cyclononatetraenide (19d) could be isolated by metathesis of 19a with tetraethylammonium chloride.[44,61]

The spectroscopic properties of these salts are in accord with the view that cyclononatetraenyl anion is an all-cis, planar system. In the NMR spectrum the salts show only one resonance signal at low field (19a, $\tau3.10$,[43] 3.28[44]; 19b, $\tau2.96$[43]; 19c, $\tau3.18$[44]; 19d, $\tau3.10$[44]), the chemical shift indicating that the molecule has an induced diamagnetic ring current. Assuming that the proton chemical shifts are proportional to the charge densities, $\delta = K\Delta\rho$ (δ = chemical shift, $\Delta\rho$ difference in charge density, $K = 10$ ppm[62]), then the signal for the cyclononatetraenyl anion should appear at 1.4 ppm to lower field than the signal for the cyclooctatetraenyl dianion. This calculation assumes that the

[60a] M. V. Sargent, unpublished results.

[61] E. A. LaLancette and R. E. Benson, *J. Am. Chem. Soc.* 85, 2853 (1963).

[62] A number of values have been suggested for k, all ca. 10.0 ppm. See Schaeffer and Schneider[28]; G. Fraenkel, R. E. Carter, A. McLachlan, and J. H. Richards, *J. Am. Chem. Soc.* 82, 5846 (1960); B. P. Dailey, A. Gower, and W. C. Neikam, *Discussions Faraday Soc.* 34, 18 (1962).

magnitude of the ring currents in the two systems is the same. The value of ca. $\tau 3.0$ is thus in excellent agreement with the value of $\tau 2.9$ ($\tau 4.3$–1.4) predicted. The ^{13}C NMR chemical shift values are also in agreement with

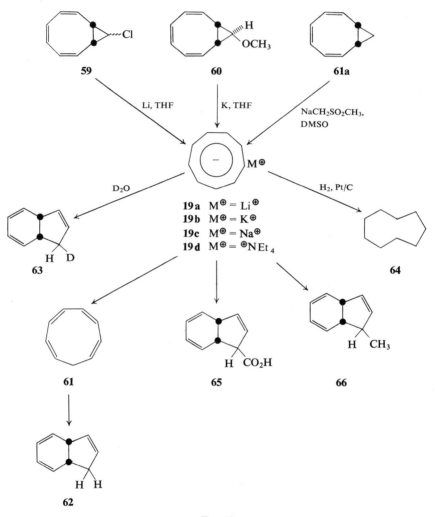

FIG. 19

simple linear correlation between chemical shift and π-electron density.[44] The UV spectrum [**19a**, THF, 251 (5.0), 319 (3.9), 324 nm (log ϵ 3.9)[43]; **19d**, CH$_3$CN, 250 (4.83), 317 (3.82), 322 nm (log ϵ 3.83)[44]] is also in accord with the planar, all-cis, nonagonal structure for the anion. The theoretical spectrum[63] identifies the high intensity band as $^1E_1 \leftarrow {}^1A_1{}'$ allowed electronic transition, and the weak bands to a vibronically allowed $^1E_4 \leftarrow {}^1A_1{}'$ transition. The simple IR spectrum of 19d is also in accord with a highly symmetric structure for the anion.

The cyclononatetraenyl anion reacts with water (or deuterium oxide), to give 8,9-dihydroindene (62) [or 1-deuterio-8,9-dihydroindene (63)],[43, 44] with *cis*-cyclononatetraene (61) as a probable intermediate.[44] The intermediacy of 61 in the hydrolysis of 19b has been confirmed, and 61 has been isolated.[64]

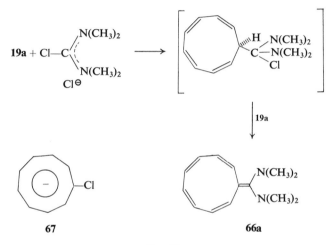

FIG. 19a

Treatment of 61 with potassium *t*-butoxide regenerates 19b.[64] Hydrogenation of the anion over Pt/C gave a mixture of products, the major compound (31.5%) being cyclononane (64).[44] Carboxylation of 19b gave 1-carboxy-8,9-dihydroindene (65),[43] and methylation with methyl iodide gave 1-methyl-8,9-dihydroindene (66).[43] The preparation of diazocyclononatetraene has been reported, but with only slight evidence for this structure.[64a]

Hafner and Tappe[64b] have recently reported that 19 reacts with N,N,N',N'-tetramethylchloroformamidinium chloride to give the nonafulvalene 66a (Fig. 19a).

[63] H. E. Simmons, D. B. Chestnut, and E. A. LaLancette, *J. Am. Chem. Soc.* 87, 982 (1965).

[64] G. Boche, H. Böhme, and D. Martens, *Angew. Chem.* 81, 565 (1969); *Angew. Chem. Intern. Ed. Engl.* 8, 594 (1969).

[64a] D. Lloyd and N. W. Preston, *Chem. & Ind.* (*London*) p. 1039 (1966).

[64b] K. Hafner and H. Tappe, *Angew. Chem.* 81, 564 (1969); *Angew. Chem. Intern. Ed. Engl.* 8, 593 (1969).

Equilibration studies of the anion with cyclopentadiene and indene suggest that some proton exchange occurs with the former, but not the latter.[44] These results indicate that the cyclononatetraenyl anion is thermodynamically more stable than the cyclopentadienyl anion, and that cyclononatetraene appears to have a *pK* between 16 and 21 on the Streitwieser scale.

Photoirradiation of **19** causes an increase in basicity of the anion, since it then reacts with acids too weak to protonate the anion in the dark.[65] Photo-irradiation of the anion in the presence of 1-hexyne gives *cis*-8,9-dihydroindene, *trans*-8,9-dihydroindene, and tricyclo[4.3.0.0.^{2,5}]nona-3,7-diene. The two latter products are also obtained on photoirradiation of *cis*-8,9-dihydroindene.

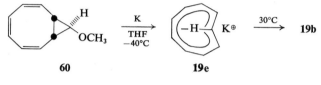

60　　　　　　**19e**

FIG. 19b

The monochlorocyclononatetraenyl anion (**67**) was prepared by treatment of 9,9-dichlorobicyclo[6.1.0]nonatriene with lithium in THF.[44]

Boche *et al.*[65a] have now shown that the cleavage of the methyl ether **60** with potassium is orbital symmetry controlled, and that at −40°C conrotatory ring opening gives the mono-*trans*-cyclononatetraenyl anion (**19e**) (Fig. 19b). In the NMR spectrum the outer protons resonate at low field ($\tau 2.73$–3.6), while the inner proton appears at high field ($\tau 13.52$), indicating that **19e** is an aromatic system. At room temperature **19e** rearranges to the *cis*-cyclo-nonatetraenyl anion (**19b**).

C. BRIDGED SYSTEMS

The nonbonded interactions between the 1 and 6 positions of di-*trans*[10]-annulene can be removed by inserting a one-carbon bridge between these positions.[66] This concept has been applied to the synthesis of both the bridged cyclononatetraenyl anion (**20**) and the bridged undecapentaenium cation (**21**).[66]

[65] J. Schwartz, *Chem. Commun.* p. 833 (1969).

[65a] G. Boche, D. Martens, and W. Danzer, *Angew. Chem.* **81**, 1003 (1969); *Angew. Chem. Intern. Ed. Engl.* **8**, 984 (1969).

[66] E. Vogel, *Chem. Soc. (London), Spec. Publ.* **21**, 113 (1967).

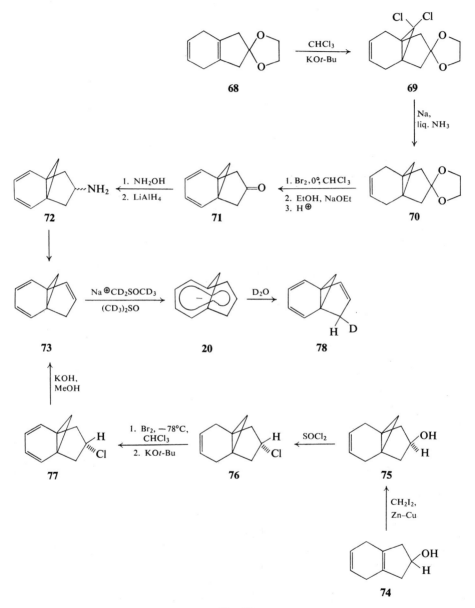

FIG. 20

1. *Bicyclo[4.3.1]decatetraenyl Anion (1,6-Methanocyclononatetraenyl Anion)* (**20**)

The bicyclo[4.3.1]decatetraenyl anion (**20**) was prepared independently by two groups in 1966.[67,68,68a] Grimme *et al.*[67] added dichlorocarbene to the ketal **68** of 4,7-dihydro-2-indanone, and then dechlorinated the resulting adduct **69** to the corresponding tricyclic ketal **70**. Bromination and dehydrobromination of the ketal **70** gave the diene **71**, which was then converted to tricyclo[4.3.1.0^{1,6}]deca-2,4,7-triene (**73**) by preparation of the amine **72**, exhaustive methylation, and Hofmann elimination.

The same tricyclic hydrocarbon **73** was prepared by Radlick and Rosen[68] through a Simmons–Smith methylene addition to 4,7-dihydro-2-indanol (**74**), conversion of the resulting alcohol **75** to the corresponding chloride **76**, and then bromination and dehydrohalogenation (Fig. 20).

Treatment of the hydrocarbon **73** with sodium methyl sulfinyl anion in DMSO-d_6 gave a brownish-red solution, which with excess deuterium oxide gave 9-deuteriotricyclo[4.3.1.0^{1,6}]deca-2,4,7-triene (**78**). The NMR spectrum of solutions of the anion **20** showed resonance signals at low field [$\tau 3.2$ (*dd*, 2 H), 4.0–4.6 (*m*, 5 H)[67]; $\tau 2.94$ (*dd*, 2 H, $J = 2.0, 5.0$ Hz), 3.98 (*m*, 5 H)[68]] and at high field [$\tau 10.7$ (*d*, 1 H, $J = 7.5$ Hz), 11.2 (*d*, 1 H, $J = 7.5$ Hz)[67]; $\tau 10.45$ (*d*, 1 H, $J = 7.5$ Hz), 10.95 (*d*, 1 H, $J = 7.5$ Hz)[68], corresponding, respectively, to the ring protons and the protons on the methylene bridge. This spectrum indicates that the anion possesses an induced diamagnetic ring current, and is consistent with the open bicyclic anion **20**, rather than the tricyclic structure **79**.

79

Radlick and Rosen[68a] found that proton removal from the hydrocarbon **73** occurred stereospecifically, the proton on the opposite side to the methylene bridge being abstracted. Protonation of the anion **20** also occurred on the side opposite to the methylene bridge, and it was suggested that this stereospecificity was due to the greater electron density on this side of the anion.

[67] W. Grimme, M. Kaufhold, U. Dettmeier, and E. Vogel, *Angew. Chem.* **78**, 643 (1966); *Angew. Chem. Intern. Ed. Engl.* **5**, 604 (1966).

[68] P. Radlick and W. Rosen, *J. Am. Chem. Soc.* **88**, 3641 (1966).

[68a] P. Radlick and W. Rosen, *J. Am. Chem. Soc.* **89**, 5308 (1967).

2. Bicyclo[5.4.1]dodecapentaenium Cation (1,6-Methanoundecapentaenium Cation) (21)

Although the macrocyclic undecapentaenium cation has not been prepared, the corresponding 1,6-bridged ion has been synthesized.[69] Addition of diazomethane to 1,6-methano[10]annulene (80) in the presence of cuprous chloride gave bicyclo[5.4.1]dodeca-1,3,5,7,10-pentaene (81) as the major product. Reaction of 81 with triphenylmethylfluoroborate in acetonitrile gave the stable orange-yellow fluoroborate salt (mp 185°C) of the 1,6-methano-undecapentaenium cation (21) (Fig. 21). The NMR spectrum of solutions of

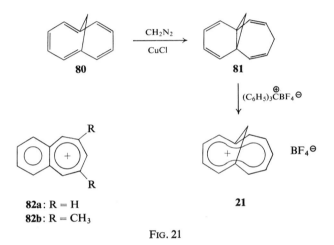

82a: R = H
82b: R = CH$_3$

FIG. 21

the cation shows proton resonance signals at low field [τ0.4–1.7 (9 H)] and an AX system at high field (τ10.3, d, 1 H, $J = 10$ Hz; τ11.8, d, 1 H, $J = 10$ Hz); corresponding, respectively, to the ring protons and the protons on the methylene bridge. The chemical shifts of the protons are similar to those of 1,6-methano[10]annulene, rather than the precursor 81, and indicate that 21 has a diamagnetic ring current. Both the electronic spectrum and chemical shifts of the ring protons of the cation 81 are quite similar to those of the benzotropylium cation (82a).

A detailed analysis of the electronic spectrum (270, 303, 435 nm) and the polarization of the observed transitions has been made for the 1,6-methano-undecapentaenium cation (21).[70] These spectra have been compared with

[69] W. Grimme, H. Hoffmann, and E. Vogel, *Angew. Chem.* 77, 348 (1965); *Angew. Chem. Intern. Ed. Engl.* 4, 354 (1965).

[70] W. Grimme, E. Heilbronner, G. Hohlneicher, E. Vogel, and J. P. Weber, *Helv. Chim. Acta* 51, 225 (1968).

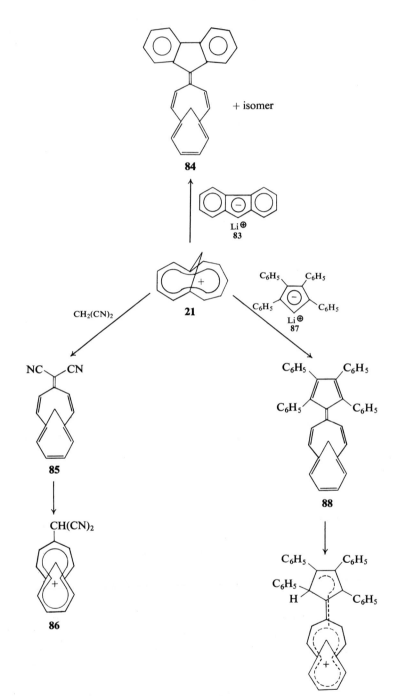

+ isomer

FIG. 22

those of the 4,6-dimethyl-1,2-benzotropylium cation (**82b**), and the polarization of the bands is found to be in agreement with the predictions of Pariser–Parr–Pople MO calculations. The inductive effect of the methylene bridge was examined, and the spectrum of the hypothetical undecapentaenylium cation was predicted with a Platt diagram to correlate bands.

Prinzbach and Knothe[71] have reported that **21** reacts with fluorenyllithium (**83**) in ether/THF to give, after subsequent oxidation with chloranil, the pentaundecafulvalene derivative **84**, together with other isomers. The cation **21** also reacts with malonitrile[71] and lithium tetraphenyl cyclopentadiene (**87**)[72] to give analogous compounds (Fig. 22). When **85** or **88**, but not **84**, are protonated with trifluoroacetic acid or perchloric acid, the corresponding undecaptenaenium cations (**86, 89**) are formed. The electronic spectra of **89** in perchloric acid/ethanol has the band [610 nm ($\log \epsilon 4.35$)] shifted to much higher wavelengths than either the parent ion **21** (423 nm, 3.60, 60% H_2SO_4) or **86** (426 nm).

IV. Monocyclic Systems with More than Ten π-Electrons

A. [16]ANNULENYL DIANION

The reduction of [16]annulene (**90**) to the corresponding anion radical **91** and dianion **92** was reported by Oth *et al.*[73] in 1968 (Fig. 23). Polarography of [16]annulene in absolute DMF indicated that two one-electron reduction processes occur to give **91** and then **92**. A solution of [16]annulene in THF-d_8 reacts with a potassium mirror, and an examination of the course of the reduction was made by recording the NMR spectrum after successive contacts of the solution with the potassium surface. After a brief contact with potassium at 0°C, the spectrum of [16]annulene, observed at −30°C, disappears, and only a broad signal is observed. This is presumably due to a rapid exchange between [16]annulene and the radical anion **91**. After further reduction new resonance signals appear which develop until, after 3 days at 0°C, the spectrum of the [16]annulenyl dianion (**92**) is observed. The spectrum shows resonance signals at τ1.17, 2.55, and 18.17, and the spectrum is invariant up to +140°C. The spectrum appears to be due to only one configurational isomer, and the change in position of the inner protons, which appear at τ−0.61 in [16]annulene and at τ18.17 in **92**, dramatically illustrates the difference in magnetic properties

[71] H. Prinzbach and L. Knothe, *Angew. Chem.* **79**, 620 (1967); *Angew. Chem. Intern. Ed. Engl.* **6**, 632 (1967).

[72] H. Prinzbach and L. Knothe, *Angew. Chem.* **80**, 698 (1968); *Angew. Chem. Intern. Ed. Engl.* **7**, 729 (1968).

[73] J. F. M. Oth, G. Anthoine, and J.-M. Gilles, *Tetrahedron Letters* p. 6265 (1968).

when going from the $4n$ to the $(4n + 2)$ π-electron system. The lack of thermal dependence of the NMR spectrum of **92** indicates that the interconversion of inner and outer protons is much slower than in [18]annulene,[74] and the resonance energy of the anion appears to be ca. 10 kcal·mole⁻¹ greater than that of [18]annulene.[75]

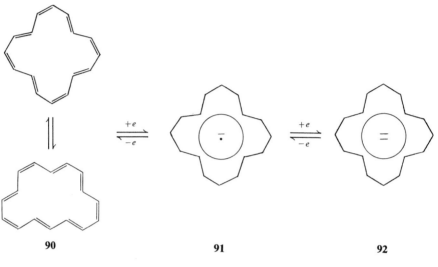

FIG. 23

B. 1-METHOXY-2,8,10-TRIDEHYDRO[17]ANNULENYL ANION

Griffiths and Sondheimer[76] have recently prepared the 1-methoxy-2,8,10-tridehydro[17]annulenyl anion (**94**) by abstraction of a proton from 17-methoxycycloheptadeca-1,3,5,11,13-pentaen-7,9,15-triyne (**93**) with methyllithium, or potassium, in THF (Fig. 23a). The reaction of the proton, rather than the ether group, with potassium indicates the pronounced acidity of the ¹⁷H proton of the cyclic ether **93**. The anion **94** has a complex NMR spectrum, the outer protons being at low field (from τ−0.47 to 2.16), and the inner protons at exceptionally high field (from τ18.15 to 19.09). The positions of these proton resonance signals, together with the low field resonance (τ5.38) of the methoxyl protons, indicate that the anion **94** has a large diamagnetic ring current. As in the case of [16]annulene, the proton resonances of the precursor have the

[74] L. M. Jackman, F. Sondheimer, Y. Amiel, D. A. Ben-Efraim, Y. Gaoni, R. Wolovsky, and A. A. Bothner-By, *J. Am. Chem. Soc.* **84**, 4307 (1962); I. C. Calder, P. J. Garratt, and F. Sondheimer, *Chem. Commun.* p. 41 (1967).

[75] I. C. Calder and P. J. Garratt, *J. Chem. Soc., B* p. 660 (1967).

[76] J. Griffiths and F. Sondheimer, *J. Am. Chem. Soc.* **91**, 7518 (1969).

inner protons at lower field (from τ1.93 to 2.77) than the outer protons (from τ3.15 to 5.48), which may indicate that a paramagnetic ring current effect is occurring in **93**. The electronic spectrum of **94** in ether had maxima at 404 (>4.48) and 657 nm (log ε > 3.90).

Treatment of solutions of the anion with water (or deuterium oxide) gave not the precursor **93** but an isomer **95a** (or the mondeuterio compound **95b**).

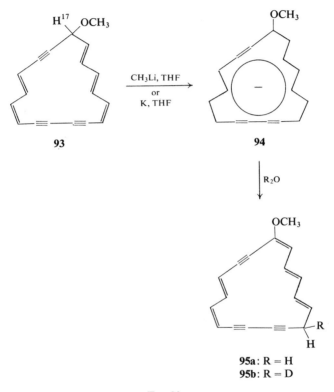

FIG. 23a

C. 1,3,7,9,13,15,19,21-OCTADEHYDRO[24]ANNULENYL DIANION

1,3,7,9,13,15,19,21-Octadehydro[24]annulene (**96**) can be reduced with potassium in THF in a manner analogous to cyclooctatetraene and [16]-annulene.[77] The reaction takes place in two one-electron processes, to give the radical anion **97** and the dianion **98** (Fig. 24). The NMR spectrum of

[77] R. M. McQuilkin, P. J. Garratt, and F. Sondheimer, *J. Am. Chem. Soc.* **92**, 6682 (1970).

solutions of **96** in THF–d_8 at −60°C after contact with a potassium mirror show the immediate disappearance of the signal due to **96**, only a very broad band being discernible, followed by the appearance of a new band at a very similar position to that of the precursor. The ESR spectrum of **97** in THF

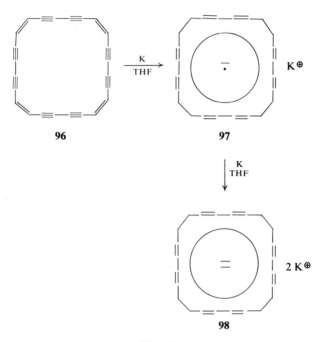

Fig. 24

showed the expected nine lines, with a value for a_H of 1.66 gauss. The electronic spectra of **97** was complex, whereas that of the dianion **98** showed only one band at 476 nm.

V. Physical Properties of Monocyclic Ions

The monocyclic ions are a group of molecules which, in the all-cis configuration, possess a high symmetry, and as such are of considerable interest for theories of bonding. Except for the bridged ions, the smaller systems have all of the carbon atoms equivalent, and thus no problems regarding charge density arise. Consequently the effects of charge on the physical properties of molecules can be readily studied in this series.

The electronic, IR and NMR spectra of these ions, including the cyclopropenium cation, the cyclopentadienyl anion, and the tropylium cation, are shown in Tables I–III.[5, 24, 27, 28, 43, 44, 58a, 67–69, 73, 78–83]

A. ELECTRONIC SPECTRA

Murrell and Longuet-Higgins,[84] using the Pariser–Parr S.C.F. theory, calculated that the tropylium cation should exhibit an allowed transition at 6.36 eV and a forbidden transition at 4.33 eV. The forbidden transition corresponds to the band at 275 nm (4.51 eV) observed by Doering and Knox,[5] and the allowed transition to the absorption at 217 nm (5.73 eV) observed by Dauben et al.[81] Longuet-Higgins and McEwen[23a] calculated, again using a semiempirical MO method, the excitation energies of a number of aromatic ions. These calculations suggest that the cyclopentadienyl anion should not show an absorption above 220 nm, in agreement with the experimental finding.[80] The cyclooctatetraenyl dianion was predicted to have two transitions at 5.80 eV (215 nm) and 3.67 eV (340 nm). The observed spectrum[27] is very different to that calculated.

Simmons et al.[63] calculated that the electronic spectrum of the cyclononatetraenyl anion should have transitions at 3.619 and 5.446 eV, and these calculations are in reasonable agreement with the observed values of 4.96 (250) and 3.88 eV (320 nm). The band of higher energy was assigned to an electronically allowed transition to the $^1E_1{}'$ state, whereas the long wavelength absorption was assigned to the electronically forbidden $^1E_4{}'$ transition. The theoretical oscillator strength of the $^1E_1{}' \leftarrow {}^1A_1{}'$ transition was found to be 1.99, which is in accord with the high intensity found for the 250 nm band.

B. INFRARED AND RAMAN SPECTRA

The analyses of the infrared and Raman spectra of a number of monocyclic ions have been carried out, and the assignments compared with those for benzene. Lippincott and his co-workers[85] analyzed the IR and Raman

[78] R. Breslow, J. T. Groves, and G. Ryan, J. Am. Chem. Soc. 89, 5048 (1967).

[79] R. Breslow and J. T. Groves, J. Am. Chem. Soc. 92, 984 (1970).

[80] W. Okamura, unpublished results; see Katz et al.[102]

[81] H. J. Dauben, F. A. Gadecki, K. M. Harmon, and D. L. Pearson, J. Am. Chem. Soc. 79, 4557 (1957).

[82] H. P. Fritz and L. Schäfer, Ber. 97, 1829 (1964).

[83] G. Herzberg, "Infrared and Raman Spectra of Polyatomic Molecules," p. 365. Van Nostrand, Princeton, New Jersey, 1945.

[84] J. N. Murrell and H. C. Longuet-Higgins, J. Chem. Phys. 23, 2347 (1955).

[85] R. D. Nelson, W. G. Fateley, and E. R. Lippincott, J. Am. Chem. Soc. 78, 4870 (1956); W. G. Fateley, B. Curnutte, and E. R. Lippincott, J. Chem. Phys. 26, 1471 (1957).

TABLE I

ELECTRONIC SPECTRA OF MONOCYCLIC $(4n + 2)$ π-ELECTRON IONS

Compound	Reference	Solvent	λ max nm	ϵ	λ max nm	ϵ	λ max nm	ϵ
$C_3H_3^{\oplus}SbCl_6^{\ominus}$	78, 79		No high intensity maximum above 205 nm					
$C_5H_5^{\ominus}Li^{\oplus}$	80	THF	No high intensity maximum above 205 nm					
$C_7H_7^{\oplus}ClO_4^{\ominus}$	81	H_2SO_4	217	41,000	273.5	4,350		
$C_7H_7^{\oplus}Br^{\ominus}$	5	Aqueous acid			275	4,360		
$C_8H_8^{\ominus}K_2^{\oplus}$	27	THF	236	740	263	1,700	407	3,240
$C_9H_9^{\ominus}Li^{\oplus}$	43	THF	251	100,000	319	7,900	324	7,900
$C_9H_9^{\ominus}Et_4N^{\oplus}$	44	CH_3CN	250	66,300	317	6,600	322	6,800
$C_{12}H_{11}^{\oplus}BF_4^{\ominus a}$	69	60% H_2SO_4	272	31,600	302	51,800	320	14,100
			385	2,300	423	4,000		
$C_{16}H_{16}^{\ominus}$	58a		333	16,000	380	50,000	413	170,000
			560	12,000	600	16,000		

^{a} Bicyclo[5.4.1]dodecapentaenium tetrafluoroborate.

TABLE II

IR SPECTRA OF MONOCYCLIC $(4n + 2)$ π-ELECTRON IONS^a

Compound	Reference	Medium	Absorption (cm^{-1})
$C_3H_3^{\oplus}SbCl_5^{\ominus}$	78, 79	Nujol	3105, 1276, 908, 736
$C_5H_5^{\ominus}K^{\oplus}$	82	Nujol	3048, 1455, 1381, 1009, 890, 731, 702, 663
C_6H_6	83	Vacuum	3099, 1485, 1037, 671
$C_7H_7^{\oplus}Br^{\ominus}$	5	KBr	3020, 1482, 678, 651
	27	^{b}	3020, 1477, 992, 633
$C_8H_8^{\ominus}K_2^{\oplus}$	27	Nujol	2994, 1431, 880, 707, 684
$C_9H_9^{\ominus}Et_3N^{\oplus}$	44	KBr	3021, 2994, 2915, 2874, 1477, 1453, 1431, 1389, 1374, 1170, 1001, 845, 782

^{a} Only strong and medium absorptions are given in this table.
^{b} Medium not stated.

spectra of ferrocene and tropylium bromide, comparing the observed spectra with calculated data and concluded that for all $(CH)_n$ planar rings with $n \geqslant 5$ similar assignments may be made, and the same selection rules apply. The IR and Raman data for potassium cyclopentadienide have more recently been

TABLE III

NMR SPECTRA OF MONOCYCLIC $(4n + 2)$ π-ELECTRON IONS

Compound	Reference	Solvent	Chemical shift (τ)
$C_3H_3^{\oplus}SbCl_6^{\ominus}$	78, 79	CH_2Cl_2	-1.1 (s)
$C_5H_5^{\ominus}Li^{\oplus}$	28	CH_3CN	4.6 $(s)^a$
$C_7H_7^{\oplus}Br^{\ominus}$	28	CH_3CN	0.82 $(s)^a$
$C_8H_8^{\ominus}K_2^{\oplus}$	24	THF-d_8	4.3 (s)
$C_8H_8^{\ominus}Na_2^{\oplus}$	28	CH_3CN	4.33 $(s)^a$
$C_9H_9^{\ominus}Li^{\oplus}$	43	THF-d_8	3.10 (s)
	44	THF	3.28 (s)
$C_9H_9^{\ominus}K^{\oplus}$	43	THF-d_8	2.96 (s)
$C_9H_9^{\ominus}Na^{\oplus}$	44	DMSO	3.10 (s)
$C_9H_9^{\ominus}Et_4N^{\oplus}$	44	DMSO-d_6	3.18 (s) (6.97, —CH_2, $J = 7$ Hz; 8.95, —CH_3, m, $J = 7$, ca. 1.5 Hz)
$C_{10}H_9^{\ominus}Na^{\oplus}$ ^{b}	67	DMSO-d_6	3.2 $(dd$, 2 H), 4.0–4.6 $(m$, 5 H), 10.7 $(d$, 1 H, $J = 7.5$ Hz), 11.2 $(d$, 1 H, $J = 7.5$ Hz)
	68	DMSO-d_6	2.94 $(dd$, 2 H, $J = 2$, 5.5 Hz), 3.98 $(m$, 5 H), 10.45 $(d$, 1 H, $J = 7.5$ Hz), 10.95 $(d$, 1 H, $J = 7.5$ Hz)
$C_{12}H_{11}^{\oplus}BF_4^{\ominus}$ ^{c}	69	CD_3CN	0.4–1.7 $(m$, 9 H), 10.3 $(d$, 1 H, $J = 10$ Hz), 11.8 $(d$, 1 H, $J = 10$ Hz)
$C_{16}H_{16}^{\ominus}$	73	THF-d_8	1.17 $(dd$, 8 H), 2.55 $(dd$, 4 H), 18.17 $(dd$, 4 H)

^{a} Converted from data giving position in ppm or Hz from benzene (assumed to be $\tau2.73$).
^{b} Sodium bicyclo[4.3.1]decatetraenide (20).
^{c} Bicyclo[5.4.1]dodecapentaenium tetrafluoroborate (21).

analyzed by Sadô et al.[86] Using a similar type of analysis West et al.[87] have also assigned modes to the observed bands in the trichlorocyclopropenylium cation, and this may be applied to the parent $C_3H_3^{\oplus}$ ion.[78, 79] The symmetries of the calculated bands, the corresponding experimentally observed absorptions, and the modes probably giving rise to these bands are tabulated in Table IV.

C. NMR SPECTRA

The effect of charge density on the NMR chemical shift of protons has been studied by a comparison of the positions of the proton resonance in the cyclopentadienyl anion, the tropylium cation, and benzene.[28, 62] The change in

[86] A. Sadô, R. West, H. P. Fritz, and L. Schäfer, Spectrochim. Acta 22, 509 (1966).
[87] R. West, A. Sadô, and S. W. Tobey, J. Am. Chem. Soc. 88, 2488 (1966).

TABLE IV

IR AND RAMAN SPECTRAL ASSIGNMENTS OF THE MONOCYCLIC $(4n + 2)$ π-ELECTRON IONS

Species, symmetry					Spectral activity	Observed bands (cm^{-1})					Description[c]
$C_3H_3^{\oplus}$ D_{3h}	$C_5H_5^{\ominus}$ D_{5h}	C_6H_6 D_{6h}	$C_7H_7^{\oplus}$ D_{7h}	$C_9H_9^{\ominus}$ D_{9h}		$C_3H_3^{\oplus}$[a]	$C_5H_5^{\ominus}$[b]	C_6H_6[c]	$C_7H_7^{\oplus}$[c]	$C_9H_9^{\ominus}$[d]	
A_1'	A_1'	A_{1g}	A_1'	A_1'	R[e]		3043	3062	3060		CH stretching
					R		983	992	868		Ring breathing
A_2''	A_2''	A_{2u}	A_2''	A_2''	IR[e]	736		671	633		CH ($\perp$)[f] bending
E''	E_1''	E_{1g}	E_1''	E_1''	R		—	—	—		CH ($\perp$) bending
E'	E_1'	E_{1u}	E_1'	E_1'	IR	3105	3039	3099	3020		CH stretching
					IR	1276	1455	1485	1477		CC stretching
					IR	908	1003	1037	992		CH ($\parallel$)[g] bending
E'	E_2'	E_{2g}	E_2'	E_2'	R		3096	3047	3075	2994	CH stretching
					R		1447	1596	1594	1477	CC stretching
					R		1020	1178	1210	1001	CH ($\parallel$) bending
					R		565	606	433		CCC ($\parallel$) bending

[a] Breslow et al.[78,79]
[b] Sadô et al.[86]
[c] Lippincott et al.[85]
[d] LaLancette and Benson.[44]

[e] R, Raman; IR, infrared.
[f] $\perp$, perpendicular to ring plane.
[g] $\parallel$, parallel to ring plane.

chemical shift, δ, was found to be proportional to the excess charge on the carbon atom attached to the proton, $\Delta\rho$, and the proportionality constant was empirically found to have a value of ca. 10 ppm/unit charge

$$\delta = K\Delta\rho$$

It is assumed for the purpose of these calculations that the deshielding effect due to the diamagnetic ring current remains constant throughout the series.

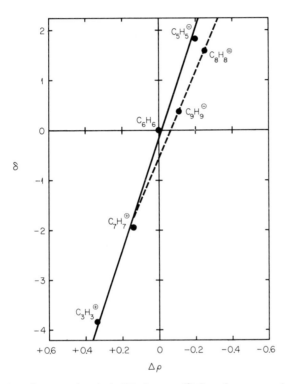

FIG. 25. Plot of proton chemical shifts in ppm (δ) from benzene against charge density ($\Delta\rho$) for monocyclic ions.

The expected value for the chemical shift of the cyclooctatetraenyl dianion using benzene as the standard is $\tau 5.23$, whereas the experimental finding ($\tau 4.33$) shows that the protons are considerably more deshielded. However, when the two 10 π-electron ions $C_8H_8^{\ominus}$ and $C_9H_9^{\ominus}$ are considered together, then, again assuming that the ring currents are constant, the predicted position of the cyclononatetraenyl anion is $\tau 2.9$, using the experimental value of $\tau 4.3$ for the cyclooctatetraenyl dianion. It should thus be possible within a series having the same number of π-electrons to predict the chemical shift of any member provided the chemical shift of one member is known.

A plot of charge density against chemical shift for the known monocyclic ions is shown in Fig. 25. Calculations using the values for the $C_8H_8^{\ominus}$ and

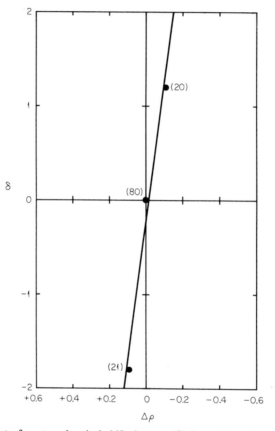

FIG. 26. Plot of proton chemical shifts in ppm (δ) from 1,6-methano[10]annulene (**80**) against charge density ($\Delta\rho$) for the bicyclic ions (**20**) and (**21**).

$C_9H_9^{\ominus}$ ions, assuming $K = 10$ ppm, suggest that planar, all-*cis*[10]annulene would have a NMR resonance at $\tau 2.0$.

In the case of the three 10 π-electron bridged systems, **20**, **21** and 1,6-methano[10]annulene, the problem of unequal charge density arises since the carbon atoms are no longer equivalent. A plot of charge density against chemical shift is shown in Fig. 26, taking a weighted average value of the chemical shift of the aromatic protons. This assumes that the charge densities on the 1,6-carbon atoms are similar to those of the carbon atoms bearing protons.

Similar correlations have been made between the charge density and ^{13}C nuclear resonance shifts.[88] The data of Spiesecke and Schneider, together with additional values for the cyclopropenium cation, the tetramethylcyclobutenium dication, and the cyclononatetraenyl anion, are tabulated in Table V.

TABLE V

^{13}C CHEMICAL SHIFTS AND ^{13}C-H COUPLING CONSTANTS OF THE MONOCYCLIC $(4n + 2)$ π-ELECTRON IONS

Compound	Reference	Solvent	^{13}C (ppm relative to benzene)	$^{13}J_{CH}$ (Hz)
$C_3H_3^{\oplus}$	78, 79, 120	SbF_5, SO_2	−47.8	265
$C_8H_{12}^{\oplus\oplus a}$	120	SbF_5, SO_2	−80.0	
$C_5H_5^{\ominus}$	88	THF	25.7	157
$C_7H_7^{\oplus}BF_4^{\ominus}$	88	H_2O, HBF_4	−27.6	171
$C_8H_8^{\ominus}$	88	THF	42.5	145
$C_9H_9^{\ominus}Li^{\oplus}$	44	THF	19.0	137

[a] Tetramethylcyclobutenium dication.

A plot of charge density against ^{13}C chemical shift is shown in Fig. 27. As can be seen from Fig. 27, the diamagnetic ring current has little effect on the position of the ^{13}C resonance, and the 2 π-, 6 π-, and 10 π-electron systems all fall on the same linear plot. The deviation in the position of the cyclopentadienyl anion found by Spiesecke and Schneider has been confirmed,[44] but the reason for this deviation is not known. These results suggest that ^{13}C nuclear resonance data may be more reliable than proton resonance data in predicting the charge densities at carbon atoms in the monocyclic ions.

VI. Polycyclic Aromatic Ions

The fusion of two monocyclic $(4n + 2)$ π-electron systems provides a new, bicyclic $(4n + 2)$ π-electron system. Such systems are extremely familiar, examples being naphthalene, conceptionally formed by the fusion of two benzene rings, and the indenyl anion, from benzene and the cyclopentadienyl anion. Two other types of fusion are also possible. Two $4n$ systems may be fused to give a new $(4n + 2)$ π-electron system, while a $4n$ plus a $4n + 2$ system gives a new $4n$ π-electron system.[89] The properties of systems formed by these

[88] H. Spiesecke and W. G. Schneider, *Tetrahedron Letters* No. 14, p. 468 (1961).
[89] R. Breslow, *Chem. Eng. News* **43**, xxvi, 90 (1965).

two latter modes are as yet little known. An alternative method of analyzing polycyclic systems is to examine the periphery, considering the transannular bonds to be perturbations. Thus naphthalene can be considered to be 1,6-

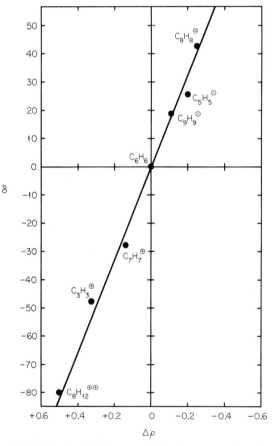

FIG. 27. Plot of ^{13}C chemical shift in ppm (δ) from benzene against charge density ($\Delta\rho$) for monocyclic ions.

bridged [10]annulene. However both of these analyses are probably mainly useful for classification purposes, and are of lesser use in predicting the properties of the ensuing system.[90] In the present section we will examine polycyclic systems containing $(4n + 2)$ π-electron rings, considering these systems in the order of the increasing size of the smallest ring.

[90] M. J. S. Dewar, *Tetrahedron* 1, Suppl. 8, 75 (1966).

A. SYSTEMS WITH FOUR-MEMBERED RINGS

Two reports have recently been published concerning the synthesis of the cyclobutadienocyclopentadienyl anion (**99**).[91, 92] The anion **99** is a $4n$ π-electron system, conceptualized by fusion of cyclobutadiene and the cyclopentadienyl anion or by the extension of a 1,4 bond across the cycloheptatrienyl anion (**100**). The anion **99** might thus be expected to exhibit the properties of a destabilized

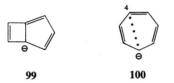

      99   **100**

cyclopentadienyl anion, or behavior similar to that of **100**, previously prepared by Dauben and Rifi.[93]

Breslow *et al.*[91] reacted 1-methoxybicyclo[3.2.0]hepta-3,6-diene (**101**) with BCl_3 at $-75°C$ and obtained the corresponding chlorodiene **102** in 63% yield.

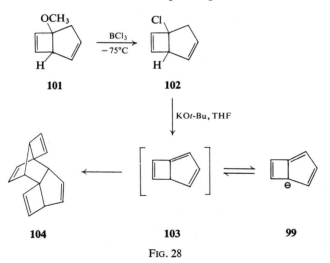

FIG. 28

Treatment of **102** with potassium *t*-butoxide in THF gave a dimer, the structure of which was considered to be possibly **104** (one possible stereoisomer shown) formed by Diels–Alder dimerization of **103** (Fig. 28). When the reaction was

[91] R. Breslow, W. Washburn, and R. G. Bergman, *J. Am. Chem. Soc.* **91**, 196 (1969).
[92] N. L. Bauld, C. E. Dahl, and Y. S. Rim, *J. Am. Chem. Soc.* **91**, 2787 (1969).
[93] H. J. Dauben and M. R. Rifi, *J. Am. Chem. Soc.* **85**, 3041 (1963).

carried out in the presence of deuterated *t*-butanol, a considerable portion of the dimer (65%) contained two atoms of deuterium, and on the addition of DMSO-d_6 to the reaction mixture 100% incorporation of two deuterium atoms into the dimer was observed. The results suggest that the hydrocarbon **103** exchanges via the anion **99** at a rate comparable to that in which it dimerizes.

However, Bauld *et al.*[92] reported that the dimer formed from bicyclo[3.2.0]-hepta-1,3,5-triene (**103**) did not have the structure **104**, but was a mixture of

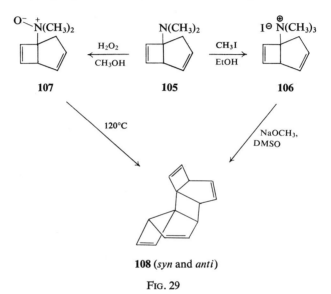

108 (*syn* and *anti*)

Fig. 29

isomers (**108**) arising from 2 + 2 addition of the strained trisubstituted double bond in **103**. The *N,N*-dimethylamine **105** was converted to the corresponding methiodide **106** with methyl iodide and to the amine oxide **107** by hydrogen peroxide in methanol (Fig. 29). The same 50:50 mixture of dimers (**108**) was obtained by either pyrolysis of the *N*-oxide **107** or base treatment of the methiodide **106**. The dimers could be separated by VPC, and have different, though extremely similar, NMR spectra. Bauld *et al.* suggest that the dimers are syn and anti isomers of structure **108**, and Breslow *et al.* have now come to the same conclusion.[94] Bauld *et al.*[92] were also unable to detect deuterium incorporation into the dimer **108** during base elimination of the methiodide **106** using potassium *t*-butoxide and a variety of deuterated solvents. At the present time the results of these two groups of workers can only be resolved by assuming that base elimination from **106** gives the hydrocarbon **103** in a

[94] See Cava *et al.*,[95] footnote 6.

higher energy state than base elimination from **102**, so that dimerization occurs much faster than deprotonation.

Cava *et al.*[95] have recently prepared the related norbiphenylene anion (**109**)

109

by the synthetic route shown in Fig. 30. The dicarboxylic acid **110** was converted to the corresponding diethyl ester, which was then transformed to the dihydrazide **111**. Reaction of **111** with nitrous acid gave the diazide **112**, which was converted to the diamine **113** by pyrolysis in benzyl alcohol to the

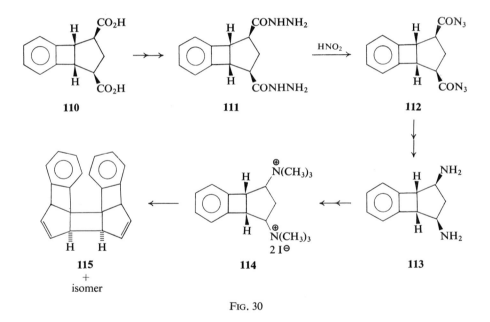

FIG. 30

benzylurethane and subsequent hydrogenolysis. Exhaustive methylation gave the dimethiodide **114**, which on treatment with potassium *t*-butoxide in *t*-butanol gave a mixture of dimers. The structure of one of these dimers was shown to be **115** by its physical properties and X-ray crystallographic analysis.

[95] M. P. Cava, K. Narasimhan, Z. Weiger, L. J. Radonovich, and M. D. Glick, *J. Am. Chem. Soc.* **91**, 2378 (1969).

When the elimination was carried out using the dimsyl anion in DMSO, a deep-brown solution was formed, which decolorized immediately on addition of water to give the same mixture of dimers. Addition of the solution to D_2O gave the dimers containing two atoms of deuterium. The brown solution shows an electronic absorption at 590 nm with a shoulder at 500 nm, and appears to be due to the anion **109**.

The increased stability of the anion **109** over that of the anion **99**, which parallels the greater stability of biphenylene over benzocyclobutadiene, is presumably due to the decreased cyclobutadienyl character of the four-membered ring in **109** compared with **99**.

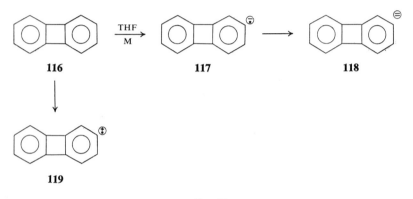

116   **117**   **118**

119

Fig. 31

Hush and Rowlands[96] have shown that the reduction of biphenylene (**116**) with alkali metals in THF gives first the monoanion **117** and then the dianion **118** (Fig. 31). The electronic spectrum of the anion **117** in dimethoxyethane shows peaks at 269, 326, 345, 360, 390, and 610 nm, whereas the dianion **118** absorbs at 255, 345, and 480 nm. The calculated energies for these bands in the spectrum of the dianion are only in moderately good agreement with the observed values.

The ESR spectra of **117** and of the cation radical **119** have been recorded.[97, 98] The spectra of both radical ions are similar showing five main lines,[97] which are further split into quintets.[98] The hyperfine splittings for **117** are 2.86 and 0.21 gauss, and for **119** are 3.69 and 0.21 gauss.[98] Hückel and semiempirical MO treatments, using the relation $a_H = Q\rho$, give calculated spectra in reasonable agreement with those experimentally observed.

[96] N. S. Hush and J. R. Rowlands, *Mol. Phys.* **6**, 317 (1963).
[97] C. A. McDowell and J. R. Rowlands, *Can. J. Chem.* **38**, 503 (1960).
[98] A. Carrington and J. Dos Santos-Veiga, *Mol. Phys.* **5**, 285 (1962).

Bauld and Banks[99] repeated these reactions, and concluded that two absorptions at 588 and 565 nm were due to the anion **117** and the dianion **118**, respectively. From the behavior of these absorptions these workers concluded that the anion **117** extensively disproportionates to the hydrocarbon and dianion, and suggested that the dianion **118** possessed aromatic character. However, Waack et al.[100] also reinvestigated this reduction and found that **117** absorbs at 270 (4.69) and 382 nm (log ε 4.72) and **118** at 345 nm (log ε 4.64), in agreement with

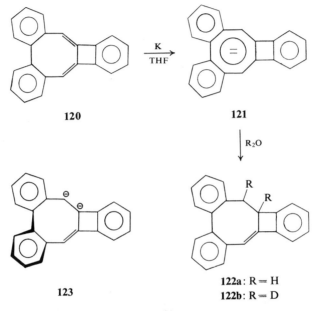

FIG. 32

the findings of Hush and Rowlands. Furthermore, these workers concluded that the visible spectral absorptions at 595 and 565 nm were both due to the monoanion. By following the progress of the reduction these authors concluded that in fact the disproportionation of the anion radical **117** was very small ($K < 1.5 \times 10^{-2}$), and that the biphenylene dianion **118** showed no special stability.

The 3,4:5,6:9,10-tribenzobicyclo[6.2.0]decapentaenyl dianion (**121**) has recently been prepared by reduction of the hydrocarbon **120** with alkali metals[101a] (Fig. 32). The dianion, the first biphenylene analog in which a

[99] N. L. Bauld and D. Banks, *J. Am. Chem. Soc.* **87**, 128 (1965).

[100] R. Waack, M. A. Doran, and P. West, *J. Am. Chem. Soc.* **87**, 5508 (1965).

[101] C. S. Baxter, P. J. Garratt, and K. P. C. Vollhardt, *J. Am. Chem. Soc.* **91**, 7783 (1969).

[101a] P. J. Garratt and K. P. C. Vollhardt, unpublished experiments.

phenyl ring has been replaced by a 10 π-electron system, is stable in solution, and the NMR spectrum (τ2.52, m, 2 H; τ3.90, m, 2 H; τ4.23, s, 2 H; τ4.51, 4.87, m, 4 H) is consistent with **121** being a planar, aromatic species. Treatment of solutions of **121** with water (or deuterium oxide) give 1,2-dihydro-3,4:5,6:9,10-tribenzobicyclo[6.2.0]decapentaene (**122a**) (or the corresponding dideuterio derivative **122b**). When the progress of the reduction of **120** is followed, it appeared that the initial reduction occurred in the nonplanar form, to give the nonplanar dianion **123** which then rehybridizes to **121**. However, subsequent experiments[101a] have established that the NMR spectrum attributed to the nonplanar dianion **123** is due to a neutral dimer which is formed from **120**. This dimer is cleaved by potassium to give the dianion **121**.

B. Systems with Five-Membered Rings

The fusion of two cyclopentadienyl anions conceptionally forms the penta-lenyl dianion (**128**), a system which can also be derived from the cyclo-octatetraenyl dianion by 1,5-transannular bonding. The synthesis of the pentalenyl dianion was described by Katz and Rosenberger[102] in 1962 (see Fig. 33). Partial hydrogenation of dicyclopentadiene gave the dihydrodicyclo-pentadiene **124**, which on oxidation with selenium dioxide in acetic acid, acetic anhydride followed by hydrolysis gave 1-hydroxy-5,6-dihydrodicyclo-pentadiene (**125**). Dehydration over alumina at 320°C gave **126**, which on pyrolysis at 570°C undergoes successive ene and retro-Diels–Alder reactions to give ethylene and dihydropentalene **127**. Dihydropentalene rapidly dimerizes at room temperature, but on treatment with n-butyllithium in n-hexane a pale-yellow solution is formed, which deposits crystals on cooling. The NMR spectrum of this solution (τ4.27, t, 2 H, $J = 3.0$ Hz; τ5.02, d, 4 H, $J = 3.0$ Hz) is that expected for the pentalenyl dianion **128**.

Treatment of solutions of **128** with neutral phosphate buffer, followed by hydrogenation of the extracted organic material, gave a 25% yield of *cis*-bicyclo[3.3.0]octane (**129**). Various MO calculations predict electronic absorptions in reasonable accord with the observed absorption at 296 nm (logϵ3.7).

Treatment of dihydropentalene **127** with n-butyllithium and ferrous chloride gives dihydropentalenyl iron (**130**) (or the double bond isomer).[103] Reaction of **130** in ether with excess t-butyllithium at −40°C gave a maroon solution, which precipitated a solid on warming to −10°C. Addition of suspensions of the solid to dilute deuteriosulfuric acid gave, after isolation, **130** containing two atoms of deuterium, indicating that the precipitate is dipenta-lenyl iron dianion (**131**). Similar treatment of the maroon solution gave **130**

[102] T. J. Katz and M. Rosenberger, *J. Am. Chem. Soc.* **84**, 865 (1962); T. J. Katz, M. Rosenberger, and R. K. O'Hara, *ibid.* **86**, 249 (1964).
[103] T. J. Katz and M. Rosenberger, *J. Am. Chem. Soc.* **85**, 2030 (1963).

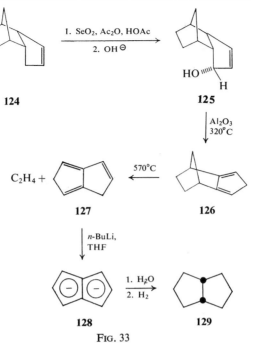

1. SeO₂, Ac₂O, HOAc
2. OH⊖

124

125

Al₂O₃
320°C

C₂H₄ +

570°C

127

126

n-BuLi,
THF

1. H₂O
2. H₂

128

129

FIG. 33

containing only one atom of deuterium, indicating that the monoanion **132** is present in this solution. Other derivatives have also been prepared.[103a] The dibenzopentalenyl dianion has been synthesized.[103b]

Both of the *sym*-indacenyl (**134**) and *as*-indacenyl (**140**) dianions have been prepared. Hafner and Sturm[104] obtained **134** by treatment of *sym*-indacene

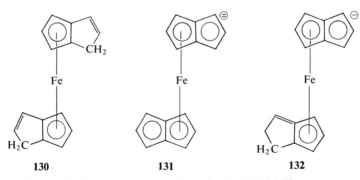

130

131

132

[103a] T. J. Katz and J. J. Mrowca, *J. Am. Chem. Soc.* **89**, 1105 (1967).

[103b] A. J. Silvestri, *Tetrahedron* **19**, 855 (1963).

[104] K. Hafner, *Angew. Chem.* **75**, 1041 (1963); *Angew. Chem. Intern. Ed. Engl.* **3**, 165 (1964).

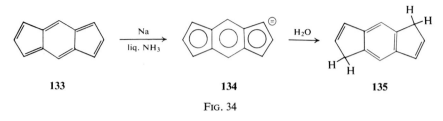

133   134   135

FIG. 34

(133) with sodium in liquid ammonia. Disodium *sym*-indacenide (134) was isolated as a thermally stable, air-sensitive, colorless compound. On hydrolysis the dianion 134 gave the known 1,5-dihydro-*sym*-indacene (135) (Fig. 34).

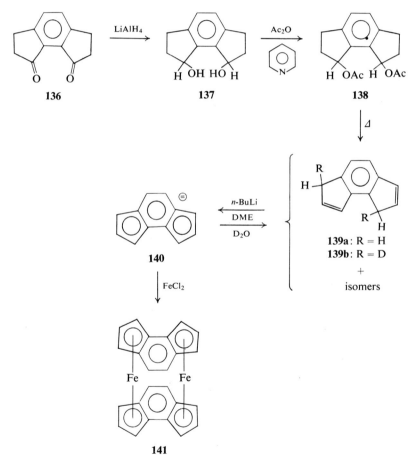

FIG. 35

The *as*-indacenyl dianion (**140**) was prepared by Katz and Schulman[105] by the method shown in Fig. 35. Reduction of 1,8-dioxo-2,3,6,7-tetrahydro-*as*-indacene (**136**) with LiAlH₄ gave a mixture of the diasterioisomeric alcohols **137**, which could be acetylated to the corresponding acetates **138**. Pyrolysis of the acetate mixture in a nitrogen stream at 595°C gave dihydro-*as*-indacene (**139a**), probably as a mixture of double bond isomers. Treatment of a solution of **139a** in dimethoxyethane with *n*-butyllithium in *n*-hexane precipitates the white, crystalline dilithium *as*-indacenide (**140**). Addition of deuterium oxide gave dihydro-*as*-indacene containing two deuterium atoms (**139b**). Reaction

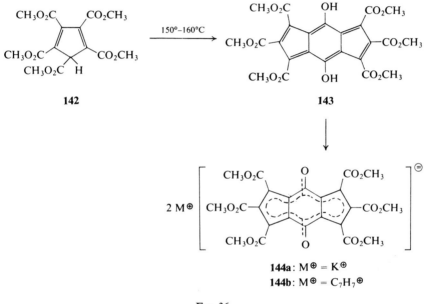

FIG. 36

of the dianion **140** with ferrous chloride gave bis(*as*-indecenyl iron) (**141**) in 87% yield as maroon crystals.[105, 106] The staggered arrangement of the *as*-indacenyl rings was indicated from the X-ray analysis, which shows that the molecules have a center of symmetry.

Le Goff and La Count[107] prepared the substituted *sym*-indacenyl dianion **144** by the route shown in Fig. 36. The diene **142** was heated at 150°–160°C, when self-condensation occurred to give the diol **143**. The diol **143** on reaction with potassium fluoride in methanol gave the dianion as the dipotassium

[105] T. J. Katz and J. Schulman, *J. Am. Chem. Soc.* **86**, 3169 (1964).
[106] T. J. Katz, V. Balogh, and J. Schulman, *J. Am. Chem. Soc.* **90**, 734 (1968).
[107] E. Le Goff and R. B. La Count, *Tetrahedron Letters* p. 1161 (1964).

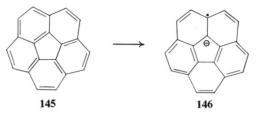

145   146

salt (**144a**). Reaction of the **144a** with tropylium fluoroborate gave the corresponding tropylium salt (**144b**).

Corannulene (**145**) is reduced by alkali metals in THF to give green solutions of the radical anion **146**.[108] The ESR spectrum shows 11 equally spaced lines

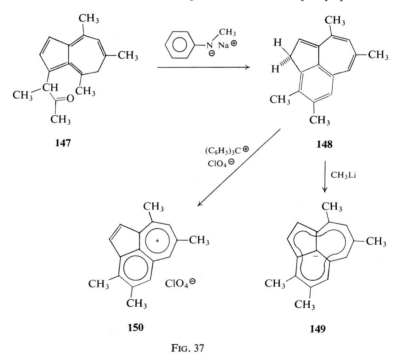

FIG. 37

$[a_{\text{H}}$ 1.560 $\pm$ 0.005 gauss) in the binomial intensity expected for 10 equivalent protons. The radical **146** absorbs in the visible region at 650 nm, and can also be prepared by polarographic reduction of **145**. The calculated nonplanar configurations are in reasonable agreement with the experimental value, but **146**, like corannulene, is probably bowl-shaped. Further reduction by alkali

[108] J. Janata, J. Gendell, G.-Y. Ling, W. Barth, L. Backes, H. B. Mark, and R. G. Lawton, *J. Am. Chem. Soc.* **89**, 3056 (1967).

metals produces a bright red species which absorbs at 500 nm, but the authors believe this is not the corannulene dianion.

Hafner and Schaum[104, 109] have prepared the tricyclic hydrocarbon **148** by cyclization of the ketone **147** (Fig. 37). When **148** is treated with methyl-lithium the anion **149** is formed–a 14 π-electron system. With tritylperchlorate a stable orange salt is formed with the charge presumably mainly localized as the tropylium cation **150**.

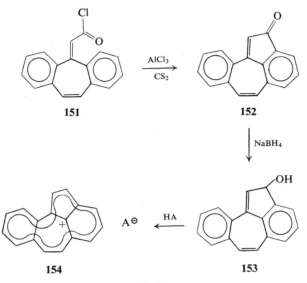

FIG. 38

Galantay et al.[110] prepared the violet dibenz[c,d,h]azulenium cation (**154**) by the route shown in Fig. 38. Cyclization of the ketone **151** with AlCl₃ in carbon disulfide at −10°C, gave maroon crystals of 2H-dibenz[c,d,h]azulen-2-one (**152**). Reduction of **152** with NaBH₄ gave the corresponding alcohol **153** which, on treatment with strong acids, gave the cation **154**.

C. SYSTEMS WITH SIX- OR MORE MEMBERED RINGS

The most extensively investigated members of this group are the ions derived from the phenalene system. Phenalene **155** can be converted into the corresponding cation **156**, radical **157**, or anion **158** by removal of a hydride ion,

[109] K. Hafner and H. Schaum, *Angew. Chem.* **75**, 90 (1963); *Angew. Chem. Intern. Ed. Engl.* **2**, 95 (1963).
[110] E. Galantay, H. Agahigian, and N. Paolella, *J. Am. Chem. Soc.* **88**, 3875 (1966).

hydrogen atom, or a proton. HMO calculations indicate that the 12 π-electrons in the cation **156** form a filled, bonding shell, and that the additional electrons in the radical and anion enter a nondegenerate, nonbonding orbital.[111] The chemistry of these and related systems has been reviewed.[112]

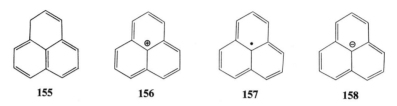

| 155 | 156 | 157 | 158 |

Heptalene (**159**) is a 12 π-electron system, and it might be expected to be easily oxidized to the 10 π-electron heptalenium dication (**160**), or reduced to the 14 π-electron heptalenyl dianion (**161**). The latter reduction appears to be less likely, from the known instability of the cycloheptatrienyl anion.

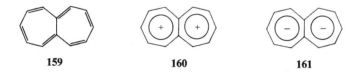

| 159 | 160 | 161 |

Dauben and Bertelli[113] prepared the heptalenium cation (**167**) by the sequence of reactions shown in Fig. 39. Reduction of 1,5-naphthalene-dicarboxylic acid (**162**) with sodium in liquid ammonia gave the tetrahydro derivative **163**. The carboxylic acid groups of **163** were then reduced with lithium aluminum hydride to give 1,5-di(hydroxymethyl)-1,4,5,8-tetra-dehydronaphthalene (**164**). Solvolysis of the ditosylate **165**, formed from **164** by treatment with tosyl chloride and pyridine, gave a mixture of 1,5- and 1,10-dihydroheptalenes (**166**). Hydride abstraction with trityl fluoroborate gave 1-heptalenium fluoroborate (**167**), as bright yellow crystals. Attempts to remove a further hydride ion from **167** were unsuccessful, but proton removal with trimethylamine gave heptalene (**159**).[114]

Some spectroscopic evidence has been found by Wilson[115] to suggest that the dication **160** is formed by treatment of iodoheptalenium iodide with silver hexafluoroantimonate in nitroethane.

[111] R. Zahradnik, J. Michl, and J. Kontecky, *Collection Czech. Chem. Commun.* **29**, 1932 (1964), and refs. therein.

[112] D. H. Reid, *Quart. Rev.* (*London*) **19**, 274 (1965.)

[113] H. J. Dauben and D. J. Bertelli, *J. Am. Chem. Soc.* **83**, 4657 (1961).

[114] H. J. Dauben and D. J. Bertelli, *J. Am. Chem. Soc.* **83**, 4659 (1961).

[115] J. D. Wilson, Ph.D. Thesis, University of Washington, 1966.

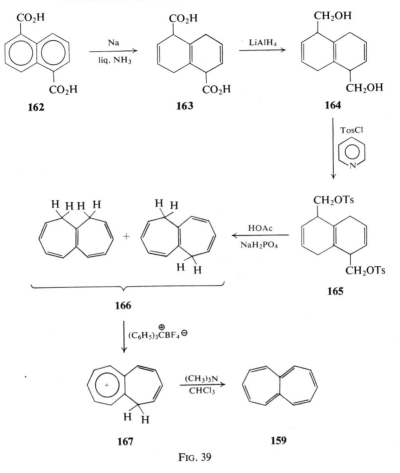

FIG. 39

VII. Homoaromatic Ions

To the best of our knowledge, the first suggestion of a stabilizing nonbonded interaction between π-systems was made by Thiele.[1a] The suggestion was put forward to explain the lack of acidity of the methylene protons of cycloheptatriene (168) compared with those of cyclopentadiene. Although it now appears that the major factor involved in the decreased acidity of these protons is the instability of the resulting cycloheptatrienyl anion.[93] the work of Doering *et al.*[115a] on the Büchner acids (169) suggests that 1,6-interactions in cycloheptatrienes may be of importance. The term "homoaromatic" to describe

[115a] W. von E. Doering, G. Laber, R. Vanderwahl, N. F. Chamberlain, and R. B. Williams, *J. Am. Chem. Soc.* **78**, 5448 (1956).

168   169

systems in which the cyclic conjugation is interrupted by a methylene group was coined by Winstein,[116] and the work in this area has been reviewed.[117,117a] The more convincing examples of "homoaromatic" stabilization are found in the ionic species,[117a] and it is these ions that will be considered in this section. Only species which have been isolated or observed spectroscopically will be considered, and for a discussion of kinetic studies in which homoaromatic intermediates have been implicated, the reader is referred to the review by Winstein.[117a]

   The extension of the Hückel rule to predict which of the more complex bishomoaromatic structures (e.g., **170**) will be aromatic has been made by Goldstein.[118] Goldstein's theory suggests that bishomo conjugates will only occur in "odd" systems. A further group of potentially aromatic systems, bicycloaromatics (e.g., **171**) was also examined, and the prediction made that only "odd" systems containing a total number of $4n$ π-electrons would be aromatic. The number of known molecules of these types is at present small.

170   171

A. HOMOCYCLOPROPENIUM CATIONS

   The cyclobutadienium (**172**) and cyclooctatetraenium (**6**) dications have proved to be elusive species, although the synthesis of the tetraphenylcyclobutadienium dication (**173**)[119] and more recently the tetramethylcyclobutadienium dication (**174**) have been reported.[120] Previous attempts to prepare the systems were frustrated by the stability of the intervening homoaromatic cations, together with the difficulty of accommodating two positive charges over the system. Katz et al.[121] examined the reaction of 1,2,3,4-tetramethyl-3,4-dichlorocyclobutene (**175**) with silver hexafluoroantimonate at −70°C in

[116] S. Winstein, J. Am. Chem. Soc. **81**, 6524 (1959).

[117] S. Winstein, E. C. Friedrich, R. Baker, and Y.-I. Lin, Tetrahedron **2**, Suppl. 8, 621 (1966).

[117a] S. Winstein, Chem. Soc. (London), Spec. Publ. **21**, 5 (1967); Quart. Rev. **23**, 141 (1969).

[118] M. J. Goldstein, J. Am. Chem. Soc. **89**, 6357 (1967).

[119] H. H. Freedman and A. E. Young, J. Am. Chem. Soc. **86**, 734 (1964); however, see Olah et al.[120]

[120] G. A. Olah, J. M. Bollinger, and A. M. White, J. Am. Chem. Soc. **91**, 3667 (1969).

[121] T. J. Katz, J. R. Hall, and W. C. Neikam, J. Am. Chem. Soc. **84**, 3199 (1962).

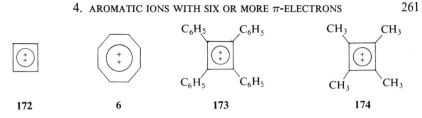

| 172 | 6 | 173 | 174 |

liquid sulfur dioxide and showed that the cation **176a** was formed (Fig. 40). Katz and Gold[122] subsequently prepared **176b** as the tetrachloroaluminate by treatment of **175** with $AlCl_3$ in methylene chloride. Other related substituted cyclobutenium cations were prepared as chloroaluminates by this method. The ultraviolet spectra of these ions is shown in Table VI.[122] The position

TABLE VI

ULTRAVIOLET SPECTRA OF CYCLOBUTENYLIUM CATIONS

Cyclobutenium cations as chloroaluminates	$\lambda_{max}^{CH_2Cl_2}$ (nm)	$\log \epsilon$
CH₃, CH₃, Cl cyclobutenyl	253	3.67
CH₃, CH₃, CH₃ cyclobutenyl	245	3.46
CH₃, CH₃, H cyclobutenyl	240	3.48
CH₃, CH₃, Br cyclobutenyl	263	3.57

of the ultraviolet maxima in these ions, ca. 250 nm, lies between that of a normal allylic cation (300 nm) and the cyclopropenium cation (ca. 185 nm). This suggests a considerable 1,3-interaction in these ions, and HMO calculations

[122] T. J. Katz and E. H. Gold, *J. Am. Chem. Soc.* **86**, 1600 (1964).

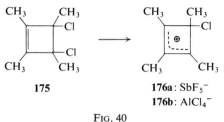

175   **176a**: SbF_5^-
           176b: $AlCl_4^-$

FIG. 40

indicate a 1,3-π-resonance integral of 0.33β, in good agreement with a previous prediction of Kiefer and Roberts.[123]

Whereas the NMR spectrum of the cation **176** showed three types of methyl resonance in SO_2 or CH_2Cl_2,[121, 122] when **175** is dissolved in sulfuric acid only one type of methyl group is observed.[124] It appears that the methyl groups are rapidly becoming equivalent in sulfuric acid, and similar findings were made for the cation **177**, which at room temperature again shows only a single sharp resonance.[125] A mechanism of intermolecular exchange was suggested to be operative in this latter system. The chemical shifts of the methyl groups in the nonequilibrating spectra support the concept that the charge is spread over the three carbon atoms by a 1,3 interaction. The term "homocyclopropenium cation" having structure **178** to describe this species seems appropriate.

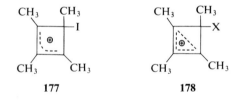

177   **178**

B. HOMOCYCLOPENTADIENYL ANIONS

1. *Bishomocyclopentadienyl Anion* (**180**)

Brown[126] and Winstein *et al.*[117a, 127] have prepared potassium bicyclo[3.2.1]-octadienide (**180**), previously postulated as an intermediate in the deuterium exchange of bicyclo[3.2.1]octa-1,2-diene (**181**),[128] by essentially the same

[123] E. F. Kiefer and J. D. Roberts, *J. Am. Chem. Soc.* **84**, 784 (1962), and references therein.

[124] C. F. Wilcox and D. L. Nealey, *J. Org. Chem.* **28**, 3446 (1963).

[125] E. H. Gold and T. J. Katz, *J. Org. Chem.* **31**, 372 (1966).

[126] J. M. Brown, *Chem. Commun.* p. 638 (1967).

[127] S. Winstein, M. Ogliaruso, M. Sakai, and J. M. Nicholson, *J. Am. Chem. Soc.* **89**, 3656 (1967).

[128] J. M. Brown and J. L. Occolowitz, *Chem. Commun.* p. 376 (1965); *J. Chem. Soc., B* p. 411 (1968).

method (Fig. 41). Treatment of *exo*-4-methoxybicyclo[3.2.1]octa-2,6-diene (**179a**) with sodium potassium alloy in THF gave a relatively stable orange solution of the anion **180**. The NMR spectrum of the anion was in accord with delocalized bishomocyclopentadienyl anion structure. The ^{6}H, ^{7}H protons

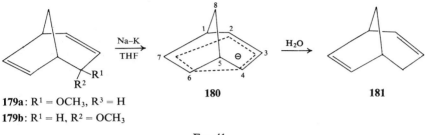

179a: R^1 = OCH$_3$, R^3 = H
179b: R^1 = H, R^2 = OCH$_3$

FIG. 41

showed a marked upfield shift from those in the hydrocarbon **181**, whereas the ^{3}H proton is at *lower* field. Treatment of solutions of **180** with water gave the hydrocarbon **181**. The endo isomer **179b** reacts much less readily with the alloy than the exo isomer, and appears to generate the 4-methoxy anion.[126]

2. *Bicyclo[3.3.2]nonatrienyl Anion* (**171**)

Grutzner and Winstein[128a] have recently prepared an example of a bicyclo-aromatic anion,[118] the bicyclo[3.2.2]nonatrienyl anion (**171**), by treatment of the methylether **181a** with sodium–potassium alloy in dimethoxyethane (DME) (Fig. 41a). Treatment of the anion with methanol gave an essentially quantitative yield of the hydrocarbon **181b**.

In the NMR spectrum, the protons on the two double bonds all appear at the same position (τ5.02). The anion thus has the delocalized structure shown in **171**, or it has a time-dependent structure in which each bridge in turn participates in bonding with the allylic anion. Since this process was not observed in the NMR spectrum at $-35°$, then the barrier to such a ring flipping, if it occurs, must be low ($\leqslant 11.8$ kcal·mole^{-1}).

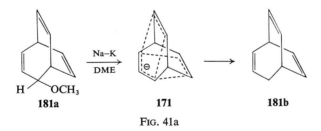

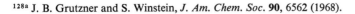

FIG. 41a

[128a] J. B. Grutzner and S. Winstein, *J. Am. Chem. Soc.* **90**, 6562 (1968).

C. HOMOTROPYLIUM CATIONS

The homotropylium cation (182) was first prepared by von Rosenberg, Mahler, and Pettit in 1962,[129] and has been one of the most extensively studied homoaromatic systems. Treatment of COT (17) with 98 % sulfuric acid, or the addition of antimony pentachloride to equimolar amounts of HCl and COT in nitromethane gave the cation 182, in the latter case as the crystalline hexa-chloroantimonate 182a (Fig. 42). The NMR spectrum of 182 showed signals at $\tau 1.4$ (5 H), 3.4 (2 H), 4.8 (1 H), and 10.6 (1 H), attributed, respectively, to

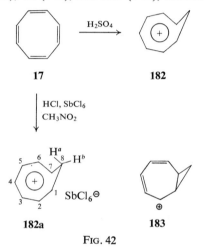

FIG. 42

the ²H–⁶H protons, the ¹H, ⁷H protons, the ᵇH proton, and the ᵃH proton. Comparison of the NMR spectrum of 182 with that of a number of related compounds,[130, 131] supports the view that the cation 182 is best represented by the delocalized homoaromatic structure, rather than the localized structure 183, with a 1,7 σ bond.

When the protonation of COT is carried out with D_2SO_4 at room tempera-ture, the deuterium is introduced in both the endo (ᵃH) and exo (ᵇH) positions.[129] However when this reaction is carried out at −10°C, then the deuterium largely enters at the endo position.[132] Warming to room temperature then allows equilibration of the deuterium between the endo and exo positions, presumably via the planar cation 184 and the energy required to compress the ring was calculated to be 22.3 kcal·mole^{-1} (Fig. 43).[132]

[129] J. L. von Rosenberg, J. E. Mahler, and R. Pettit, *J. Am. Chem. Soc.* **84**, 2842 (1962).
[130] C. E. Keller and R. Pettit, *J. Am. Chem. Soc.* **88**, 606 (1966).
[131] H. D. Kaesz, S. Winstein, and C. G. Kreiter, *J. Am. Chem. Soc.* **88**, 1319 (1966).
[132] S. Winstein, C. G. Kreiter, and J. I. Brauman, *J. Am. Chem. Soc.* **88**, 2047 (1966).

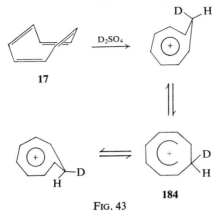

17

184

FIG. 43

The ultraviolet spectrum of **182** in sulfuric acid shows absorptions at 232.5 (4.52) and 313 nm (log ε 3.48), and a 1,7 interaction of 0.73β was calculated, considerably greater than that found for the homocyclopropenium cation.

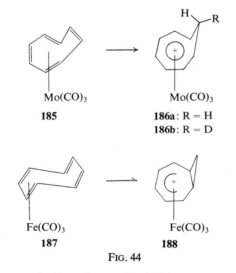

185

186a: R = H
186b: R = D

187

188

FIG. 44

The molybdenum tricarbonyl complex **186a** was prepared by treatment of the complex **185** with sulfuric acid.[133] In this case when the reaction is carried out with D_2SO_4 the deuterium largely enters the exo and not the endo position (Fig. 44).[131,132] The corresponding iron tricarbonyl complex **187** however protonates to the bicyclic cation **188**,[134] whereas the tungsten tricarbonyl complex protonates to give the homoaromatic cation analogous to **186**.[117a]

[133] S. Winstein, H. D. Kaesz, C. G. Kreiter, and E. C. Friedrich, *J. Am. Chem. Soc.* **87**, 3267 (1965).

[134] A. Davison, W. McFarlane, L. Pratt, and G. Wilkinson, *J. Chem. Soc.* p. 4821 (1962).

A number of substituted homotropylium ions have been prepared. Treatment of methyl (**189**) or phenyl (**191**) cyclooctatetraene with sulfuric acid gave the corresponding 1-substituted homotropylium cations, **190, 192** (Fig. 45).[135] The deuterated cations, prepared by treatment of the cyclooctatetraene with D_2SO_4, had the deuterium atom mainly in the endo position. The ions **190** and **192** appear to have larger barriers to ring inversion than the unsubstituted ion **182**. Homotropone (**193**) on treatment with $HSbCl_6$ in benzene, methylene

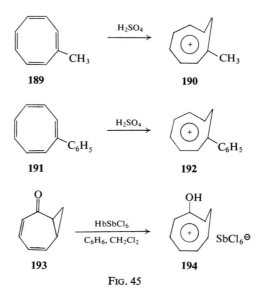

FIG. 45

chloride, gave the 2-hydroxyhomotropylium cation (**194**) as pale-yellow crystals.[136] The 1-hydroxyhomotropylium cation (**196**) was prepared by treatment of cyclooctatrienone (**195**) with fluorosulfonic acid in SO_2 or SbF_5, SO_2 (Fig. 46).[137] Solutions of **196** are stable at low temperatures, but at room temperature the cation **196** irreversibly rearranges to protonated acetophenone (**197**).

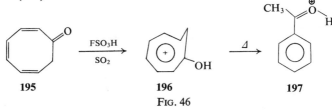

FIG. 46

[135] C. E. Keller and R. Pettit, *J. Am. Chem. Soc.* **88**, 604 (1966).
[136] J. D. Holmes and R. Pettit, *J. Am. Chem. Soc.* **85**, 2531 (1963).
[137] M. Brookhart, M. Ogliaruso, and S. Winstein, *J. Am. Chem. Soc.* **89**, 1965 (1967).

The 8-halohomotropylium cations were discussed in Section II,B.

The dibenzohomotropylium ion (199) was prepared by treatment of 198 with strong acids (Fig. 47).[138] The NMR spectra shows the characteristic large chemical shift difference between the ^{a}H and ^{b}H protons. Quenching a solution of 199 in methanol gave the methyl ether 200, mainly as the cis

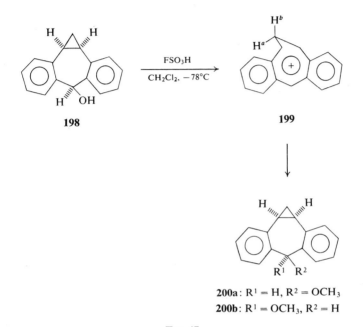

200a: R^1 = H, R^2 = OCH$_3$
200b: R^1 = OCH$_3$, R^2 = H

FIG. 47

isomer 200a. The isomeric cation 204 can be prepared by treatment of *sym*-dibenzocyclooctatetraene (201) or the alcohol 202 with fluorosulfonic acid, or by treatment of the halides 203a,b with antimony pentafluoride in SO$_2$ (Fig. 48).[139] Quenching the cation 204 with methanol gave the methyl ether 205. The NMR spectrum of 204 again shows the characteristic chemical shift difference of the methylene protons.

Protonation of benzocyclooctatetraene (206) with sulfuric acid gave the benzohomotropylium cation 207a (Fig. 49).[140] The protonation of 4,5,6,7-tetradeuteriobenzocyclooctatetraene (208) gave the corresponding tetradeuterio cation 207b, the NMR spectrum of which was completely analyzed,

[138] R. F. Childs and S. Winstein, *J. Am. Chem. Soc.* 89, 6348 (1967).
[139] G. D. Mateescu, C. D. Nenitzescu, and G. A. Olah, *J. Am. Chem. Soc.* 90, 6235 (1968).
[140] W. Merk and R. Pettit, *J. Am. Chem. Soc.* 90, 814 (1968).

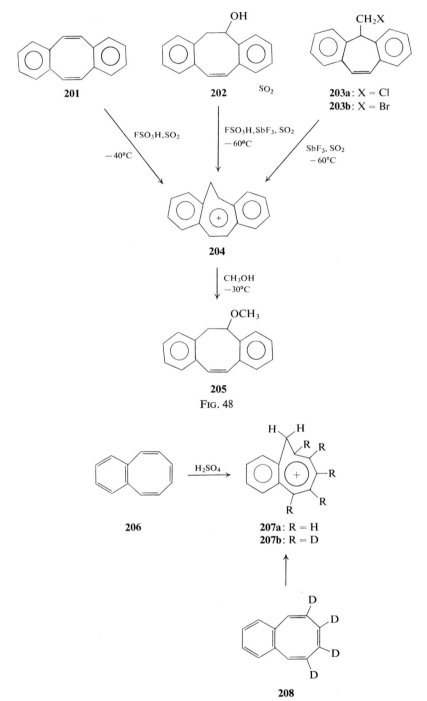

201

202 SO₂

203a: X = Cl
203b: X = Br

FSO₃H, SO₂
−40°C

FSO₃H, SbF₅, SO₂
−60°C

SbF₅, SO₂
−60°C

204

CH₃OH
−30°C

OCH₃

205

Fig. 48

206

H₂SO₄

H H
R R
R
R
R

207a: R = H
207b: R = D

D
D
D
D

208

Fig. 49

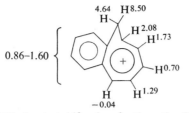

FIG. 50. The NMR chemical shift values for the cation **207a** in τ units.

allowing a complete assignment of the chemical shifts of the protons in **207a** (Fig. 50). The protonation of **206** at 0°C again occurs on the inside, giving the endo-deuterio isomer with D_2SO_4. The exo and endo protons rapidly equilibrate at room temperature. The electronic spectrum of **207a** shows absorption at 233 (4.25), 280 (3.91), 315 (3.38), and 400 nm ($\log \epsilon$ 3.43), and is very similar to that of the benzotropylium cation.

D. MONOHOMOCYCLOOCTATETRAENYL ANION AND DIANION

The monohomocyclooctatetraenyl anion radical (**209**) can be prepared by treatment of *cis*-bicyclo[6.1.0]nonatriene (**61**) with potassium in DME,[117a, 141] or by electrolysis of **61** in liquid ammonia saturated with tetramethylammonium iodide (Fig. 51).[142] The ESR spectrum of the anion radical has the hyperfine

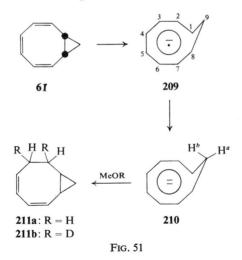

61   **209**

211a: R = H   **210**
211b: R = D

FIG. 51

splitting constants shown in Table VII[143] with a Q value of 26.4,[141] 27.1[142] gauss. The original discrepancy between the reported coupling constants has

[141] R. Rieke, M. Ogliaruso, R. McClung, and S. Winstein, *J. Am. Chem. Soc.* **88**, 4729 (1966).
[142] T. J. Katz and C. Talcott, *J. Am. Chem. Soc.* **88**, 4732 (1966).
[143] F. J. Smentowski, R. M. Owens, and B. D. Faubion, *J. Am. Chem. Soc.* **90**, 1537 (1968).

been resolved[143] in favor of those of Katz and Talcott.[142] The nonequivalence of the methylene protons at C-9 indicates that the ring is nonplanar, and the large value of the splitting constants of the C-9 protons favor the nonclassical ion **209** rather than the classical ion with a C-1, C-8 σ-bond.

TABLE VII

ESR HYPERFINE COUPLING CONSTANTS OF THE
MONOHOMOCYCLOOCTATETRAENYL ANION
RADICAL (**209**)[a]

Position	No. of protons	a_H (gauss)[b,c]
1,8	2	5.36
2,7	2	0.92
3,6	2	5.13
4,5	2	2.08
9	1	4.78
9	1	12.24

[a] Values from Smentowski et al.[143]
[b] Spectrum of Li$^\oplus$ or Na$^\oplus$ C$_9$H$_{10}$$^\ominus$ in DME at −90°C.
[c] Error ± 0.05 gauss.

The monohomocyclooctatetraenyl dianion (**210**) is prepared by allowing a dilute solution of **61** in THF (or DME) to react with excess potassium at −80°C (Fig. 51).[144] The NMR spectrum, obtained on dilute solutions in DME by a time-averaging computer, shows signals at ca. τ4.8 (6.1 H), τ6.1 (2.2 H), τ8.0 (1 H), and τ10.0 (1 H). When the anion is prepared from 9,9-dideuterio-bicyclo[6.1.0]nonatriene, the high field signals are absent. The NMR spectrum is best accommodated by the nonclassical homocyclooctatetraenyl structure. Quenching the dianion **210** with methanol gave mainly (85%) bicyclo[6.1.0]-nona-2,4-diene (**211a**), and quenching with deuteriomethanol gave the corresponding hydrocarbon **211b** containing two atoms of deuterium.[145]

E. 1-METHYLSULFINYLMETHYL-1,6-METHANOCYCLODECATETRAENYL ANION

Böll[145a] has reported that treatment of 1,6-methano[10]annulene (**212**) with the sodium methylsulfinyl carbanion in dimethyl sulfoxide gave the adduct **213**, the 1-methylsulfinylmethyl-1,6-methanocyclodecatetraenyl anion (Fig.

[144] M. Ogliaruso, R. Rieke, and S. Winstein, J. Am. Chem. Soc. **88**, 4731 (1966).
[145] M. Ogliaruso and S. Winstein, J. Am. Chem. Soc. **89**, 5290 (1967).
[145a] W. A. Böll, Tetrahedron Letters p. 5531 (1968).

52). Treatment of **213** with water gave a mixture of isomers from which the tetraene **214** was isolated. This compound could be reconverted to **213** by treatment with potassium *t*-butoxide. The NMR spectrum of **213** indicates that the molecule is symmetrical, since protons which would be equivalent in such a structure show coincident chemical shifts. The anion **213** is thus best considered a monohomoaromatic cyclononatetraenyl anion.

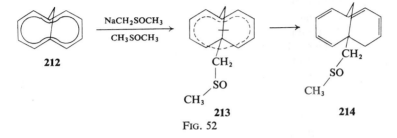

FIG. 52

VIII. Conclusions

Despite, or even because of, its limitations, the Hückel Molecular Orbital (HMO) method has stimulated a large body of chemical research on the synthesis and properties of cyclic-conjugated π-electron systems. It is quite clear that, notwithstanding the incorrect conclusions of the theory with regard to the delocalization energies of such systems, the theory does predict the difference in properties between those molecules with closed and those with partially filled electron shells. In fact, the simple application of Hückel's rule, without any considerations of its theoretical justification,[90] satisfactorily predicts the relative stabilities of the monocyclic 2,4, and 6 π-electron species. Computations in which the oversimplified assumptions of the HMO theory are removed can provide more satisfactory estimates of delocalization energies, and the use of perturbation treatments, which again have the advantage that the organic chemist will be able to use "shirt cuffs"[146] rather than a computer, are currently finding favor with our theoretical colleagues.[147]

In the case of the medium ring ions, factors other than the number of π-electrons become important. It thus appears likely that the undecapenta-enium cation will be destabilized by nonbonded hydrogen interactions, or, in the all-cis configurations, by bond angle strain. Accounts of attempts to prepare this cation, and the related [12]annulenium dication, will be awaited with interest. In the 14 π-electron series, of which the uncharged hydrocarbon-[14]annulene has been prepared,[148] the ionic species are so far unknown.

[146] O. Wilde, "The Importance of Being Earnest," Act 1, Scene 1.
[147] R. Hoffmann, *J. Am. Chem. Soc.* **91**, 4614 (1969).
[148] F. Sondheimer, *Proc. Roy. Soc.* **A297**, 173 (1967).

The synthesis of the tridecahexaenyl anion and the pentadecaheptaenium cation appear more feasible than the corresponding 10 π-electron analogs, though the [12]annulenyl dianion will have to overcome considerable strain energy. The study of larger ions is likely to expose, as in the case of the higher annulenes, a chemistry very different from that of the smaller homologous ions.

The properties of the polycyclic systems seem to be dominated by their composite parts rather than by the overall electron count. In those systems composed of $(4n + 2)$ π-electron rings the structure of the component parts is in harmony with the overall count of $(4n + 2)$ π-electrons. The formation of the relatively stable pentalenyl dianion reflects this situation. In the case of systems which overall have $(4n + 2)$ π-electrons, but are composed of $4n$ rings, the properties of the individual rings are different from those expressed by the overall number of π-electrons. Whether the heptalenyl dianion will show the properties of a 14 π-electron system, or that of two fused cycloheptatrienyl anions is not known, but one suspects that it will be the latter. However, if stabilization of such systems is to be found, it will probably occur with the ions rather than the uncharged hydrocarbons. The synthesis of a variety of such polycyclic ions composed of both $4n$ and $4n + (4n + 2)$ π-electron rings is an intriguing challenge to the synthetic chemist.

Much of the strongest evidence for homoaromatic perturbation has come from the study of the ions. It is quite clear that the homocyclopropenium and homotropylium cations have properties not expected for simple allylic cations, and the spectroscopic evidence indicates a considerable nonbonded interaction between the termini of the conjugated system. The quantitative extent of this interaction is more problematical, and may be clarified by further experimental and theoretical investigations of these and related systems.

The synthesis of further bicycloaromatic ions can be expected, together with their nonaromatic counterparts, and it will then be possible to make some estimate of the importance of this type of stabilization.

We may expect a continuation of interest in the synthesis and properties of the monocyclic, polycyclic, and homoaromatic ions in the next few years, and a variety of new compounds and new synthetic reactions will undoubtedly be discovered.

Acknowledgments

We thank Professors R. Breslow, K. M. Harmon, and J. F. M. Oth for making unpublished results available to us.

5

Chemical Binding and Delocalization

in Phosphonitrilic Derivatives

D. P. Craig and N. L. Paddock

I. Introduction

Phosphonitrilic derivatives, containing the repeating unit NPX_2, resemble typical organic compounds in many ways. They exist as both cyclic and linear molecules in which the skeletal bonds are essentially covalent; the cyclic molecules undergo substitution and, to a smaller extent, addition reactions, and they exhibit various types of structural and stereoisomerism. In short, they are characterized by a stable structural core which retains its individuality through a series of reactions of the exocyclic groups. Our concern in this chapter is with the electronic structure of the core and with the structural and chemical features which depend on it.

There are many other series based on a molecular framework $(AB)_n$, in which A can be one of the elements silicon, phosphorus, or sulfur, and B is either nitrogen or oxygen. They include the siloxanes $(R_2SiO)_n$, the silazanes $(R_2SiNR')_n$, metaphosphates $(OPO_2{}^-)_n$, and sulfur trioxide $(SO_3)_n$. In all these cases the A–B bond can be written formally as a single bond, though it is often supposed to be supplemented by the use of the formally unshared electrons of oxygen or nitrogen in $p\pi$–$d\pi$ bonds to the second-row element. The potential use of d-orbitals is a central concern in the chemistry of the second-row elements, and a study of these homologous series can be expected to be informative on this point. This is partly because in cyclic molecules (and in high polymers) end effects are avoided, and also because theory is more reliably applied to the differential effects arising, for example, from a change in ring size or of a substituent than to the interpretation of the properties of a single compound.

FIG. 1. Trimeric phosphonitrilic chloride.

In this connection, phosphonitrilic derivatives have a special importance. They can be obtained in a large range of ring sizes [all the fluorides in the range $(NPF_2)_{3-17}$ have been prepared], and the smaller cyclic compounds have an elaborately developed chemistry. Unlike the other series mentioned above, they are formally unsaturated. The best known derivative is the trimeric chloride, and from earliest days the analogy of the simple bonding scheme (Fig. 1) with that of benzene has been stressed. The similarity in structural formula conceals important points of difference in bond type, as is evident from the behavior of the larger rings; the chemistry of homogeneously substituted tetrameric phosphonitriles $(NPX_2)_4$ is broadly similar to that of their trimeric analogs, and the lengths of the ring bonds do not alternate, as they do in cyclooctatetraene. The thermochemical stabilities of the larger phosphonitrilic rings increase steadily with increase of ring size; the larger unsaturated organic rings, by contrast, are much more difficult to obtain and are very reactive.

Early work on the phosphonitrilic chlorides had shown them to be much less reactive than the noncyclic compounds of pentavalent phosphorus, such as the oxyhalides to which the chemical analogy seemed closest. This led to the first definite suggestion, a decade ago,[1] that underlying the comparative inertness of these compounds there was a cyclic delocalization of π-electrons, which

[1] D. P. Craig and N. L. Paddock, *Nature* **181**, 1052 (1958).

might be connected with chemical inertness in the way familiar in aromatic hydrocarbons and their derivatives. The concept of aromaticity had already grown out of its original attachment to benzenoid substances and had been applied first to nitrogen-containing heterocyclics and, then, in the inorganic field, to borazines. The further development to rings including elements of the second row, such as PN rings, raised more fundamental issues. The first concerns what may be called the general valency problem of elements in the second row, namely: What are the conditions in which phosphorus displays five-covalency, and hence has available an electron capable of being delocalized; and what determines the degree of availability? The second is a matter of symmetry. The indicated valence configuration of pentavalent phosphorus is $(3s)(3p)^3(3d)$, and since the single bonds to the neighboring ring nitrogen atoms and to the two exocyclic atoms are more naturally treated as sp^3 hybrids, a $3d$-electron is left to participate in a π system. The early work made clear that the properties of $2p\pi$–$3d\pi$ delocalization could, depending on which d-orbital was involved, be qualitatively different from the familiar $2p\pi$–$2p\pi$ systems of benzene (e.g., s-triazine and borazine) and that, in particular, Huckel's rule confining aromatic stability to cyclic systems of $4n + 2$ electrons might not hold. We need to ask if the d-orbitals differ in their importance, and to find those principally concerned in bonding.

Such questions can be usefully studied in the context of phosphonitrilic chemistry. Many derivatives have been synthesized, some, such as the chlorides and the methyl and phenyl derivatives, by ammonolysis of halogen derivatives of pentavalent phosphorus, others by substitution reactions, usually of the chlorides, with organic or inorganic nucleophiles. Some of these are illustrated in Fig. 2. Many structures have been determined, and their significance will be discussed below; briefly, it is possible to see in them the influence of directional effects arising from both π-electron and steric interactions. The rates of some reactions have also been determined, reactivity depending on both the attacking reagent and on the nature and ring size of the phosphonitrile; electronic substitutional effects can be investigated as in carbocyclic chemistry, and it is to be expected that quantitative work of this sort will increase. The chemistry and the structure determinations and the other molecular properties to be referred to later together make it clear that, although within a particular series $(NPX_2)_n$ there is a general similarity of chemistry, phosphonitrilic derivatives cannot be regarded as formed by simple linkage of a series of noninteracting units. In later sections we shall examine the possible types of interaction, and discuss the experimental evidence for them. We may expect the conclusions to have some bearing on the chemistry of the other $(AB)_n$ series, in which however the effects are less strongly marked.

The general questions of bonding to second-row elements span a wide range, and we shall restrict our attention to the nature and extent of electronic delocalization in phosphonitrilic derivatives, together with some closely related matters such as conjugation. We shall not deal systematically with general and preparative phosphonitrilic chemistry, of which several reviews

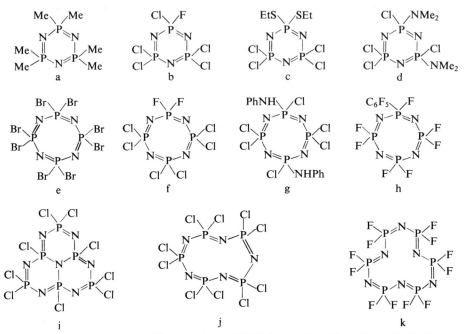

FIG. 2. Typical phosphonitrilic derivatives: (*a*) H. T. Searle, *Proc. Chem. Soc.* p. 7 (1959); (*b*) J. Emsley and N. L. Paddock, *J. Chem. Soc.*, *A* p. 2590 (1968); (*c*) A. P. Carroll and R. A. Shaw, *J. Chem. Soc.*, *A* p. 914 (1966); (*d*) M. Becke-Goehring, K. John, and E. Fluck, *Z. Anorg. Allgem. Chem.* **302**, 103 (1959); (*e*) H. Bode, *Z. Anorg. Allgem. Chem.* **252**, 113 (1943); (*f*) Reference *b*; (*g*) K. John, T. Moeller, and L. F. Audrieth, *J. Am. Chem. Soc.* **82**, 5616 (1960); (*h*) T. Chivers and N. L. Paddock, *Chem. Commun.* p. 704 (1968); (*i*,*j*) H. N. Stokes, *Am. Chem. J.* **19**, 782 (1897); (*k*) A. C. Chapman, N. L. Paddock, D. H. Paine, H. T. Searle, and D. R. Smith, *J. Chem. Soc.* p. 3608 (1960).

are available.[2,2a] The next section deals with general considerations affecting the use of *d*-orbitals in bonding, and is followed by an account of the theoretical background, which is limited to those aspects that have a clear relevance to current factual knowledge. Finally, the various types of experimental evidence are considered in detail in relation to theory, and their significance discussed.

[2] L. F. Audrieth, R. Steinman, and A. D. F. Toy, *Chem. Rev.* **32**, 109 (1943); N. L. Paddock and H. T. Searle, *Advan. Inorg. Chem.* **1**, 347 (1959); R. A. Shaw, B. W. Fitzsimmons, and B. C. Smith, *Chem. Rev.* **62**, 247 (1962); C. D. Schmulbach, *Progr. Inorg. Chem.* **4**, 275 (1962); I. A. Gribova and U. Ban-yuań, *Russ. Chem. Rev.* (*English Transl.*) **30**, 1 (1961); T. Yvernault and G. Casteignau, *Bull. Soc. Chim. France* p. 1469 (1966).

[2a] N. L. Paddock, *Quart. Rev.* (*London*) **18**, 168 (1964).

II. General Considerations Affecting the Use of d-Orbitals in Bonding

In the neutral phosphorus,[3,4] sulfur,[5] and chlorine atoms[6] the $4s$ level is below $3d$, so that the valence state $3sp^3\,4s$ is the lowest pentavalent state of phosphorus, lying about 1 eV below the $3sp^3\,d$ state. In the presence of electronegative ligands, however, the $3sp^3\,d$ valence state becomes much lower, since, in this region of the Periodic Table the $3d$-orbital sizes and energies are sensitive to effective nuclear charge. There is now wide support for the view that participation of d-orbitals in bonds to phosphorus depends on the presence of electronegative ligands, the d-orbitals being thereby contracted, so that overlap takes place in a strong nuclear field. $d\pi$-Interaction is likely to be more important than $d\sigma$, because the contraction need not be so great.

Detailed studies of the free phosphorus atom[7] in the valence configuration $sp^3\,d$ show that the $3d$-orbital is so diffuse that it could play no significant part in chemical bonds. A convenient measure of orbital size, and thus of diffuseness, is the orbital exponent factor, α, in the hydrogenic radial wave function

$$f(r) = N(\alpha)\,r^2\,e^{-\alpha r} \tag{1}$$

The value of this constant is sometimes set by reference to Slater's rules, but is more accurately found by minimization of the atomic energy with respect to variations of α. The optimum value in atomic phosphorus is about 0.34 units. The radial distance r_m at which the radial function reaches its maximum is, in atomic units of distance, $3/\alpha$; thus in this case the maximum is at about 9 au or 4.5 Å. This is so much greater than a bond distance that no contribution to bonding is expected, a conclusion that is confirmed from the appropriate overlap integral. When the same calculation of optimum exponent is made for the atom in the same valence state but in the presence of the electrostatic field of fluorine atoms at their positions in phosphorus pentafluoride, the value is about 1.2 units, giving a contracted orbital with a radial maximum at about 1.3 Å. Participation of such an orbital in bonding orbitals raises no special difficulties, and our account of the bonds in PN ring systems largely takes this for granted, in view of the substantial theoretical support available.

Although a considerable body of fact also supports this interpretation now, it is not surprising, in view of the high energies of $3d$ states in the neutral free atom, that the use of $3d$-orbitals in bonding has hitherto often been denied and, indeed, some features of phosphonitrilic structures can be interpreted without

[3] K. A. R. Mitchell, *J. Chem. Soc.*, A p. 2676 (1968).
[4] R. G. A. R. Maclagan, Ph.D. Thesis, Australian National University, 1969.
[5] F. Bernardi and C. Zauli, *J. Chem. Soc.*, A p. 2633 (1968).
[6] D. P. Craig and R. G. A. R. Maclagan, *J. Chem. Soc.*, A p. 1431 (1970).
[7] G. S. Chandler and T. Thirunamachandran, *J. Chem. Phys.* **49**, 3640 (1968).

them. The argument is a general one, and not restricted to phosphonitrilic derivatives, or even to compounds of second-row elements. Lindqvist, for instance, has related an impressively large number of bond length variations in terms of the single variable of bond polarity,[8] and Rundle has explained many structures of nonmetallic compounds on the basis of s- and p-orbital interactions only.[9] The nature of the difficulty is exemplified by the structure of the ethyl sulfate anion[10] (Fig. 3). The difference between the lengths of the bridging and terminal S—O bonds can be explained, as Cruickshank has shown,[11] in terms of the use by the different types of oxygen atom of one and two p-orbitals, respectively, but the same conclusion as to relative bond lengths follows from the ionic model in which σ-bonds to all four oxygen atoms are supplemented by electrostatic reinforcement of the terminal bonds. Such a simplified picture is inadequate, however, to explain the finer details found in

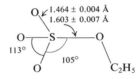

FIG. 3. The ethyl sulfate anion.

a comparison of different structures, and it is not obvious why, from the point of view of electrostatics, the S—O bond in the sulfate ion[12] (1.49 Å) should be longer than the terminal bonds in the ethyl sulfate anion (1.46 Å). For such comparisons, an interpretation in terms of the sharing of a pair of strongly interacting d-orbitals is more generally useful.[11]

Similarly, some properties of phosphonitrilic derivatives are understandable without using d-orbitals. If they are to be avoided, formal charges at phosphorus and nitrogen are required as shown (Fig. 4). It then follows that the P—N bond lengths should be equal within a particular molecule, and should be independent of ring size; further, there would be no resistance to torsional motion about the ring bonds, and the larger molecules should consequently be flexible. In agreement, it is found that in molecules which vary only in ring size, the ring bonds are all approximately equal in length, and the infrared spectra of several derivatives, especially the phosphonitrilic fluorides $(NPF_2)_n$, appear to indicate a high molecular flexibility.[13] The P—N bonds should also be

[8] I. Lindqvist, *Nova Acta Regiae Soc. Sci. Upsaliensis* **17** (11), pp. 1–19 (1960).

[9] R. E. Rundle, *Surv. Progr. Chem.* **1**, 81 (1963).

[10] J. A. J. Jarvis, *Acta Cryst.* **6**, 327 (1953); M. R. Truter, *ibid.* **11**, 680 (1958).

[11] D. W. J. Cruickshank, *J. Chem. Soc.* p. 5486 (1961).

[12] M. R. Atoji and E. Rundle, *J. Chem. Phys.* **29**, 1306 (1958).

[13] A. C. Chapman and N. L. Paddock, *J. Chem. Soc.* p. 635 (1962).

highly polar, and though there is no direct measurement of bond polarity, the intensities of infrared bands involving P—N stretching are high. As mentioned earlier, most phosphonitrilic reactions are nucleophilic displacements at phosphorus, and while the reactions are not rapid it seems likely that the phosphorus atoms carry at least a partial positive charge. The effects of fluorine substitution (both in its geminal orientation[14] and in its acceleration of further substitution[15]) and of amine substitution (in nongeminal orientation and in retardation of successive substitution[16]) are both consistent with electrostatic interactions of the substituent groups with phosphorus.

Fig. 4. Hypothetical singly bonded phosphonitrilic structure.

Nevertheless, such a simple picture of the bonding has serious defects. On the basis of Fig. 4, ring angles near 109° would be expected at both phosphorus and nitrogen, whereas they are always greater and in the trimeric derivatives usually near 120°, so that few such molecules deviate far from planarity. The aminoborine (Fig. 5) forms an interesting contrast. In this compound,[17] formal

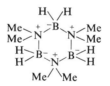

Fig. 5. Dimethylaminoborine trimer.

charges as shown are required for the formation of the ring σ-bonds, and the molecule is chair shaped, the average ring angles at boron and nitrogen being 114° and 113°. Not only are the angles larger in phosphonitrilic derivatives, but both bond angles and bond lengths depend on ring size, the angle at nitrogen varying within the range 120°–150°, and the ring bond length between 1.50 and 1.60 Å. These variations are unexpected on the basis of simple polarity, as

[14] A. C. Chapman, D. H. Paine, H. T. Searle, D. R. Smith, and R. F. M. White, *J. Chem. Soc.* p. 1768 (1961).

[15] J. Emsley and N. L. Paddock, *J. Chem. Soc.* p. 2590 (1968).

[16] M. Becke-Goehring, K. John, and E. Fluck, *Z. Anorg. Allgem. Chem.* 302, 103 (1959); S. K. Ray and R. A. Shaw, *J. Chem. Soc.* p. 872 (1961).

[17] L. M. Trefonas, F. S. Mathews, and W. N. Lipscomb, *Acta Cryst.* 14, 273 (1961).

are the frequent geminal substitution by amines[18] and the details of the relative rates of the reactions of $N_3P_3Cl_6$ and $N_4P_4Cl_8$ with fluorinating agents.[15] As will be seen below, not only are these phenomena explicable if d-orbitals are included, but an estimate of the relative importance of the orbitals is sometimes possible.

The frequent difficulty of distinguishing the effects of polarity from those of $p\pi$–$d\pi$ bonding has two causes. First, such bonding is often weak, compared with $p\pi$–$p\pi$ bonding in the first row, and the chemical effects are thus to a large extent those of a highly polar σ-bond; also, because of the different shapes of p- and d-orbitals, the center of electron density in a $p\pi$–$d\pi$ bond is always close to the p-center, even for equally electronegative orbitals, thus giving a polar distribution of charge in a formally nonionic bond. Comparative freedom of motion about the P—N bond is possible partly for these reasons, and partly because $p\pi$–$d\pi$ bonding is less restricted geometrically than $p\pi$–$p\pi$ bonding in, e.g., borazines, in which only one π-orbital is available at each atom. For these reasons, the certain recognition and characterization of a π-electron system in phosphonitriles requires the consideration of diverse types of evidence; a single fact is often ambiguous. It must also be remembered that the distinction between descriptions using d-orbitals, and those using single bonds with charge separation is by no means as sharp as might be imagined. The purpose is to describe the electron distribution in a complex molecule. It is not impossible that two different bases might allow roughly equally good accounts to be given of the energetics, or that one might be only marginally better than the other. In that case, the choice is made in favor of the basis that allows the more comprehensive and systematic account of the chemistry and molecular properties. The use of outer d-orbitals in bonding has been considered in more detail by Mitchell.[19]

These comments can be illustrated by reference to the chemistry and structures of the phosphorus ylides, often written as $R_3P^+ \cdot C^-R_2'$, a formulation which accounts very well for their reactions with, e.g., ketones in a 4-center reaction to give olefins (the Wittig reaction)[20]:

$$R_3P^+ \cdot C^-R_2' + R_2''CO \rightarrow R_3PO + R_2'C{:}CR_2'' \qquad (2)$$

These reactions are explicable on the basis of bond polarity alone, but, apart from the evidence of the more detailed chemistry, the occurrence of $p\pi$–$d\pi$ bonding in such compounds is shown by the structure of methylenetriphenylphosphorane,[21] $Ph_3P{:}CH_2$, in which the P=C bond is particularly short

[18] Y. Kobayashi, L. A. Chasin, and L. B. Clapp, *Inorg. Chem.* **2**, 212 (1963); S. K. Das, R. Keat, R. A. Shaw, and B. C. Smith, *J. Chem. Soc.* p. 5032 (1965).
[19] K. A. R. Mitchell, *Chem. Rev.* **69**, 157 (1969).
[20] S. Trippett, *Quart. Rev. (London)* **17**, 406 (1963).
[21] J. C. J. Bart, *J. Chem. Soc., B* p. 350 (1969).

(1.661 Å) [the mean P—C(Ph) length is 1.823 Å], and the methylene carbon atoms are sp^2 hybridized. In the ethyl anion in LiC_2H_5, on the other hand, the C—C—H angles to the methylene group are 108.1° and 104.7°, so that here there is a close approach to sp^3 hybridization.[22] The chemical behavior of the methylenetriphenylphosphorane can still be that of a highly polar compound, and, as the local symmetry at phosphorus approaches C_3, the d-orbitals involved in the π-bonding become a degenerate pair, in the limit offering (by themselves) no resistance to rotation about the P—C bond. Consistently, the radical $Ph_3P^+ \cdot CH_2$ (in which π-bonding would in any case be expected to be weak) rotates rapidly about the P—C bond at room temperature, even though the PCH_2 group is essentially planar.[23] The d-orbital basis therefore refines the polar concept without contradicting it, and for phosphonitrilic compounds as well as phosphorus ylides it allows the more coherent discussion, suggests useful analogies with organic aromatic compounds, and helps to rationalize their stereochemical characteristics.

The occurrence in phosphonitrilic molecules of a π system in some respects analogous to that in benzenoid compounds is suggested by many structural features, especially the equality and comparative shortness of the ring bonds, and the near-planarity of many derivatives, though the preference of $p\pi$–$d\pi$ systems for planar conformations is not strong. Detailed molecular structures and their significance will be discussed later; meanwhile it is useful to note certain points of difference from organic aromatics. Carbon bond orbitals are not very sensitive to the atoms to which they are attached by bonds; thus, bond lengths can be calculated to a useful precision by adding bond radii, the value for carbon (in a given hybridization) being largely independent of the attached groups. The same statement applies to other elements of the first row, fluorine being exceptional. For instance, the C—B distances in the planar molecules Me_3B, Me_2BF, and $MeBF_2$ are all equal[24] $(1.56 \pm 0.01$ Å) whereas in the compounds PF_5, $MePF_4$, and Me_2PF_3,[25] (Fig. 6) the introduction of the comparatively electropositive methyl groups lengthens all the bonds, especially the axial bonds to fluorine. The difference between the equatorial P—F distances in PF_5 and Me_2PF_3 (0.019 Å) is less than that between the C—F distances in CF_4 and CF_2H_2 (0.035 Å),[26] but it is outweighed by the difference in the axial P—F bonds (0.066 Å), and ^{19}F-coupling constants suggest that

[22] H. Dietrich, *Acta Cryst.* **16**, 681 (1963).

[23] E. A. C. Lucken and C. Mazeline, *J. Chem. Soc., A* p. 439 (1967).

[24] H. A. Levy and L. O. Brockway, *J. Am. Chem. Soc.* **59**, 2085 (1937); S. H. Bauer and J. M. Hastings, *ibid*. **64**, 2686 (1942).

[25] K. W. Hansen and L. S. Bartell, *Inorg. Chem.* **4**, 1775 (1965); L. S. Bartell and K. W. Hansen, *ibid*. p. 1777.

[26] C. G. Thornton, Publ. No. 7746. University Microfilms, Ann Arbor, Michigan; see C. G. Thornton, *Dissertation Abstr.* **14**, 604 (1954); see also *Chem. Soc. (London), Spec. Publ.* **11**, M108 (1958); D. R. Lide, *J. Am. Chem. Soc.* **74**, 3548 (1952).

further replacement, in Me_3PF_2, increases the axial-equatorial differentiation still more.[27] It is not necessary to consider the specific orbitals involved to recognize that the bonding electrons are more polarizable in compounds of second row elements than in those of the first row. $Me_4P^+F^-$ is ionic, the concentration of electrons at the electronegative ligand having then reached the limit. The same sensitivity to ligand is evident in the fact that there are few molecules which contain pentavalent phosphorus joined to atoms or groups, all of which are of low electronegativity. One such example is pentaphenylphosphorane, Ph_5P, which has[28] a bipyramidal structure similar to the compounds shown in Fig. 6, having very long axial bonds (1.987 Å) compared to the equa-

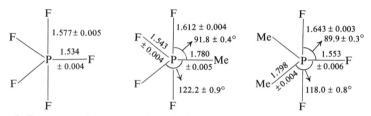

FIG. 6. Structures of pentacoordinated phosphorus compounds [K. W. Hansen and L. S. Bartell, *Inorg. Chem.* **4**, 1775 (1965); L. S. Bartell and K. W. Hansen, *ibid.* p. 1777]. (Bond lengths in angstroms.)

torial bonds (1.850 Å), the latter only marginally longer than the P—C bonds in triphenylphosphine, Ph_3P (1.828 Å).[29] The phosphorus–hydrogen bond in these circumstances is unstable unless accompanied by electronegative ligands, as it is in phosphorous acid $HP(O)(OH)_2$, HPF_4, and H_2PF_3[30]; PH_5 is not yet known. In the context of ring compounds with both σ- and π-bonds, it is obviously necessary to interpret bond length variations with such points in mind; indeed the concept of a skeleton of σ-bonds, common to a series of molecules, has much less force where elements of the second row are concerned than it does in organic chemistry, and attention will be subsequently drawn to many situations where uncertainties arise in this way.

III. Theoretical Background

A. Local Symmetry and Topological Type

Aromaticity can appear in molecular situations where cyclic delocalization of electrons is possible. In carbon compounds the conditions for aromaticity

[27] E. L. Muetterties, W. Mahler, and R. Schmutzler, *Inorg. Chem.* **2**, 613 (1963).
[28] P. J. Wheatley, *J. Chem. Soc.* p. 2206 (1964).
[29] J. J. Daly, *J. Chem. Soc.* p. 3799 (1964).
[30] R. R. Holmes and R. N. Storey, *Inorg. Chem.* **5**, 2146 (1966).

are well understood, both in terms of symmetry and geometrical structures. When the concept is extended to inorganic ring systems, such as the phosphonitrilics, a broader discussion is called for, first in terms of the wider symmetry conditions, and second in the more detailed energetics. It has first to be noted that the overlap of $p\pi$ orbitals (Fig. 7a) occurring typically in aromatic carbon compounds, and the overlap of alternating $p\pi$ and $d\pi$ orbitals (Fig. 7c) in in cyclic inorganic systems are characteristic of systems in which the consequences of delocalization may be different. There is a similar dualism in the

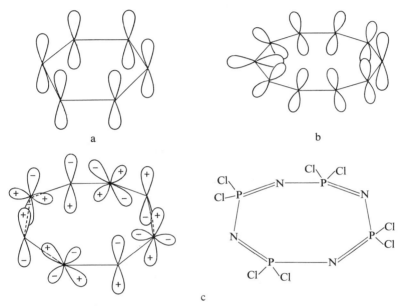

FIG. 7. Normal connection of $2p\pi$ orbitals (a); and Möbius connection of hypothetical 10-atom system (b). Orbitals of alternating $3d\pi(xz)$–$2p\pi$ orbitals (c).

orbitals (lone-pair electrons at N, d_{xy} and $d_{x^2-y^2}$ at P) capable of in-plane delocalization. Thus a preliminary problem is to classify systems according to atomic orbital types. It will be shown that only two types are possible; but there is an additional freedom, in that a chain of π-orbitals may be joined end-to-end to form a cycle in different ways, defined in relation to the local symmetry planes of successive orbitals. In benzene, and in the $p\pi$–$d\pi$ system of $(NPCl_2)_3$, the atoms of the ring are in one plane, and this plane is a common symmetry element of all π-orbitals. Thus the ring is formed from an open chain by joining the local symmetry planes of the π-orbitals to give a simple surface. Alternatively,[31,32] the local symmetry planes may be connected to form a

[31] E. Heilbronner, *Tetrahedron Letters* p. 1923 (1964).
[32] S. F. Mason, *Nature* **205**, 495 (1965).

single-sided surface (Fig. 7b) in the form of the Möbius strip so that the local π-orbital axes are twisted through 180° in making one complete turn around the closed cycle of atoms.

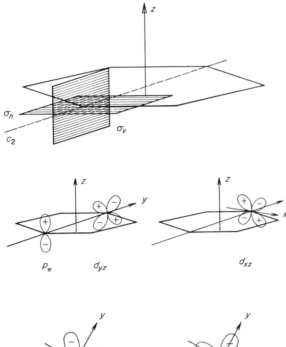

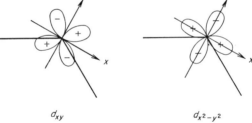

FIG. 8. Axis system and site symmetry operations.

There are thus at least two topologically distinct types of delocalized systems. At first sight the Möbius type seems of little chemical relevance, but an example will indicate a possible interest. The skeleton of phosphabenzene made of five carbons and one phosphorus has a delocalized π system of adjacent $2p\pi$ orbitals on the carbons together with a $3d\pi$ orbital on phosphorus. If this is a d_{xz} orbital (Fig. 8) the connected set of π-orbitals has a single change of relative sign round the ring, a property also possessed by the orbitals in Fig. 7b.

B. Classification by Local Symmetry

Delocalization in the planar phosphonitrilics depends on the use by phosphorus of its $3d$ orbitals, in particular $3d_{xz}$, $3d_{yz}$ and $3d_{xy}$, $3d_{x^2-y^2}$ in the axes of Fig. 8. Nitrogen uses a $2p_z$ and a hybrid $2s-2p_y$ orbital. There are two systems of delocalization: in one the participating orbitals are antisymmetric to reflection in the molecular plane (π_a system), and in the other they are symmetric (π_s system). The electrons of each of the ring atoms are in a potential field imposed by all atoms of the molecule. Where the ring is planar the local field belongs to the symmetry group C_{2v} and each atomic orbital can be classified according to a representation of this group. A classification is shown in Table I.

TABLE I

SYMMETRY SPECIES OF s-, p-, AND d-ORBITALS

		E	C_2	σ_h	σ_v
$s,p_y,d_{x^2-y^2}$	A_1	1	1	1	1
d_{xz}	A_2	1	1	-1	-1
d_{xy}	B_1	1	-1	1	-1
p_z,d_{yz}	B_2	1	-1	-1	1

In a nonplanar ring the field imposed on an atom *by nearest neighbors* remains C_{2v}, and classification under this "site" group retains an approximate significance. If two orbitals, as for example d_{yz} and p_z, belong to the same representation of the site symmetry group C_{2v}, no symmetry distinction is possible between the ways in which they participate in a delocalized system. If orbitals belong to different representations, as d_{xz} and p_z, their behavior in delocalized systems is expected to be different. We may investigate the differences by examining the properties of molecular orbitals constructed from the different types of atomic orbitals.

We begin with a discussion of the planar system, for which the molecular orbitals follow from symmetry arguments. In a planar monocyclic $(AB)_n$ molecule with ring bond lengths and ring angles fixed to conform to the molecular symmetry group D_{nh} molecular orbitals are formed by combining the atomic orbitals according to representations of the group. Let us number the A sites by $p = 1,2\ldots n$, and the B sites by $p = 1\frac{1}{2}, 2\frac{1}{2}\ldots n + \frac{1}{2}$. The linear combinations of atomic orbitals then are as follows in Eq. (3)

$$\phi_l^A = n^{-1/2} \sum_{p=1}^{n} e^{2\pi i l p/n}\, A_p$$

$$\phi_l^B = n^{-1/2} \sum_{p=1}^{n} e^{2\pi i l (p+1/2)/n}\, B_{p+1/2}$$

(3)

The orbitals are classified by the ring quantum number l, with allowed values:

$$l = 0, \pm 1 \ldots (n-1)/2 \quad (n \text{ odd})$$
$$l = 0, \pm 1 \ldots n/2 \quad (n \text{ even})$$

The symmetry classifications for the several combinations of ring size and orbital type are given in Table II. In π_a delocalization the atomic orbitals belong either to a_2 or b_2 of C_{2v}, and the molecular orbitals transform according to the second and fourth species in each entry of Table II. In π_s delocalization

TABLE II

MOLECULAR ORBITAL SPECIES (IN D_{nh}) FOR VARIOUS SITE SPECIES[a]

l \\ n	3		4		5		6	
0	a_1'	a_1''	a_{1g}	a_{1u}	a_1'	a_1''	a_{1g}	a_{1u}
	a_2'	a_2''	a_{2g}	a_{2u}	a_2'	a_2''	a_{2g}	a_{2u}
± 1	e'	e''	e_u	e_g	e_1'	e_1''	e_{1u}	e_{1g}
	e'	e''	e_u	e_g	e_1'	e_1''	e_{1u}	e_{1g}
± 2			b_{1g}	b_{1u}	e_2'	e_2''	e_{2g}	e_{2u}
			b_{2g}	b_{2u}	e_2'	e_2''	e_{2g}	e_{2u}
± 3							b_{1u}	b_{1g}
							b_{2u}	b_{2g}

[a] The four entries in each box refer (in order: top row, bottom row) to linear combinations of orbitals belonging to site species a_1, a_2, b_1, and b_2 of group C_{2v}. See Table I.

the first and third entries apply. It is well known that in $(AB)_n$ rings of alternating atoms, in particular in ground states, the essential distinction is between homomorphic delocalization (participating orbitals of the same species) and heteromorphic (different species)[1,33]. It is apparent that this same classification may be made within π_a and π_s delocalization types, and that in both the proper molecular orbitals are linear combinations of the symmetry-adapted functions found by solving a secular equation. In a π_a system both $3d_{xz}$ and $3d_{yz}$ orbitals in phosphorus may be expected to take part, together with the $2p_z$ orbital of nitrogen. For a π_s system the $3d_{z^2}$, $3d_{x^2-y^2}$, and $3d_{xy}$ are all allowed to participate, together with the s and p_y nitrogen orbitals. For example, for the π_a system of a planar $(NPX_2)_n$ ring with a phosphorus ring angle of $120°$ the secular equation is, in Hückel approximation,

$$\begin{vmatrix} \alpha - E & 2i\beta(xz)\cos(\pi/6)\sin(1\pi/n) & 2\beta(yz)\sin(\pi/6)\cos(1\pi/n) \\ -2i\beta(xz)\cos(\pi/6)\sin(1\pi/n) & \alpha(xz) - E & 0 \\ 2\beta(yz)\sin(\pi/6)\cos(1\pi/n) & 0 & \alpha(yz) - E \end{vmatrix} = 0$$

(4)

In this equation the Coulomb integrals α, $\alpha(xz)$, and $\alpha(yz)$ apply to the $2p\pi$ orbital at nitrogen and the d-orbitals at phosphorus. The resonance integrals $\beta(xz)$ and $\beta(yz)$ are defined for optimum orientation toward the neighbor $2p$ nitrogen orbital. By inserting appropriate values for the electronegativities α and the resonance integral and solving Eq. (4), the energies of the molecular orbitals are found belonging to each allowed value of the ring quantum number l. The secular equation for the π_s system has the same form, the basis orbitals then being the s-p_y hybrid at nitrogen, and the d_{xy} and $d_{x^2-y^2}$ orbitals at phosphorus.

Many of the phosphonitriles to be discussed in later sections have lower symmetry than the ideal D_{nh}, some being nonplanar, others planar, but with ring structures with reentrant angles. It is now necessary to assess the applicability of the theory to these systems. The first requirement is that the cyclic system should have no reentrant contacts, or near approaches, by atoms that are nonneighbors. Where this condition is met the approximation of neglecting all but nearest neighbors can be retained. Other cases, with reentrant contacts, such as $N_5P_5Cl_{10}$ and $N_8P_8(OMe)_{16}$ in their crystals and perhaps some examples in the gas phase, require a theory enlarged to include nonneighbor interactions, which may enhance the delocalization. Next, if the system is planar, and the neighbor π overlaps are equal around the ring, the secular equation is identical in form with that for D_{nh} symmetry, and the theory already outlined holds without change.

In nonplanar systems, the classification into π_a and π_s retains only its local significance with respect to the plane defined by an atom and its two nearest neighbors. Thus an orbital with π_a properties at one center will overlap orbitals that have the π_s property with respect to the local plane of its neighbor

[33] D. P. Craig, *Proceedings of the Kekulé Symposium*, p. 20. Butterworth, London and Washington, D.C., 1958.

as well as that with π_a symmetry. Where the dihedral angle is small, the $\pi_a-\pi_s$ overlap might be neglected, allowing the simple theory to be used; otherwise the secular problem within a basis set including all the orbitals must be solved.

C. Classification by Topological Type

Molecules connected as in Fig. 7b are not yet known. However, a possible application has already been sketched in phosphabenzene, and in view also of the tendency of $(NPX_2)_n$ systems to form large rings in which the twisting of the local planes required to give a total torsion of π in a complete circuit could easily be accommodated with small dihedral angles between neighbors, a brief survey of possibilities is appropriate. Such molecules need have no symmetry, even with coplanar ring bonds. Nevertheless, some useful results can be found by beginning with a linearly connected polyene of P—N units and simulating the effect of closing into a cyclic system by applying cyclic boundary conditions. Evidently nearest neighbor interactions alone can be allowed, insofar as the linear and cyclic structures are assumed to differ only in the constraints on the wave function at the terminal atoms, which become neighbors in the cyclic systems.

The application of this argument will first be given for normal aromatic molecules. We think of an infinite linear polyene, and adapt it to an n-cyclic aromatic by taking a segment of the polyene, folding it into a circle, and joining the ends. Only those wave functions of the infinite polyene that are equal at atoms n apart (period n) possess the symmetry required of solutions to the cyclic problem. Thus if $\psi(p)$ is the wave function at the pth atom in the chain, we require the property [Eq. (5)],

$$\psi(p + n) = \psi(p) \tag{5}$$

A solution can be written in the form of Eq. (6),

$$\psi(p) = e^{2\pi i l p/n} \phi_p \tag{6}$$

where ϕ_p is a site atomic wave function at the pth atom and l is a quantum number chosen to satisfy Eq. (5); and $l = 0, \pm1, \pm2 \ldots$. Distinct solutions are limited to the values in Eq. (7),

$$l = 0, \pm1, \pm2 \ldots n/2 \tag{7}$$

The result [Eq. (7)] is usually found by noting that the molecule has its cyclic symmetry covered by the group C_n with characters $X_l = e^{2\pi i l/n}$, l being given by Eq. (7). Where $n/2$ is an odd number (e.g., benzene) the n electrons are accommodated pairwise in molecular orbitals, each of which transforms like a representation of C_n. The electron configuration is then closed. Where $n/2$ is even, two electrons half-fill a doubly degenerate pair of orbitals, and the configuration is open.

In systems with the Möbius topology, the hypothetical linear polyene to be closed on itself is uniformly twisted about its axis with period $2n$. By applying boundary conditions at points n atoms apart one may simulate the symmetry of the Möbius molecules. The characters of the cyclic group of period $2n$ are given in Eq. (8),

$$\chi_l = e^{2\pi i l/2n} \tag{8}$$

and the boundary conditions require that only those representations are realized for which there is a sign change for the translation through n atoms

TABLE III

TOTAL MO ENERGIES ACCORDING TO DELOCALIZATION TYPE

Local symmetry type \ Topological type	Normal	Möbius
Homomorphic $n/2$ odd	cosec	cot
$n/2$ even	cot	cosec
Heteromorphic	cot	cosec

along the axis. Thus the allowed values of the quantum number l are $l = \pm1, \pm3 \ldots \pm(n-1)$. The orbitals in this case all fall in degenerate pairs, and it is evident that closed configurations are attained for even $n/2$ and not, as in the normal case, for the Hückel $4n + 2$ systems.

In a heteromorphic molecule of n atoms [formula $(AB)_{n/2}$] the structural unit is a pair of adjacent atoms. The boundary condition for a molecule with normal topology is that only representations occur which have the same sign after $n/2$ units (for $n/2$ odd) or only representations with opposite sign (for $n/2$ even). The permitted values of the ring quantum number l are the following,

$$n/2 \text{ odd}: \quad l = 0, \pm2, \pm4 \ldots$$
$$n/2 \text{ even}: \quad l = \pm1, \pm3 \ldots$$

In molecules with Möbius topology the association of l values with odd and even $n/2$ is reversed.

The symmetry relationships within the four possible delocalization types are summarized in Table III. In the special case that the electronegativities of all participating atoms are the same, with a common resonance integral β, it may readily be shown that the total π-electron energy for an n-electron system is given by one of the two following expressions (9a) or (9b),

$$n\alpha + 4\beta \csc (\pi/n) \qquad (9a)$$

$$n\alpha + 4\beta \cot (\pi/n) \qquad (9b)$$

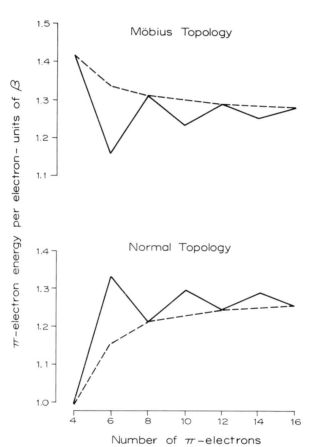

FIG. 9. π-Electron energy per electron relative to the Coulomb integral $[(1/n)(E_\pi - n\alpha)$, Eqs. (9a) and (9b)] for two possible topologies and for homomorphic (full-line) and hetero-morphic (dashed-line) interactions.

The simplest examples of the application of these formulas are benzene and cyclooctatetraene for $n = 6$ and $n = 8$, respectively. Table III shows how these energy formulas apply in the several cases; they are illustrated in Fig. 9.

D. The Island Model

We have described the electron wave functions for phosphonitrilic ring systems by emphasizing their connection with benzenoid aromatics, using a model in which all π-electrons are free to circulate over the ring system, compatibly with local conditions at each atom defined by the electronegativities of the available orbitals. Both d_{xz} and d_{yz} orbitals participate in the delocalization in π_a systems, and d_{xy} and $d_{x^2-y^2}$ in π_s. The range of possible types of behavior is very wide, when all degrees of importance of the several d-orbitals

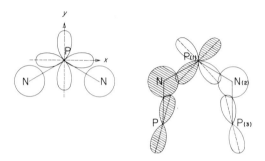

FIG. 10. π-Orbitals at nitrogen and phosphorus projected on to the local PNP plane [after D. P. Craig and K. A. R. Mitchell, *J. Chem. Soc., A* p. 4682 (1965)].

are allowed. In all cases the cyclic delocalization picture is acceptable. However, there is a span of theoretically possible situations in which the d_{xz} and d_{yz} electronegativities are roughly equal, and then the eigenfunctions of the 3×3 secular equation for the π_a system may be interpreted according to an alternative picture.[34] Let us take equally electronegative d-orbitals. The alternative description is that, instead of full cyclic delocalization over both d_{xz} and d_{yz} orbitals there is limited delocalization over sets or "islands" of three adjacent centers as shown in Fig. 10. The recognition of the importance of π_s delocalization which came later leads to other choices of model (including equal d_{xy} and $d_{x^2-y^2}$ orbitals) that can be appropriate to special combinations of d-orbital properties, but only the π_a island model justifies detailed discussion.

The situation to be described can be most readily understood where the d_{xz} and d_{yz} electronegativities are equal. Then the (x,y) quantization shown in Fig. 8 is not necessary, and the orbitals may be rotated about the z axis by any desired angle convenient for bond formation. By making a 45° rotation the scheme shown in the right-hand section of Fig. 10 is obtained. One $d\pi$ orbital overlaps its right-hand neighbor strongly and the other the left-hand, giving

[34] M. J. S. Dewar, E. A. C. Lucken, and M. A. Whitehead, *J. Chem. Soc.* p. 2423 (1960).

the illustrated three-center regions of overlap. If the N—P—N angle were precisely 90° there would be zero overlap between the shaded d-orbital at P(1) in Fig. 10 and the orbital at N(2), and the three-center molecular orbitals would be accurately orthogonal. In the actual phosphonitrilic ring systems the angle at phosphorus is near to 120°, and there is a degree of nonorthogonality between adjacent three-center islands, which inevitably superimposes some cyclic delocalization on the island model. We shall see in detail that the energy correction for this deviation from the island pattern is in some cases quite small, and in such cases (nearly equal d-orbital electronegativities) the two models are more or less equivalent. Where the d-orbitals have different electronegativities the island model becomes difficult to define and does not in general give a satisfactory account of the energies. A full comparison has been published,[35] and should be consulted for details. The essential features are seen by doing the calculation of the energies in the island model, first assuming the d-orbital electronegativities to be equal to one another. We first give in Eq. (10) the transformed d-orbitals found by rotating d_{xz} and d_{yz} by 45°.

$$d^+(k) = 2^{-1/2}[-d(xz, k) - d(yz, k)] \tag{10}$$
$$d^-(k) = 2^{-1/2}[d(xz, k) - d(yz, k)]$$

The index k numbers the atoms in the ring. The three-center molecular orbital is given by expression (11)

$$2^{-1/2} \cos \xi [d^-(k) + d^+(k + 2)] + \sin \xi \cdot p\pi(k + 1) \tag{11}$$

in which the appropriate d-orbitals overlapping a given nitrogen orbital appear with equal coefficients, and the angle parameter ξ determines the relative importance of nitrogen and phosphorus orbitals. The energy of the orbital is minimized when ξ is chosen to satisfy Eq. (12), $\gamma\beta$ being the difference of electronegativity between the d-orbitals and the p-orbital, i.e., $\alpha(xz) = \alpha(yz) = \alpha - \gamma\beta$,

$$\tan 2\xi = -2\sqrt{2} \cos 15°/\gamma \tag{12}$$

The energy after eliminating γ is given by Eq. (13),

$$E(\xi) = \alpha + \sqrt{2}\beta \cos 15° \cot \xi \tag{13}$$

We are now ready to compare the π-electron energies for the two models. For cyclic delocalization we take the molecular orbital energies from the solutions to secular Eq. (4), and for the island model note that all electrons can be accommodated, two from each island, in orbitals of energy $E(\xi)$ given in Eq. (13), with a total energy $nE(\xi)$. The essential comparison is between the delocalization energies in the two models, calculated by subtracting from the

[35] D. P. Craig and K. A. R. Mitchell, *J. Chem. Soc.* p. 4682 (1965).

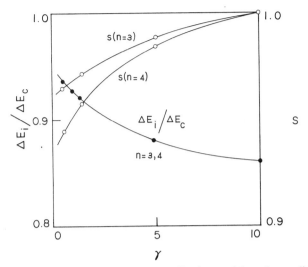

FIG. 11. Comparison of cyclic and island delocalization models under conditions of equal d_{xz} and d_{yz} electronegativities. The lower curve gives the delocalization energy ratio for six- and eight-membered ring systems. The upper curves give the overlap integrals for the two molecular wave functions [after *J. Chem. Soc.* p. 4682 (1965), by permission].

totals the energies of electrons confined to two-center π bonds between phosphorus and nitrogen. The two-center orbital is found by optimizing a linear combination of d_{xz} and d_{yz} orbitals at P and a $2p\pi$ orbital at N. The result is given in Eq. (14),

$$\epsilon(\zeta) = \alpha + \beta \cot \zeta; \qquad \tan 2\zeta = -2/\gamma \qquad (14)$$

Next, by subtracting $n\epsilon(\zeta)$ from the respective total energies we arrive at E_c and E_i, the delocalization energies for equal d-electronegativities in the cyclic and island models, respectively. The values are compared in Fig. 11 by plotting the ratio $\Delta E_i/\Delta E_c$ as a function of γ. The island model accounts for about 90% of the delocalization energy under these conditions. It is also illuminating to compare the wave functions in the two models by calculating their overlap, namely expression (15),

$$\int \ldots \int \psi_i^* \psi_c \, d\tau_1 \, d\tau_2 \ldots d\tau_n \qquad (15)$$

An overlap of unity, as on the right-hand side of Fig. 11, means that the wave functions in the two models have become identical in the limit where the p-orbital centers are very much more electronegative than d-orbital centers. In this (unphysical) limit the π-electrons concentrate exclusively at the nitrogen atoms in both models, giving the same wave functions. Otherwise the overlap of the wave functions is in the range 0.9–1.0.

We conclude that, insofar as the d-orbitals are taken to be equally electronegative, the island model can account for most of the delocalization energy in the π_a system. The remainder is accounted for by the nonorthogonality of neighbor islands.

E. Extension of the Island Concept to Inequivalent d-Orbitals

The assumption on which the foregoing is built, namely equally electronegative d_{xz} and d_{yz} orbitals, has no firm basis. The island model in its original and simplest form has an application only within this limiting assumption, but it is interesting to try to extend it to a wider framework, and to continue the comparison with the cyclic model. It is first to be noted that in the local field at the phosphorus sites the orbitals d_{xz} and d_{yz} are not equivalent: they belong to different representations of the site group C_{2v}. The relative electronegativities depend on the local fields acting on each, deriving from the phosphorus atom potential and from the molecular environment. Calculations have been made[36] leading to the conclusion that the electronegativities may be widely different, that they are sensitive to environment, and that the d_{xz} orbital is the more electronegative in the π_s system, and $d_{x^2-y^2}$ in the π_s, but we make no use of these detailed arguments here.

d-Orbitals of unequal orbital exponents cannot be transformed in the manner of Fig. 10 to give new orbitals rotated through 45°. However, it is still convenient in a three-center model to retain the linear combinations d^+ and d^- of Eq. (10). They are orthogonal, and their electronegativities are the same, equal to the mean of those of d_{xz} and d_{yz}. The common Coulomb parameter is given by (16)

$$\alpha(d^+) = \alpha(d^-) = [\alpha(xz) + \alpha(yz)]/2 \qquad (16)$$

d^+ and d^- are not diagonal in the one-electron Hamiltonian h, the off-diagonal element being given by Eq. (17),

$$\langle d^+|h|\, d^-\rangle = [\alpha(xz) - \alpha(yz)]/2 \qquad (17)$$

For convenience in presenting results we adopt an energy stated in units of $\beta \equiv \beta(xz)$, viz., the resonance parameter for interaction of a $2p\pi$ orbital at nitrogen with a d_{xz} orbital at a neighbor phosphorus, the orbital being *directed for maximum overlap*. For the 120° N—P—N angle assumed in the model the $2p\pi$–$3d_{xz}$ resonance integral is $\beta\cos(\pi/6)$. Variation parameters γ, δ, and σ enable the essential definitions to be completed, as follows,

$$\alpha(xz) = \alpha - \gamma\beta; \quad \alpha(yz) = \alpha - \delta\beta; \quad \beta(yz) = \sigma\beta$$

The energies are measured relative to α, the Coulomb integral for the nitrogen

[36] K. A. R. Mitchell, *J. Chem. Soc., A* p. 2683 (1968).

$2p\pi$ orbital. The necessary matrix elements of h are given in expressions (18),

$$\langle d^+|h|\,d^+\rangle = \alpha(d^+) = \alpha(d^-) = \alpha - (\gamma + \delta)\beta/2$$
$$\langle d^+|h|\,d^-\rangle = (\gamma - \delta)\beta/2$$
$$\rho = \beta(d^-, p\pi) = (\beta/\sqrt{2})[\cos(\pi/6) + \sigma\sin(\pi/6)]$$
$$\epsilon = \beta^+(d^+, p\pi) = -(\beta/\sqrt{2})[\cos(\pi/6) - \sigma\sin(\pi/6)]$$

(18)

It is to be noted that the last two expressions, both resonance integrals, refer to the same pair of atoms. $\beta(d^-, p\pi)$ is the larger ($\sigma > 0$), and refers to a nitrogen orbital overlapping with a d-orbital of the same island; β^+ refers to the overlap with the d-orbital forming part of an adjacent island.

The framework is now complete for the calculation of energies in both models. The energy of the lowest-energy island molecular orbital is $\alpha + \sqrt{2}\rho\cot\xi$, for $\tan 2\xi = -(4\sqrt{2}\rho/\beta)/(\gamma + \delta)$, and this result after subtraction of the localized π-bond energy [Eq. (14)] gives the island delocalization energy ΔE_i for any combination of energy parameters. To find ΔE_c, namely the best value for orbitals d_{xy}, d_{yz}, and $p\pi$, it is necessary, as before, to solve Eq. (4), sum the energies of the occupied molecular orbitals, and subtract the energies appropriate to localized bonding.

The values of parameters for particular molecules cannot be chosen with confidence. It is possible only to display the general character of the comparison between the models for wide ranges of the parameters. Here we give one set of comparisons only, but note that they are representative of the more complete survey.[35] Figure 12 gives values of the ratio of delocalization energies $\Delta E_i/\Delta E_c$ as a function of the ratio of resonance parameters, in terms of $\eta = (1 - \sigma)/(1 + \sigma)$. The d_{xz} orbital is here assumed less electronegative (smaller Coulomb integral) than the $p\pi$ of nitrogen by the fixed amount of 1.5β. The d_{yz} orbital is assigned various Coulomb parameters including a value ($\delta = 1.5$) equal to that of d_{xz}, and values both greater and less. For equal electronegativities, $\gamma = \delta = 1.5$, one expects equal resonance parameters, namely $\eta = 0$, and we then find that the island model gives about 80% of the total delocalization energy, as anticipated in Fig. 11. Where $\delta > \gamma$, giving a d_{yz} orbital less electronegative and, therefore, spatially more diffuse than d_{xz}, one expects that its resonance integral with $2p\pi$ will be less ($\eta > 0$). Thus the left-hand side of the diagram applies. For less diffuse d_{yz} orbitals ($\delta < \gamma$) the right side applies. It is evident that by setting $\sigma = 0$ we uncouple d_{yz} altogether from the delocalized system, and so have at the extreme left the condition of a purely heteromorphic delocalization. At the extreme right d_{xz} is relatively uncoupled, and the delocalization is homomorphic.

This review makes evident how the island model is related to the cyclic delocalization model in a formal sense, showing it to be a special case appropriate to the physical situation where the two $d\pi$ orbitals are of equal electronegativity. From a theoretical point of view this seems improbable, as explained, because the calculations of the orbital properties in the molecular field strongly suggest differences. However, the subsequent analysis allows the matter to be explained from the standpoint of experimental results, in terms of observed molecular properties contrasted with those expected according to the two models. The main point is that the island model in its simplest form (no allowance for delocalization) describes the π-electron contribution to all molecular properties as strictly additive, so that the contribution

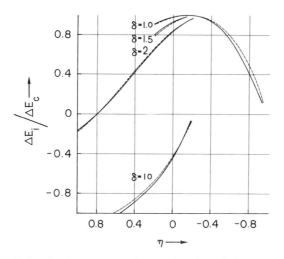

FIG. 12. Delocalization energy ratio as a function of the resonance parameter η. The Coulomb parameter of the d_{xz} orbital is set at $\alpha(xz) = \alpha - 1.5 \beta(xz)$. Full lines refer to six-membered rings, and dotted lines to eight-membered rings. [After *J. Chem. Soc.* p. 4682 (1965) by permission.]

per π-electron is independent of ring size. Among lesser points we note that π-electron effects cannot be transmitted at distances beyond a single island, and that the stereochemical relationships between adjacent islands are strictly determined in relation to the phosphorus atom common to them. Comparisons of the models will be made on these points in Section IV,B.

F. FURTHER APPLICATION OF THE π-ELECTRON THEORY

When the problem of the π-electron energy levels and electron distribution in any system has been solved, a study can be made of the changes in these quantities induced by perturbations, represented as changes in the Coulomb parameters of the ring atoms, as changes in the resonance parameters, or a combination of the two. One can then deal for example with the orientation of substitution by an initial substituent, or with changes in π bond order in a partially substituted system. Such studies are familiar in π-electron treatments of organic aromatics, where the possibilities have been widely explored[37] and approximate methods of calculation developed. In the phosphonitrilic rings one cannot at present expect to go beyond qualitative conclusions reflecting the symmetry differences between homo- and heteromorphic modes of delocalization, and variations of ring size.

G. PERTURBATIONS EXPRESSED AS ELECTRONEGATIVITY CHANGES

A phosphonitrilic $(NPX_2)_n$ into which a new atom, say F, is substituted is affected in the π_a and π_s systems mainly by the change in the Coulomb parameter, measuring the electronegativity, for the $3d$ orbitals on the phosphorus atom concerned, and in a secondary way, here neglected, by the change in resonance parameters between the atom and its neighbors. The polarization of a molecule by an attacking reagent might be simulated in the same way. Protonation of the nitrogen atom could be represented as an electronegativity increase in the p_z orbital of nitrogen insofar as effects on the π_a system are concerned, and as a very large increase in the electronegativity of the lone-pair orbital, sufficient to remove the electrons from the delocalized π_s system altogether. Thus calculations made to determine changes in electron distributions and energies caused by altering the Coulomb parameter of one ring atom might be valuable, in the first place to explore correlations with chemical behavior.

The essential problem is to calculate the changes produced in the energy levels and electron distributions by perturbations at one or more of the ring atoms. Let us suppose at first that the unperturbed system is a cyclic molecule in which each atom has the same electonegativity, corresponding to a Coulomb parameter α. If a perturbation is applied to one of the atoms, increasing the Coulomb parameter to $\alpha + \delta$, we may find the new energy levels by using the method first given by Koster and Slater[38] for electron-trapping crystals, and later developed[39] for excitons in mixed crystals. In the simplest situation of a ring of $2N$ identical atoms, with π-electron energies $\epsilon(k)$, k being a ring quantum number, the energies of the perturbed electron levels are given by the roots E of

$$1 = \frac{\delta}{2N} \sum_k \frac{1}{\epsilon(k) - E} \tag{19}$$

[37] See, e.g., C. A. Coulson, "Valence," Chapter IX. Oxford Univ. Press, London and New York, 1961.

[38] G. F. Koster and J. C. Slater, *Phys. Rev.* **95**, 1167 (1954).

[39] D. P. Craig and M. R. Philpott, *Proc. Roy. Soc.* **A290**, 583 (1966).

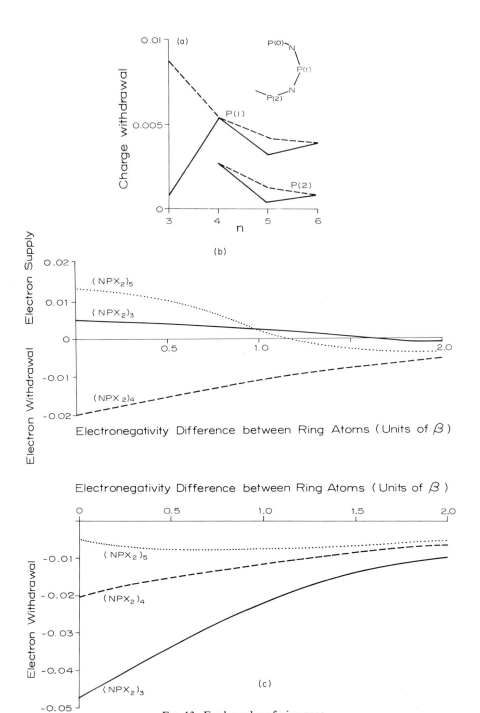

FIG. 13. For legend see facing page.

The perturbed π energy is the sum of the energies of electrons occupying the lowest available orbitals. The calculation is exact, so that, having found a new set of levels, we can repeat the calculation for a second perturbation and continue in a sequence of steps, each based on the previous one without loss of accuracy. We do, however, need a generalized expression for any case where the initial electron distribution is nonuniform. This is given by Eq. (20):

$$1 = \frac{\delta_s}{N} \sum_k \frac{p_{ks}^2}{\epsilon(k) - E} \qquad (20)$$

where s labels the perturbed site and p_{ks} is the amplitude at the site s for the orbital k. Thus to apply the method one must know the energy levels and molecular orbital coefficients for the parent problem. Solution of Eq. (20) then leads to the new energy levels while the perturbed molecular orbital coefficients are found from Eq. (21)

$$c_{i\lambda} = c_{s\lambda} \frac{\delta}{N} \sum_k \frac{p_{ki} p_{ks}}{\epsilon(k) - E(\lambda)} \qquad (21)$$

where $c_{i\lambda}$ is the amplitude at the ith center for the molecular orbital of energy $E(\lambda)$.

Let us suppose that we have a phosphonitrilic ring system defined for the π-electrons by the usual electronegativities $\alpha(xz)$, $\alpha(yz)$ and resonance parameters as in the text following Eq. (17). Then, for any chosen set, we can find the energy levels and electron densities for π_a, likewise for π_s systems, and have the complete starting data for Eq. (21). As discussed in Section IV,B studies of substitution behavior show that in some examples (e.g., chloride replacement by fluoride) the second substituent enters geminally, but in others (chloride replacement by dimethylamide) the second enters at a different phosphorus atom. Use of Eqs. (20) and (21) allows straightforward calculation of the energies and electron densities for the singly substituted molecule. A second application would allow a calculation of the energies of the two types of doubly substituted molecules, and so to see which gave the more stable π-electron system. However even at the first stage of calculation (for mono-

---

FIG. 13. (*a*) Charge withdrawal (as a function of ring size) from nearest phosphorus P(1) and next nearest P(2) produced by a perturbation 0.5β at P(0) in ring systems $(NPX_2)_n$ for which $\alpha_N = \alpha_P + 2\beta$. Full-line corresponds to homomorphic; dashed-line to heteromorphic systems. (*b*) Charge withdrawal from the nearest phosphorus in homomorphic systems caused by a perturbation 0.5β at P(0) for rings with various electronegativity differences between nitrogen and phosphorus. (*c*) Same as (*b*) in heteromorphic systems.

substitution) the electron densities will indicate the position of second substitution in the manner familiar in the organic aromatics.

The use of these methods will be illustrated by two sets of calculations, both being applied to examples in Section IV,B. The first calculations (Fig. 13a) are for PN rings of varying ring size and fixed electronegativity difference such that $\alpha_P = \alpha_N - 2\beta$. One phosphorus atom, P(0), is perturbed by substitution so that $\alpha_{P(0)} = \alpha_P + 0.5\beta$. One sees that heteromorphic systems show a smooth variation of electron withdrawal with ring size while homomorphic systems alternate. Also, of the homomorphic systems the tetramer exhibits the largest electron withdrawal from the phosphorus vicinal to the point of substitution, a fact which will be related in a later section to observations of reactivity, and orientation of successive substitution.

The second calculations (Fig. 13b and 13c) are related to the following question. For a given perturbation how does the electronegativity difference between the ring atoms influence the transmission of the perturbation to non-neighbor centers? The perturbation is represented by a change in the Coulomb parameter of one phosphorus atom, P(0) from α_P to $\alpha_P + 0.5\beta$. The relative electronegativities of the ring atoms are allowed to vary according to $\alpha_P = \alpha_N - \gamma\beta, \gamma = 0, 0.5, 1.0, 1.5,$ and 2.0. The degree to which the influence of the perturbation is transmitted from P(0) to other centers is measured by its influence at the nearest phosphorus atom, viz., P(1). Figure 13b applies to homomorphic systems, and shows that for Hückel $4n + 2$ rings the neighbor phosphorus atoms are electron-rich for small electronegativity differences in the ring, and electron-deficient for larger differences. The $4n$ systems, and all heteromorphic systems (Fig. 13c) are electron-deficient for all electronegativity differences. The main point however is that the extent to which the influence is propagated is strongly dependent on the delocalization present in the unperturbed system, and when this is reduced by increased electronegativity differences within the ring, the perturbation becomes more and more localized. We refer later (Section IV,B,4) to this "insulation" of perturbations in connection with evidence of the directional effects of substituents, which, as shown by these arguments, must be less where the electronegativity difference between N and PX_2 is greater.

H. Bond Length Alternation

There has been much discussion of the variation of bond lengths in organic π systems especially in pseudoaromatic systems.[40] Whether or not molecules like benzene and the larger cyclic polyenes (annulenes) adopt structures with equal bond lengths depends on a balance between the π-energy gain in structures

[40] For a review, see T. Nakajima, in "Molecular Orbitals in Chemistry, Physics and Biology" (P.-O. Löwdin and B. Pullman, eds.), p. 451. Academic Press, New York, 1964.

with equal length, and the loss by σ-bond stretching and compression energy involved in maintaining the length equality. The scale of these terms can be measured by the resonance integral ratio $k = \beta_s/\beta_d$, β_s and β_d being the resonance integrals for single and double bond lengths, and by the stretching force constant for the σ component of the bond. The same general principles can be applied to P—N bonds in a qualitative sense.[41, 42] Evidence of the σ contribution to the stretching force constant in this case is unavailable and in any case the σ–π separation is less well based than in benzene. The force constants for the P—N bond stretching in some phosphonitriles have been recorded by Chapman and Carroll.[43] The value for $(NPCl_2)_3$ is ~8.2 mdynes $Å^{-1}$, greater than the corresponding C—C force constant in benzene (7.6 mdynes $Å^{-1}$). This is surprising. However, the variation of the resonance integral β with bond length is certainly very much slower than in organic aromatics if it be assumed that there is a proportionality between overlap integral and resonance integral: the resonance integral varies with bond length at half or less the rate in P—N than in C—C. Thus we do not necessarily expect that partial loss of delocalization in P—N systems (e.g., by out-of-plane displacements) will show itself by the appearance of bond length inequalities, although at some stage of diminishing delocalization it should do so.

IV. Experimental Evidence Relating to π-Bonding

In the foregoing sections, the general evidence for the availability of d-orbitals in phosphorus has been reviewed, and the models for the description of electronic delocalization briefly summarized. We now consider the experimental evidence bearing on the three major questions that arise in a discussion of aromatic character in the rings, namely: (i) Do $d\pi$ electrons contribute to the bonds? (ii) Are the molecules stabilized by delocalization of the π-bonds? (iii) Are the finer details of the delocalization in line with homomorphic or heteromorphic character, or with an island model?

The problems are more difficult than in benzene, and the evidence has to be scrutinized closely if conclusions about aromaticity are to be well based. In benzene and other aromatics the reality of delocalization is accepted on the basis of the cumulative support of a number of lines of evidence. The primary one is the physical evidence from thermochemical measurement of heats of combustion or hydrogenation and comparison with reference substances coupled with assumed additivity relationships. Then there is the evidence of

[41] C. W. Haigh and L. Salem, *Nature* **196**, 1307 (1962).
[42] D. W. Davies, *Nature* **196**, 1309 (1962).
[43] A. C. Chapman and D. F. Carroll, *J. Chem. Soc.* p. 5005 (1963).

ring currents measured by proton resonance shifts and diamagnetic susceptibility, and structural evidence of the variation of bond lengths and their
interpretation in terms of theoretical models including delocalization. There
are also chemical indications in the transmission of the influence of substituents
in aromatic rings.

Phosphonitrilic compounds show important differences. That their π-
bonding scheme is not simply homomorphic is strongly suggested by the
different types of product obtained by ammonolysis of boron trichloride and
phosphorus pentachloride. In the first reaction

$$NH_4Cl + BCl_3 \rightarrow \tfrac{1}{3}(NH \cdot BCl)_3 + 3HCl \tag{22}$$

the cyclic product is exclusively as shown.[44] The π-bonding in borazines is
exclusively homomorphic, and the absence of bigger rings corresponds to the
fact that the π-electron energy (per electron) for a homomorphic system is
greatest for a six-membered ring (Fig. 9). There seems no geometric reason
why larger rings should not be formed, if they were electronically stable, and
the condition for the formation of the eight-membered borazocine family is
that the nitrogen atom should carry a bulky alkyl group.[45] Formation of the
six-membered ring is then sterically hindered, and the eight-membered ring is
formed by default:

$$\textit{tert-}BuNH_2 + BCl_3 \rightarrow \tfrac{1}{4}(\textit{tert-}BuN \cdot BCl)_4 + 2\,HCl \tag{23}$$

In the related preparative method for the phosphonitrilic chlorides, by contrast,
a large range of ring sizes is found. In the reaction

$$NH_4Cl + PCl_5 \rightarrow \frac{1}{n}(NPCl_2)_n + 4\,HCl \tag{24}$$

about 25 % of the product consists of the bigger rings, $(n > 6)$,[46] and, although
the possibility of delocalization of lone pairs on nitrogen into the d-orbitals of
phosphorus provides a flexibility which is absent in the $p\pi$–$p\pi$ system of the
borazines, the difference from the borazine synthesis suggests that a different
type of bonding is involved. Thermochemical measurements confirm this, but
also show that the scale of π-interactions is smaller than in benzene and its
homologs; the evidence from ring currents is inconclusive. For these reasons,
and because the molecules are more polarizable than carbocyclic compounds
and structural changes larger, more use must be made of structural and

[44] C. A. Brown and A. W. Laubengayer, *J. Am. Chem. Soc.* **77**, 3699 (1955).

[45] H. S. Turner and R. J. Warne, *Proc. Chem. Soc.* p. 69 (1962).

[46] L. G. Lund, N. L. Paddock, J. E. Proctor, and H. T. Searle, *J. Chem. Soc.* p. 2542 (1960);
H. N. Stokes, *Am. Chem. J.* **19**, 782 (1897).

chemical evidence than of physical properties in delineating the nature and extent of delocalization.

A. PRIMARY EVIDENCE FOR $p\pi$–$d\pi$ BONDING IN PHOSPHONITRILIC DERIVATIVES

Some of the general features of $p\pi$–$d\pi$ bonds were considered in Section II, mainly by reference to other than phosphonitrilic derivatives. In this section we present more detailed evidence for $p\pi$–$d\pi$ bonding, estimate the relative magnitudes of the σ- and π-interactions, and illustrate some of the ways in which they are affected by variations in ligand and in ring size. In later sections we shall attempt to discriminate some of the different types of $d\pi$-interaction described in Section III.

1. *Thermochemistry*

The heats of formation of three trimeric and two tetrameric phosphonitrilic derivatives have been determined by combustion,[47,48] and of the chlorides $(NPCl_2)_{5-7}$ through their heats of polymerization.[49] For benzene and its derivatives, it is customary to express the effects of delocalized π-bonding in terms of a resonance energy, but this is not possible for phosphonitrilic derivatives, because the appropriate value of the bond energy term $E(P{=}N)$ is especially uncertain. As an alternative way of expressing the results, Table IV shows the amounts by which the measured $E(P{-}N) + E(P{-}X)$ exceeds the sum of their estimated single-bond equivalents, together with comparable information for benzene and cyclooctatetraene. These deviations can be taken to be primarily due to π-electron effects, either localized or delocalized, but also to differences in σ-bond strength, where the electronegativities of the substituents differ from those of the reference compounds. The relative importance of σ- and π-effects will be discussed in connection with structure, for which the data are more abundant, though the same difficulties recur.

The tabulated figures require several comments. The first is that, partly because of uncertainties in the chemistry of the combustion process, the heats of formation of the phosphonitrilic derivatives are much less accurately known than those of the hydrocarbons; this results in probable errors in bond energy pairs of the order of ±1.5 kcal in, e.g., $N_4P_4Ph_8$. Another sort of inaccuracy arises from the general difficulty of establishing single-bond energies to second-row elements. These elements interact, through their acceptor orbitals, with neighboring atoms carrying unshared electrons, in a way which is not possible for carbon, and the resulting contributions to the measured

[47] A. F. Bedford and C. T. Mortimer, *J. Chem. Soc.* p. 4649 (1960).
[48] S. B. Hartley, N. L. Paddock, and H. T. Searle, *J. Chem. Soc.* p. 430 (1961).
[49] J. K. Jacques, M. F. Mole, and N. L. Paddock, *J. Chem. Soc.* p. 2112 (1965).

bond energy can only be assessed indirectly. Their reality is shown by the existence of the donor–acceptor complex $Me_3N \cdot PCl_3$, in which they form the only bond between the two molecules[50]; the dissociation energy of the complex

TABLE IV

BOND ENERGY TERM INCREMENTS OVER SINGLE-BOND VALUES FOR PHOSPHONITRILIC DERIVATIVES, BENZENE, AND CYCLOOCTATETRAENE[a]

	$N_3P_3Cl_6$	$N_4P_4Cl_8$	$N_3P_3Me_6$	$N_4P_4Ph_8$	$N_3P_3(OC_6H_{11})_6$
$\Delta E(P-N) + \Delta E(P-X)^b$					
kcal	9.3	9.7^c	8.7	15.8	16.1
Reference	d	d, e	f	f	f

	C_6H_6	C_8H_8
$\Delta E(C-C) + \Delta E(C-H)^g$		
kcal	38.5 (21.3)	33.0 (15.8)
Reference	h	i

[a] So expressed because the ring and exocyclic bonds are equally numerous.

[b] Reference single-bond energies assumed to be: $E(P-N) = 66.8$ kcal [P. A. Fowell and C. T. Mortimer, *J. Chem. Soc.* p. 2913 (1959)]; $E(P-O) = 92.0$ kcal, $E(P-Cl) = 76.2$ kcal [based on E. Neale, L. T. D. Williams, and V. T. Moores, *J. Chem. Soc.* p. 422 (1956) and E. Neale and L. T. D. Williams, *J. Chem. Soc.* p. 2156 (1954)]; $E(P-C)(Me_3P) = 65.3$ kcal [L. H. Long and J. F. Sackman, *Trans. Faraday Soc.* **53**, 1606 (1957)]; $E(P-C)(Ph_3P) = 70.6$ kcal [A. F. Bedford and C. T. Mortimer, *J. Chem. Soc.* p. 1622 (1960)].

[c] For $(NPCl_2)_{5-7}$, $\Delta E(P-N) + \Delta E(P-Cl) = 9.8, 10.0, 10.1$ kcal (calculated from Reference d).

[d] S. B. Hartley, N. L. Paddock, and H. T. Searle, *J. Chem. Soc.* p. 430 (1961).

[e] J. K. Jacques, M. F. Mole, and N. L. Paddock, *J. Chem. Soc.* p. 2112 (1965).

[f] A. F. Bedford and C. T. Mortimer, *J. Chem. Soc.* p. 4649 (1960).

[g] The first value given for each compound was obtained by using the values of $E(C-C)$ and $E(C-H)$ recommended by T. L. Cottrell ("The Strengths of Chemical Bonds." 2nd ed. Butterworth, London and Washington, D.C., 1968) for general use. The second value (in parentheses) is based on values of $E(C-C)$ and $E(C-H)$ appropriate to sp^2 hybridization [J. D. Cox, *Tetrahedron* **18**, 1337 (1962)], derived by assuming that the conjugation energy in 1:3-butadiene is zero.

[h] E. J. Prosen, R. Gilmont, and F. D. Rossini, *J. Res. Natl. Bur. Std.* **34**, 65 (1945).

[i] H. D. Springall, T. R. White, and R. C. Cass, *Trans. Faraday Soc.* **50**, 815 (1954).

is 6.4 kcal mole^{-1}. It is possible, therefore, that the P—N bond in $P(NEt_2)_3$, used as the reference compound for $E(P-N)$, may have a component arising from the delocalization of lone pairs on nitrogen. If so, the deviations from

[50] R. R. Holmes, *J. Phys. Chem.* **64**, 1295 (1960); *J. Am. Chem. Soc.* **82**, 5285 (1960).

estimated single bond energies quoted in Table IV would be lower limits to the π-electron contributions. Similarly, in phosphonitrilic derivatives, formally unshared electrons on both the exocyclic atoms and the ring nitrogen atoms contribute to the bonding, and influence each other's contribution competitively[51]; attempts to divide the deviation into parts characteristic of the individual bonds are therefore not very helpful.

We also expect bond energies to be dependent on the hybridization state. This is certainly an important effect for carbon,[52,52a] $E(C—C)$ varying from 85.0 to about 98.0 kcal[52a] as the hybridization changes from sp^3 to sp^2; bond energy deviations for the hydrocarbons calculated for both states are given in Table IV, the second (in parentheses) being the more appropriate. There are too few thermochemical data for phosphorus compounds to allow properly for the energetic effects of hybridization changes, but the absolute effect in phosphonitrilic derivatives is not likely to be so big, because the geometrical changes are smaller; the relative stereochemical effects within a series of phosphonitrilic molecules are considered in Section IV,A,3.

With these reservations in mind, it seems possible to draw the following conclusions from Table IV. (1) π-Electron effects are smaller in the phosphonitrilic series than in the benzene series. (2) The energetic difference between benzene and cyclooctatetraene, some of which would persist if the latter molecule were planar, is not paralleled in the phosphonitrilic series. The mean bond energy is greater in $N_4P_4Cl_8$ than in $N_3P_3Cl_6$, even though the tetramer, in one of its forms, has a similar shape to cyclooctatetraene (Section IV,A,3), and the strengthening continues in $(NPCl_2)_{5-7}$. Although π-bonding is involved in both the carbocyclic and the phosphonitrilic series, its thermochemical consequences depend differently on ring size, and the π-interactions are therefore of different types. (3) The effect of ligand variation is not certain. It is possible that conjugative effects may contribute to the stability of the phenyl derivatives, and that exocyclic π-bonding through delocalization of lone pairs may also occur in the cyclohexyloxy compound. Such matters cannot be decided by thermochemistry alone; for phosphoryl compounds, where similar difficulties occur, individual bond energies have been allocated by taking account of spectroscopic evidence.[51]

2. Vibrational Spectra and Base Strength

The stretching frequency of a nominally single P—N bond occurs at 620, 640 cm^{-1} in $P(NCS)_3$, $OP(NCS)_3$,[53] in the range 640–653 cm^{-1} in the series

[51] S. B. Hartley, W. S. Holmes, J. K. Jacques, M. F. Mole, and J. C. McCoubrey, *Quart. Rev. (London)* **17**, 204 (1963).

[52] M. J. S. Dewar and H. N. Schmeising, *Tetrahedron* **5**, 166 (1959); **11**, 96 (1960); J. D. Cox, *ibid.* **18**, 1337 (1962).

[52a] J. D. Cox, *Tetrahedron* **19**, 1175 (1963).

[53] K. Oba, F. Watari, and K. Aida, *Spectrochim. Acta* **23A**, 1515 (1967).

$N_nP_nF_{2n-1}(NCS)$, $(n = 3-6)$,[54] at 645 cm^{-1} in $N_3P_3(NCS)_6$,[55] and at 707–718 cm^{-1} in $N_3P_3F_5NMe_2$,[54] and a series of dimethylamino derivatives $N_3P_3Y_x(NMe_2)_{6-x}$, $(Y = Cl, Br)$.[56] Although they are more nearly related to the properties of particular bonds than are the thermochemical measurements, none of the frequencies is characteristic of a simple local stretching motion, and they cannot be taken as a direct measure of the stretching force constants; we attach significance only to the order of magnitude of the frequencies, 620–720 cm^{-1}.[57]

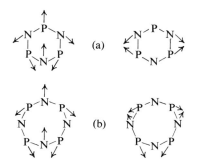

Fig. 14. Approximate atomic motions in degenerate P—N stretching modes. (a) E' in N_3P_3; (b) E_u in N_4P_4.

Infrared spectroscopy suggests that the ring bonds in phosphonitrilic derivatives are stronger than single bonds. The spectra of all the compounds have a strong band (usually the strongest) in the range 1180–1450 cm^{-1}, to which the stretching of the ring P—N bonds makes a major contribution. The approximate atomic motions in the corresponding degenerate modes are shown in Fig. 14 for planar six- and eight-membered rings. The vibrations are qualitatively similar in exhibiting approximate antisymmetric PNP motion [and are hence often collectively referred to as ν_{as}(PNP)], but they differ quantitatively, and their relationship to bond stretching constants is more remote than for the single-bond frequencies referred to above. It is all the more so because the remaining frequencies in the set are widely distributed; the lowest is the totally symmetric (ring-breathing) frequency, which occurs

[54] T. Chivers, R. T. Oakley, and N. L. Paddock, *J. Chem. Soc., A* p. 2324 (1970).

[55] R. Stahlberg and E. Steger, *Spectrochim. Acta* **23A**, 2185 (1967).

[56] R. Stahlberg and E. Steger, *Spectrochim. Acta* **23A**, 2005 (1967).

[57] In a comparative study of a large number of phosphorus–nitrogen compounds, R. A. Chittenden and L. C. Thomas [*Spectrochim. Acta* **22**, 1449 (1966)] find that a band near 750 cm^{-1} frequently, but not invariably, occurs. They assign ν(P—N) to the range 873–1053 cm^{-1}, and ν(P=N) to 1325–1400 cm^{-1}. The values of ν_{as}(PNP) found in phosphonitrilic derivatives are close to this latter group of frequencies.

at 746–756 cm^{-1} in $N_3P_3Br_6$,[58-60] 782–785 cm^{-1} in $N_3P_3Cl_6$,[13,58,61,63] and at 689 cm^{-1},[13] 743 cm^{-1},[64] or 567–570 cm^{-1} in $N_3P_3F_6$.[13,65] Since these frequencies are greater than some assignments of the single-bond frequencies, and the other frequencies of the set are much greater, there is evidently a net strengthening of the ring bonds, but for the reasons given above, direct comparison of corresponding frequencies of molecules of different ring size cannot be used to assess relative bond stretching constants.

TABLE V

ν_{as}(PNP)(cm^{-1}) FOR TRIMERIC AND TETRAMERIC PHOSPHONITRILIC DERIVATIVES

	Me[a]	Ph[b]	Br[c-g]	NMe$_2$[h]	Cl[d-f, i-m]	OMe[h]	F[f,j,n-p]
$N_3P_3X_6$ (E or E')	1180	1190	1171–1179	1195	1210–1220	1275	1297–1305
$N_4P_4X_8$ (E or E_u)	1180	1213	1272–1280	1265	1295–1325	1337	1425–1445

[a] H. T. Searle, *Proc. Chem. Soc.* p. 7 (1959).
[b] A. J. Bilbo, *Z. Naturforsch.* **15b**, 330 (1960).
[c] E. Steger and R. Stahlberg, *J. Inorg. & Nucl. Chem.* **28**, 688 (1966).
[d] T. R. Manley and D. A. Williams, *Spectrochim. Acta* **23A**, 149 (1967).
[e] U. Stahlberg and E. Steger, *Spectrochim. Acta* **23A**, 672 (1967).
[f] E. Steger and R. Stahlberg, *Z. Naturforsch.* **17b**, 780 (1962).
[g] K. John and T. Moeller, *J. Inorg. & Nucl. Chem.* **22**, 199 (1961).
[h] R. A. Shaw, *Chem. & Ind. (London)* **54** (1959).
[i] L. W. Daasch, *J. Am. Chem. Soc.* **76**, 3403 (1954).
[j] A. C. Chapman and N. L. Paddock, *J. Chem. Soc.* p. 635 (1962).
[k] S. Califano, *J. Inorg. & Nucl. Chem.* **24**, 483 (1962).
[l] I. C. Hisatsune, *Spectrochim. Acta* **21**, 1899 (1965).
[m] I. C. Hisatsune, *Spectrochim. Acta* **25A**, 301 (1969).
[n] F. Seel and J. Langer, *Z. Anorg. Allgem. Chem.* **295**, 316 (1958).
[o] H. J. Becher and F. Seel, *Z. Anorg. Allgem. Chem.* **305**, 148 (1960).
[p] A. C. Chapman, N. L. Paddock, D. H. Paine, H. T. Searle, and D. R. Smith, *J. Chem. Soc.* p. 3608 (1960).

Comparisons within the same ring size are more useful, because the frequencies are distributed similarly, and the common practice of taking ν_{as}(PNP) as a rough measure of ring bond strength is partly justified by the calculation

[58] U. Stahlberg and E. Steger, *Spectrochim. Acta* **23A**, 627 (1967).
[59] G. E. Coxon and D. B. Sowerby, *Inorg. Chim. Acta* **1**, 381 (1967).
[60] T. R. Manley and D. A. Williams, *Spectrochim. Acta* **23A**, 149 (1967).
[61] L. W. Daasch, *J. Am. Chem. Soc.* **76**, 3403 (1954).
[62] S. Califano, *J. Inorg. & Nucl. Chem.* **24**, 483 (1962); S. Califano and A. Ripamonti, *ibid.* p. 491.
[63] I. C. Hisatsune, *Spectrochim. Acta* **21**, 1899 (1965).
[64] H. J. Becher and F. Seel, *Z. Anorg. Allgem. Chem.* **305**, 148 (1960).
[65] E. Steger and R. Stahlberg, *Z. Naturforsch.* **17b**, 780 (1962).

of P—N bond stretching constants in $N_3P_3Cl_6$ and $N_3P_3F_6$ (8.2, 8.9 mdynes Å^{-1}),[43] which are in the order of their $\nu_{as}(PNP)$. Frequencies of this type are shown in Table V for a variety of trimeric and tetrameric derivatives. For each series, $\nu_{as}(PNP)$ increases with the electronegativity of the ligand on Pauling's scale, and we can take this as evidence that the ring bonds, all stronger than single bonds, are further strengthened by electronegative substituents. These results are as expected from the general properties of d-orbital bonding, as described in Section II.

Those substituents which decrease $\nu_{as}(PNP)$ also increase the base strength of phosphonitrilic derivatives, and it is therefore convenient to discuss the spectroscopic results and the base strength measurements together. Compounds in which a P—O or a P—N bond is essentially single are highly polar and susceptible to hydrolysis; dialkyl phosphinic acids R_2POOH are slightly basic,[66] and the phosphoramidate esters $R_2N \cdot PO(OR')_2$ (especially when R is an alkyl group) are easily split by hydrogen chloride[67]; the mechanism probably involves the formation of the conjugate acid $[R_2NH \cdot PO(OR')_2]^+$. The formally unsaturated phosphonitrilic derivatives, especially those carrying the more electronegative substituents, are not so readily decomposed, and reactions with nucleophiles are usually more important than their base properties. Nevertheless, even $N_3P_3Cl_6$ forms complexes with SO_3[68] and $AlCl_3$[69]; both $N_3P_3Cl_6$ and $N_4P_4Cl_8$ form salts with perchloric acid,[70] and all the chlorides $(NPCl_2)_n$ $(n = 3-8)$ are reversibly protonated in sulfuric acid.[71] The base strengths of several phosphonitrilic derivatives have been determined in solution in nitrobenzene,[72-75] and the results considered in terms of electron distribution. Aminophosphonitriles also react with Lewis acids such as iodine.[76]

The primary problem, whether the base strengths show the lowering expected from participation of the nitrogen orbitals in a π system, is a difficult one because of the absence of information on the single-bonded structures. The molecular structure of 2,4,6-trimethoxy-1,3,5-trimethyl-2,4,6-trioxocyclo-triphosphazane $N_3Me_3P_3O_3(OMe)_3$ shows, by its comparatively short bond

[66] P. C. Crofts and G. M. Kosolapoff, *J. Am. Chem. Soc.* **75**, 3379 (1953).
[67] Z. Skrowaczewska and P. Mastalerz, *Roczniki Chem.* **29**, 415 (1955).
[68] M. Goehring, H. Hohenschutz, and R. Appel, *Z. Naturforsch.* **9b**, 678 (1954).
[69] H. Bode and H. Bach, *Ber.* **75B**, 215 (1942).
[70] H. Bode, K. Bütow, and G. Lienau, *Ber.* **81**, 547 (1948).
[71] D. R. Smith, unpublished work (1960).
[72] D. Feakins, W. A. Last, and R. A. Shaw, *J. Chem. Soc.* p. 2387 (1964).
[73] D. Feakins, W. A. Last, and R. A. Shaw, *J. Chem. Soc.* p. 4464 (1964).
[74] D. Feakins, W. A. Last, N. Neemuchwala, and R. A. Shaw, *J. Chem. Soc.* p. 2804 (1965).
[75] D. Feakins, W. A. Last, S. N. Nabi, and R. A. Shaw, *J. Chem. Soc., A* p. 1831 (1966).
[76] S. K. Das, R. A. Shaw, B. C. Smith, W. A. Last, and F. B. G. Wells, *Chem. & Ind.* (*London*) p. 866 (1963).

lengths within the ring and the essentially planar configuration at nitrogen, that the σ-bonds are supplemented by appreciable π-bonding (Fig. 15).[77] In spite of this, and of the large inductive effect of the phosphoryl group, the related ethyl compound is sufficiently strongly basic to form a stable hydrochloride,[78] whereas its phosphonitrilic counterpart $N_3P_3(OEt)_6$ is not. This example suggests that a moderate decrease in base strength accompanies formal π-bonding, presumably because protonation of such a structure entails a loss of delocalization energy.

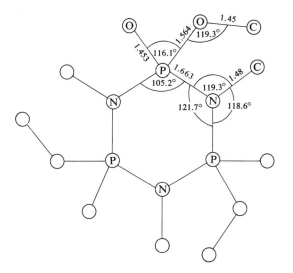

FIG. 15. The structure 2,4,6-trimethoxy-1,3,5-trimethyl-2,4,6-trioxocyclotriphosphazane. Averaged dimensions from G. B. Ansell and G. J. Bullen, *J. Chem. Soc.*, *A* p. 3026 (1968), by permission.

Although direct evidence of the relative base strengths of σ- and π-bonded structures is at present inadequate, some light is thrown on the problem by a study of the effect of substituents on base strength, especially those which can release electrons to the ring by delocalization of lone pairs, and a comparison of their effects on base strength and on $\nu_{as}(PNP)$. The main factors which affect the base strength of a phosphonitrilic derivative are (1) the hybridization state of the nitrogen atom; (2) the extent to which the nitrogen lone pair participates in π-bonding; and (3) differences in the degree of solvation of the protonated species. We shall discuss the first two effects, but the third is specific, and in the absence of detailed information on particular cases, its possibility does no more

[77] G. B. Ansell and G. J. Bullen, *J. Chem. Soc.*, *A* p. 3026 (1968).
[78] R. Rätz and Hess, *Ber.* **84**, 889 (1951).

than reduce the force of conclusions drawn from a consideration of hybridization and π-bonding.

Experimentally, the effect of ligand variation is far greater than that expected or found for variation of ring size. By comparison with the base strengths of 2-substituted pyridines, and after making allowance for the difference in reaction media, it is found that the effect of a particular substituent on base strength is similar in both series, and depends on both "multiple-bond" and inductive effects.[74] The inductive effects are apparent (Table VI) in the large

TABLE VI

BASE STRENGTHS OF TRIMERIC AND TETRAMERIC PHOSPHONITRILIC DERIVATIVES[a,b]

	NMe_2^c	Et	Ph	OEt	SEt	OPh	SPh	Cl^d
$N_3P_3X_6$	7.6	6.4	1.6	−0.2	−2.8	−5.8	−4.8	<−6.0 (−8.2)
$N_4P_4X_8$	8.3	7.6	2.2	+0.6	—	−6.0	—	<−6.0 (−8.6)

[a] pK_a', determined in nitrobenzene, except where stated.
[b] Results from D. Feakins, W. A. Last, N. Neemuchwala, and R. A. Shaw [*J. Chem. Soc.* p. 2804 (1965)] except where stated.
[c] D. Feakins, W. A. Last, and R. A. Shaw, *J. Chem. Soc.* p. 4464 (1964).
[d] Values in parentheses from equilibria in sulfuric acid, and therefore not directly comparable with the other values [D. R. Smith, unpublished work (1960)].

difference in pK_a' between $N_3P_3Et_6$ and $N_3P_3Cl_6$. Exocyclic π-bonding is established by structural studies, which demonstrate the shortness of the exocyclic P—N bonds and the near-planarity of the exocyclic groups in $N_4P_4(NMe_2)_8$[79] and in $N_6P_6(NMe_2)_{12}$,[80] and related features are found in the structures of $N_4P_4(OMe)_8$,[81] $N_6P_6(OMe)_{12}$,[82] and $N_8P_8(OMe)_{16}$.[83] Such exocyclic π-bonding results in electron transfer to the ring, so that protonation of aminophosphonitriles occurs there rather than on the substituent. The inference that it does so, arrived at from a study of base strengths,[73] has been confirmed crystallographically, with the additional result[84] that the added proton in $N_3P_3Cl_2(NHPr^i)_4H^+Cl^-$ is found to lie in the local PNP plane.[85]

The results in Table VI show that amino- and alkoxyphosphonitriles are

[79] G. J. Bullen, *J. Chem. Soc.* p. 3193 (1962).
[80] A. J. Wagner and A. Vos, *Acta Cryst*. **B24**, 1423 (1968).
[81] G. B. Ansell and G. J. Bullen, *Chem. Commun.* p. 430 (1966).
[82] M. W. Dougill, unpublished work (1969).
[83] N. L. Paddock, J. Trotter, and S. H. Whitlow, *J. Chem. Soc., A* p. 2227 (1968).
[84] N. V. Mani and A. J. Wagner, *Chem. Commun.* p. 658 (1968).
[85] For further discussion of this structure, see Sections IV,B,2 and IV,C.

stronger bases than purely inductive considerations would suggest; both are more strongly basic than the chlorides, even though the electronegativities of nitrogen and chlorine are usually taken to be equal. The stretching frequencies are also affected, those of amino derivatives being somewhat lower than those of the chlorides, (Table V) but since ν_{as}(PNP) is greater for the alkoxy than for the chloro derivatives, the effect of electron-releasing ligands on bond stretching is evidently less than on base strength. The results are explicable if there is a competition between the ligands and the ring nitrogen atoms for the phosphorus π-orbitals, and the larger effect on base strength suggests that, insofar as π-interactions are concerned, the π_s system is chiefly involved, whereas bond stretching involves both π_s and π_a. Confirmation of the occurrence of conjugative interactions between the ligand and the ring is provided by the base strengths of the thioethoxides and thiophenoxides. The former are weaker bases than the ethoxides, because sulfur is less able than oxygen to conjugate in the sense $-S^+{=}P-N^-{-}$; phenoxides are weaker bases than alkoxides, on account of such interactions as $-O^+{=}C_6H_5{}^-$, which would reduce conjugation with phosphorus. Thiophenoxides are therefore the stronger bases, because the competitive interaction with the phenyl group is reduced.[74]

The base strengths of individual molecules are expected to depend in part on the hybridization state of the lone pair. As the ring size increases from N_3P_3 to N_4P_4, the angle at nitrogen invariably increases, and, insofar as this indicates a change in σ-hybridization, the base strengths of the eight-membered rings are, from this cause alone, expected to be the higher.[74] Such a change is offset by the delocalization of the lone pairs into the π-orbitals of phosphorus, but the relative effects in the six and eight-membered rings depend on the symmetry type of the π_s-interactions. If heteromorphic, N_3P_3 rings would be more strongly basic than N_4P_4 rings, and the near-equality of the base strengths of corresponding trimers and tetramers has been attributed to a compensation of hybridization changes and π delocalization of this type.[74] There is usually, however, a slight increase of base strength from trimer to tetramer, compatible either with a dominance of the hybridization effect, or, if π-electron charge is a suitable index of base strength, with some degree of homomorphic interaction (Table VII). π-Contributions to base strength would be independent of ring size for partial delocalization of the "island" type.

Since the hybridization effect cannot be estimated accurately, the foregoing results, although demonstrating the importance of π-electron effects, do not clearly indicate their type. Study of the larger rings is in principle more informative (since further increases in the angle at nitrogen are likely to be smaller), but so far have been carried out only for the chlorides, which are so weakly basic that indirect methods of determinations have to be used, and comparability with the other results is not certain. All the chlorides reduce the O—H stretching frequency of phenol, their base strengths, by this criterion,

geneously substituted phosphonitriles. The significance to be attached to bond length inequalities arising from inhomogeneous substitution is discussed in Section IV,B, and conjugation with exocyclic groups, which has some structural consequences, in Section IV,C.

 a. Conformations and Bond Lengths. Structural data for the compounds of immediate interest are summarized in Table VIII. Two conclusions are immediately apparent. The first is that the lengths of the ring bonds fall in the range 1.50–1.60 Å, and, although they are different in different molecules, all are appreciably shorter than the single bond (1.77 Å)[89] in the phosphorami-

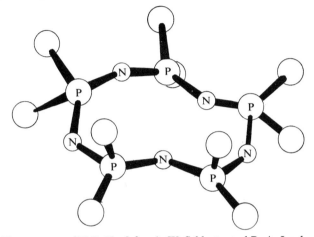

FIG. 16. The structure of $N_5P_5Cl_{10}$ [after A. W. Schlueter and R. A. Jacobson, *J. Chem. Soc., A* p. 2317 (1968), by permission].

date ion $(NH_3 \cdot PO_3)^-$. It should be said at once that this length has not the status of the C—C single bond length of 1.54 Å, in that it will depend on the electronegativity of the ligands, and may be increased by the net negative charge on the ion. While correction for these and other effects, which are discussed in Section IV,A,3b, would reduce the standard length, appreciable differences from typical phosphonitrilic bond lengths still remain, and can be ascribed to direct or indirect *d*-orbital interactions. Second, and with the exception of $N_5P_5Cl_{10}$,[90] which is discussed in Section IV,A,3b, all the ring bonds with a particular molecule are equal in length within experimental error, even though the molecular conformations are widely different.

In only two molecules $(N_3P_3F_6,$[91] $N_4P_4F_8$[92]$)$ are the departures of the PN

[89] E. Hobbs, D. E. C. Corbridge, and B. Raistrick, *Acta Cryst.* **6**, 621 (1953); D. W. J. Cruickshank, *ibid.* **17**, 671 (1964).
[90] A. W. Schlueter and R. A. Jacobson, *J. Chem. Soc., A* p. 2317 (1968).
[91] M. W. Dougill, *J. Chem. Soc.* p. 3211 (1963).
[92] H. McD. McGeachin and F. R. Tromans, *J. Chem. Soc.* p. 4777 (1961).

TABLE VIII

Summarized Structural Information for Homogeneously Substituted Phosphonitrilic Derivatives[a]

Compound	(P—N) (Å)	(P—X) (Å)	PN̂P (deg)	NP̂N (deg)	XP̂X (deg)	Conformation and approx. symmetry	Ref.
$N_3P_3F_6$	1.560 (10)	1.521 (10)	120.6 (8)	119.4 (9)	99.3 (6)	Planar D_{3h}	b
$N_3P_3Cl_6$	1.593 (15)	1.975 (10)	119.7 (10)	120.0 (12)	102.0 (4)	Flat chair $C_{3v} \sim D_{3h}$	c
$N_3P_3Cl_6$	1.561 (11)	1.976 (11)	122.3 (7)	117.2 (7)	101.8 (7)	Flat chair $C_{3v} \sim D_{3h}$	d
$N_3P_3Cl_6$	1.585 (10)	2.006 (7)	119.7 (3)	119.7 (3)	—	Flat chair $C_{3v} \sim D_{3h}$	e
$N_3P_3Br_6$	1.583 (38)	2.175 (15)	121.3 (25)	116.6 (20)	102.8 (4)	Flat chair $C_{3v} \sim D_{3h}$	f
$N_3P_3Ph_6$	1.597 (6)	1.804 (7)	122.1 (4)	117.8 (3)	103.8 (3)	Flat chair $C_{3v} \sim D_{3h}$	g
$N_3P_3(O_2C_6H_4)_3$	1.575 (2)	1.594	122.5	117.5	97.0	D_{3h}	h
$N_4P_4F_8$	1.507 (16)	1.515 (15)	147.2 (14)	122.8 (10)	99.9 (9)	Planar D_{4h}	i
$N_4P_4Cl_8(K)$	1.570 (9)	1.989 (4)	131.3 (6)	121.2 (5)	102.8 (2)	Tub S_4	j
$N_4P_4Cl_8(T)$	1.559 (12)	1.989 (4)	135.6 (8)	120.5 (7)	103.1 (2)	Chair C_{2h}	k
$N_4P_4(OMe)_8$	1.57	1.58	132.1	121.0	105	Saddle $S_4 \sim D_{2d}$	l
$N_4P_4(NMe_2)_8$	1.578 (10)	1.678 (10)	133.0 (6)	120.0 (5)	103.8 (5)	Saddle $S_4 \sim D_{2d}$	m
$N_4P_4Me_8$	1.596 (5)	1.805 (8)	132.0 (3)	119.8 (2)	104.1 (2)	Saddle $S_4 \sim D_{2d}$	n
$N_5P_5Cl_{10}$	1.521 (13)	1.961 (8)	148.6 (11)	118.4 (8)	102.0 (4)	Planar C_{2v} (see text)	o
$N_6P_6(NMe_2)_{12}$	1.563 (10)	1.669 (10)	147.5 (7)	120.1 (5)	102.9 (5)	Related to Tub-S_6	p
$N_6P_6(OMe)_{12}$	1.567 (8)	1.584 (6)	134.4 (5)	118.6 (4)	103.3 (3)	Double Tub C_i	q
$N_8P_8(OMe)_{16}$	1.561 (14)	1.576 (13)	136.7 (10)	116.7 (7)	101.3 (7)	Chair C_i (see text)	r

[a] All equivalent bond lengths and angles averaged, even where [as in $N_4P_4Cl_8(T)$ ($\tilde{N}$ angles) and $N_5P_5Cl_{10}$ ($\tilde{N}$ angles)] they differ significantly. To avoid accumulation of errors, lengths and angles are normally given to one more decimal place than the accuracy of the structure justifies. Errors (in units of the last place) are quoted as the standard deviation of an individual bond length or angle, averaged if necessary. They therefore have no strict meaning, but give a general idea of the accuracy of the structure determination; exceptionally, the standard deviation quoted for $N_3P_3Ph_6$ (footnote g) is an RMS value.

[b] M. W. Dougill, *J. Chem. Soc.* p. 3211 (1963).

[c] A. Wilson and D. F. Carroll, *J. Chem. Soc.* p. 2548 (1960).

[d] E. Giglio, *Ric. Sci.* Suppl. **30**, 721 (1960). Molecular parameters recalculated from atomic coordinates. Apparently significant inequalities in ring bond lengths disregarded, in view of later electron diffraction results (footnote e).

[e] M. I. Davis and I. C. Paul, *Acta Cryst.* **22**, 304 (1967).

[f] E. Giglio and R. Puliti, *Acta Cryst.* A**25**, S116 (1969).

[g] F. R. Ahmed, P. Singh, and W. H. Barnes, *Acta Cryst.* A**25**, 316 (1969).

[h] L. A. Siegel and J. H. Van den Hende, *J. Chem. Soc.*, A p. 817 (1967).

[i] H. McD. McGeachin and F. R. Tromans, *J. Chem. Soc.* p. 4777 (1961).

[j] R. Hazekamp, T. Migchelsen, and A. Vos, *Acta Cryst.* **15**, 539 (1962).

[k] A. J. Wagner and A. Vos, *Acta Cryst.* B**24**, 707 (1968).

[l] G. B. Ansell and G. J. Bullen, *Chem. Commun.*, p. 430 (1966); personal communication (1969).

[m] G. J. Bullen, *J. Chem. Soc.* p. 3193 (1962).

[n] M. W. Dougill, *J. Chem. Soc.* p. 5471 (1961).

[o] A. W. Schlueter and R. A. Jacobson, *J. Chem. Soc.*, A p. 2317 (1968).

[p] A. J. Wagner and A. Vos, *Acta Cryst.* B**24**, 1423 (1968).

[q] M. W. Dougill, unpublished work (1969).

[r] N. L. Paddock, J. Trotter, and S. H. Whitlow, *J. Chem. Soc.*, A p. 2227 (1968).

framework from planarity insignificant, and the comparability with benzene obvious. For most of the other compounds, the observed conformations approach symmetries which are high enough (C_{3v}, D_{2d}) to imply equal bond lengths, and this can perhaps be taken as an indication of the tendency toward planarity in cases where the natural bond angles prevent its attainment. In $N_5P_5Cl_{10}$ (Fig. 16) the PN framework is planar within 0.1 Å, in spite of the low symmetry. All these results are compatible with the occurrence of out-of-plane π_a-bonding. The structures of some other nonplanar derivatives present a problem, in that the ring bond lengths are equal, even though not required to be so by symmetry, and a consideration of their structures in terms of π_a-bonding is now found to be inadequate.

Many of the relevant structures are of eight-membered rings, and the conformations which have been found so far are illustrated in Fig. 17. The pure "saddle" form, in which the phosphorus atoms are coplanar, has been observed so far in only one molecule, $N_4P_4Me_4F_4$, which has a twofold axis through

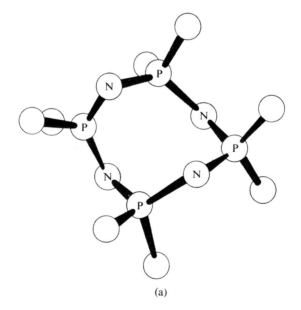

(a)

FIG. 17. Conformations of eight-membered rings: (a) saddle; (b) tub; (c) chair.

opposite phosphorus atoms.[93] The "saddle" can be deformed continuously into the "tub" form without change in ring angles, and the molecules

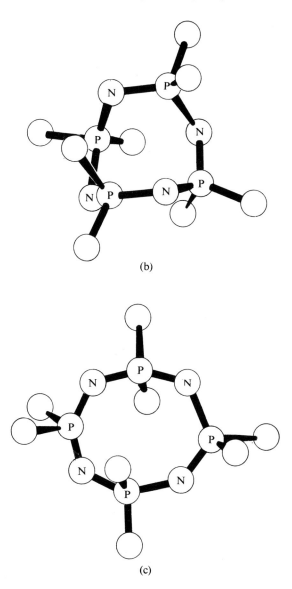

(b)

(c)

[93] W. C. Marsh, T. N. Ranganathan, J. Trotter, and N. L. Paddock, *Chem. Commun.* p. 815 (1970).

$N_4P_4(OMe)_8$,[94] (close to "saddle") $N_4P_4(NMe_2)_8$,[79,95] $N_4P_4Me_8$,[96] and $N_4P_4Cl_8(K)$[97] (close to "tub") fall in this conformational series. The chair form is represented by another polymorphic form of $N_4P_4Cl_8(T)$[98] and by two nongeminally substituted derivatives $N_4P_4Ph_4Cl_4$ and $N_4P_4Ph_4(NHMe)_4$.[99]

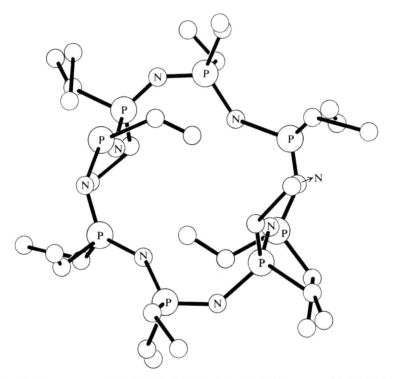

FIG. 18. The structure of $N_8P_8(OMe)_{16}$ [after N. L. Paddock, J. Trotter, and S. H. Whitlow, *J. Chem. Soc., A* p. 2227 (1968), by permission].

Another saddle form is possible, in which the nitrogen, rather than the phosphorus, atoms lie in a plane (as in S_4N_4), and a second chair form, in which, in the limiting symmetry C_{2h}, the twofold axis passes through two nitrogen atoms. In both of these, and in the "crown" conformation similar to that of e.g., S_8 and $S_4(NH)_4$, repulsive steric interactions outweigh any possible gain in

[94] See Ansell and Bullen[81]; personal communication (1969).
[95] N. L. Paddock, unpublished work (1968).
[96] M. W. Dougill, *J. Chem. Soc.* p. 5471 (1961).
[97] R. Hazekamp, T. Migchelsen, and A. Vos, *Acta Cryst*, **15**, 539 (1962).
[98] A. J. Wagner and A. Vos, *Acta Cryst.* **B24**, 707 (1968).
[99] G. J. Bullen, P. R. Mallinson, and A. H. Burr, *Chem. Commun.* p. 691 (1969).

π-electron stability, and these conformations are not found (but see Reference 99a).

It is to be expected that nonbonded interactions will have an important bearing on the detailed shapes of even the sterically acceptable conformations, and, if the symmetry is lowered in consequence, π_a-interactions may no longer be equal. The tetrameric chloride structures provide an illustration. The angles between the normals to successive PNP, NPN planes (the axes of successive atomic π-orbitals) are 15.6° and 57.3° in the K form (S_4) and 45.4°, 75.9°, 59.4° and 19.9° in the T form (C_i). Similarly, in $N_8P_8(OMe)_{16}$ (Fig. 18), in which compound, also, the ring bonds are all equal within close limits, the dihedral angles vary over the range 16°–79°.

The overlaps of the two $d\pi_a$ orbitals at phosphorus, d_{xz}, and d_{yz}, are proportional to $\cos\tau$, the dihedral angle, and are hence similarly reduced by out-of-plane deformation. So are those of their linear combinations $1/\sqrt{2}(d_{xz} \pm d_{yz})$, as appropriate for the formation of PNP three-center bonds. Neither for limited "island" delocalization nor for full homomorphic or heteromorphic delocalization is equality of bond lengths to be expected from π_a-interactions alone, unless the molecular symmetry is high, e.g., D_{2d} for an eight-membered ring; C_4 or S_4 is not enough.

Equality of bond lengths and inequality of dihedral angles are compatible if π_s-interactions are included, since $d\pi_s–p_z\pi$ overlaps are proportional to $\sin\tau$, and therefore tend to compensate the decreased overlap of the π_a-orbitals, and the observed structures have been interpreted in these terms[2a,100] (though the qualitative arguments need refinement). The problem of determining the types of π-electronic interaction therefore has two stages: (1) the estimation of the relative importance of the π_a and π_s systems; and (2) the assessment of the relative contributions of the individual atomic orbitals within the π_a, π_s classification. In Section IV,A,3b bond length changes are divided, on a structural basis, into estimated σ and (total) π components. The questions of symmetry type and individual orbital contributions are discussed in Section IV,B in connection with delocalization.

b. σ and π Contributions to Bond Lengths. It can be seen from Table VIII that the ring bonds in phosphonitrilic derivatives are, on the average, about 0.2 Å shorter than the "standard" single bond in $(NH_3 \cdot PO_3)^-$. This contraction is greater than that from the C—C single bond length of 1.54 Å to the 1.39 Å characteristic of benzene, but in neither case can the comparison be taken at its face value, because changes can be expected in the σ-bonds as well as the π-bonds. If, as seems likely on theoretical grounds (Section II) the σ framework involves mainly s- and p-orbitals, the d-orbitals contributing mainly to the

[99a] G. J. Bullen and P. A. Tucker, *Chem. Commun.* p. 1185 (1970).
[100] D. P. Craig and N. L. Paddock, *J. Chem. Soc.* p. 4118 (1962).

π-bonds, then we can expect ligand electronegativity to affect mainly the π-bonds.

Some support for its smaller effect on σ-bonds is given by the comparative structures of PF_5, $MePF_4$, and Me_2PF_3 (Fig. 6); the greater change in P—F bond length (0.06 Å) takes place in the axial direction, in which d-orbital interactions are likely to be concentrated. The effect on the length of an equatorial P—F bond of changing its partners from PF_2 to PMe_2 is smaller (0.02 Å), and this expansion can be regarded as the extreme likely in phosphonitrilic compounds. In view of the variability of the angles at nitrogen, and, to a smaller extent, at phosphorus, somewhat larger changes in bond length are to be expected from changes in σ-hybridization, corresponding to the change in the C—C bond length from 1.54 Å to 1.483 Å in the central sp^2–sp^2 σ-bond in 1,4-butadiene.[101]

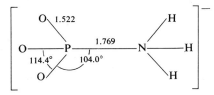

FIG. 19. The structure of the phosphoramidate ion [E. Hobbs, D. E. C. Corbridge, and B. Raistrick, *Acta Cryst.* **6**, 621 (1953); D. W. J. Cruickshank, *ibid.* **17**, 671 (1964)]. The esd's of the P—N, P—O bond lengths are 0.019 and 0.011 Å, respectively.

The structure of the phosphoramidate ion is shown in Fig. 19, and its geometry reflects both these and other factors. Since it is a zwitterion, there is no contribution to bonding from delocalization of an unshared pair of electrons. This is an important point, as can be seen by comparing the structures of sulfamic acid,[102] $H_3N \cdot SO_3$, also a zwitterion, in which S—N = 1.76 Å, with that of the sulfamate ion[103] $(SO_3 \cdot NH_2)^-$, in which S—N = 1.70 Å. The net negative charge on the phosphoramidate ion would lengthen the P—N bond slightly, but, although the P—N stretching force constant is not accurately known, probably by not more than 0.02 Å, which would be reduced by interaction with the cation lattice. We ignore the resultant change, and take 1.770 Å, close to the Schomaker–Stevenson estimate[104] of 1.76 Å, to be the single P—N bond length appropriate to sp^3-hybridization at each atom, and to the particular ligand electronegativities. The study of the structures of other reference compounds, especially phosphoramidic acid itself, would be valuable in

[101] A. Almenningen, O. Bastiansen, and M. Traetteberg, *Acta Chem. Scand.* **21**, 1221 (1958).

[102] R. L. Sass, *Acta Cryst.* **13**, 320 (1960).

[103] G. A. Jeffrey and H. P. Stadler, *J. Chem. Soc.* p. 1467 (1951).

[104] V. Schomaker and D. P. Stevenson, *J. Am. Chem. Soc.* **63**, 37 (1941).

estimating the electronegativity effects on bond lengths, so refining the arguments below.

Inspection of Table VIII shows that the ring bond lengths in $N_3P_3F_6$ are significantly shorter than in the other trimeric derivatives, and since the ring angles do not vary greatly among these compounds, the bond length variations can be regarded as resulting from changes in ligand electronegativity. The variations are more marked in the tetrameric series, there being an increase of bond length from 1.51 Å in $N_4P_4F_8$ to 1.60 in $N_4P_4Me_8$, but now the change is accompanied by an appreciable decrease in the ring angle at nitrogen, from 147.2° to 132.0°; both angles are very different from their counterparts in the six-membered rings. It is probable, therefore, that the bond length depends both directly on π-interactions and on changes in σ-hybridization, as indicated by the angular changes. The likely importance of σ-hybridization changes is shown by the structure of $N_5P_5Cl_{10}$ (Fig. 16), there being a correlation between the bond angle at nitrogen and the mean length of the two bonds from it. Increasing angle, which would decrease the p-character of the σ-bonds, is associated with decreasing bond length, as shown in the following tabulation:

PNP angle:	133.6°	143.5°	149.8°	157.2°	159.0°
Mean bond length (Å): (to nearest 0.005 Å)	1.545	1.525	1.515	1.510	1.510

We shall attempt to assess the σ and π effects in phosphonitrilic derivatives on a structural basis. Other phenomena, especially the coupling of the nuclear spins of phosphorus atoms joined through a nitrogen atom in cyclic[105] and (in part, at least) in acyclic compounds,[106] have suggested the importance of interaction through the σ framework, and with the refinement of NMR theory, an independent check in the relative magnitudes of σ-hybridization changes and π-electron effects can be expected.

In carbon compounds, bond angles are often close to 109.5° or 120°, and change little in a series of chemically related molecules. When second-row elements are involved, the constancy is lost, and the range of variation increased, although the primary stereochemistry is still determined by the σ-bond structure, which is modified in detail by the π-interactions. Thus, if only s- and p-orbitals are involved, exocyclic and endocyclic angles at phosphorus should vary in opposite senses, the relationship between the angles being expressed by $\cot^2(N\hat{P}N/2) + \cot^2(X\hat{P}X/2) = 1$. If $N\hat{P}N = 120°$, $X\hat{P}X$, on this basis, should be 101.5°, a pair of angles typical of many phosphonitrilic derivatives. For five such compounds, the above angular function is: $N_4P_4Me_8$ (0.94); $N_4P_4(NMe_2)_8$, (0.95); $N_4P_4Cl_8$ (0.96); $N_4P_4(OMe)_8$ (0.91); $N_4P_4F_8$ (1.00)

[105] E. G. Finer, J. Mol. Spectry. 23, 104 (1967).
[106] J. F. Nixon, J. Chem. Soc., A p. 1087 (1969).

The deviations from unity are significant, and in part reflect conjugative interactions with the ligands, but the figures suggest strongly that the effect of π-bonding within the ring is to widen the angle at phosphorus from $109.5°$ to $\sim 120°$ as a direct effect, and redistribute the p-orbitals accordingly. The effect of $p\pi$–$d\pi$ bonding is thus to modify an existing σ structure, rather than to produce a new qualitative arrangement of its own, and the consequent changes in bond length are the sum of two parts, (1) the direct contraction resulting from π-bonding, and (2) the indirect effect on σ-hybridization. Both are the result of d-orbital interactions, and an approximate separation of them can be made as follows.

According to Dewar and Schmeising[107] the length of a single C—C bond is linearly related to its fractional s-character. Since we are concerned with two different atoms, we have made the following similar but arbitrary assumptions. (1) The hybridization and the direct π effects on the P—N bond lengths are additive. (2) The standard single P—N bond length is the sum of the effective radii of P (1.069 Å) and N (0.701 Å). (3) These radii are changed by the same fraction as a C—C bond would be on changing the angle at either P or N, the fractions being taken to be those found by Dewar and Schmeising.[107] (4) The modified radii are added to give a new calculated bond length, its difference from 1.77 Å being attributed to changes in σ-hybridization at phosphorus and nitrogen. (5) The difference between the calculated length and the observed length is regarded as a direct π effect.

The results of the analysis of the averaged dimensions of the homogeneously substituted phosphonitriles are shown in Table IX. In so far as the assumptions are valid, the following conclusions are possible. (1) Changes in σ-hybridization, assumed to involve s- and p-orbitals only, account for about 40% of the total contraction. (2) The hybridization and direct π-contractions run roughly parallel. (3) The π-contraction, especially, is not only large, but shows a significant increase in the series $N_4P_4Me_8 < N_4P_4Cl_8 = N_4P_4(OMe)_8 < N_4P_4F_8$, and the same type of variation is found in the shorter trimeric series. To the extent that the assumptions are justified, the direct strengthening effect of an electronegative ligand can be regarded as established. From previous argument, this is regarded mainly as a π effect, but a more extensive range of reference compounds may allow later detection of a σ component, which would diminish the effect of ligand electronegativity on the π-contraction. (4) There is no convincing evidence of an effect of ring size on the π-contraction. The significance of the especially short bonds in $N_4P_4F_8$ compared with those in $N_3P_3F_6$ is much reduced after allowance for hybridization changes, and the apparent slight strengthening in the ring bonds from $N_4P_4(NMe_2)_8$ to $N_6P_6(NMe_2)_{12}$ and in the series $N_4P_4(OMe)_{12}$, $N_8P_8(OMe)_{16}$ cannot be attributed, on the

[107] M. J. S. Dewar and H. N. Schmeising, *Tetrahedron* **11**, 96 (1960).

TABLE IX

DIVISION OF BOND LENGTH CHANGES (Å) into σ and π EFFECTS[a]

	$N_3P_3F_6$	$N_3P_3Cl_6$	$N_3P_3Br_6$	$N_3P_3Ph_6$
Total Contraction	0.21	0.19	0.18	0.17
σ-Hybridization	0.06	0.06	0.05	0.06
π-Bonding	0.15	0.13	0.13	0.11

	$N_4P_4F_8$	$N_4P_4Cl_8(K)$	$N_4P_4Cl_8(T)$	$N_4P_4Me_8$	$N_4P_4(OMe)_8$
Total Contraction	0.26	0.20	0.21	0.17	0.19
σ-Hybridization	0.10	0.08	0.09	0.08	0.09
π-Bonding	0.16	0.12	0.12	0.09	0.10

	$N_4P_4(NMe_2)_8$	$N_5P_5Cl_{10}$	$N_6P_6(NMe_2)_{12}$	$N_6P_6(OMe)_{12}$	$N_8P_8(OMe)_{16}$
Total Contraction	0.19	0.25	0.21	0.20	0.21
σ-Hybridization	0.08	0.10	0.10	0.08	0.08
π-Bonding	0.11	0.15	0.11	0.12	0.13

[a] See text.

basis we have discussed, to a direct π-electron effect. On the other hand, although there is no change in π-contraction from $N_3P_3Cl_6$ to $N_4P_4Cl_8$, the shortness of the bonds in $N_5P_5Cl_{10}$ cannot wholly be attributed to hybridization changes; the large π-contraction which remains after correction for them seems genuine. Crystallographic study of other pentameric and of, especially, heptameric derivatives is very desirable.

B. DELOCALIZATION

Evidence on the extent of delocalization of the π-bonds is now presented. The analysis is based on the qualitative results of mass spectrometry, bond length inequalities in neutral and in protonated molecules, and ionization potentials in relation to both the π_a and the π_s systems. The conclusions are applied to the interpretation of chemical reactivity and other properties.

1. Mass Spectra

Our discussion should be read in the context of the quasiequilibrium theory,[108] which asserts that the molecular ion produced by electron impact dissociates by a series of competing and consecutive unimolecular decompositions, of which the rate constants are calculable, at least in principle, by absolute reaction rate theory. Accordingly, the most abundant fragments in the mass spectrum are expected to be those with a high chemical stability. In organic systems there are confirmatory findings.[109] As would be expected from such general considerations, the mass spectra of phosphonitrilic derivatives and those of organic compounds have some common features, such as the high relative abundance of even-electron ions. In aliphatic compounds, the parent molecular ion P^+ usually appears in low abundance; 3.2% in n-hexane,[110] decreasing to 0.3% in n-octacosane $C_{28}H_{58}$.[111] Cyclization improves stability, but even so the yield of P^+ decreases from 17.2% in cyclohexane[112] to 5.4% in cyclooctane[113]; in all cases skeletal fragmentation is the main process. Fluorine

[108] A. L. Wahrhaftig, in "Mass Spectrometry" (R. I. Reed, ed.), p. 137. Academic Press, New York, 1965.

[109] F. W. McLafferty, in "Mass Spectrometry of Organic Ions" (F. W. McLafferty, ed.), p. 309. Academic Press, New York, 1963.

[110] "Mass Spectral Data," Res. Proj. No. 44, Spectrum No. 147. Am. Petrol. Inst., Pittsburgh.

[111] "Mass Spectral Data," Res. Proj. No. 44, Spectrum No. 886, Am. Petrol. Inst., Pittsburgh.

[112] "Mass Spectral Data," Res. Proj. No. 44, Spectrum Nos. 1589 and 1605. Am. Petrol. Inst., Pittsburgh.

[113] "Mass Spectral Data," Res. Proj. No. 44, Spectrum No. 842. Am. Petrol. Inst., Pittsburgh.

has a destabilizing effect, C_6 fragments constituting 0.6% and 1.2% of the mass spectra of C_6F_{14} and cyclo-C_6F_{12}, respectively.[114]

Molecular features which allow the charge to divide over two or more centers, such as unsaturated centers or conjugated carbonyl groups, stabilize the molecule as a whole[115] and, especially, the (odd-electron) parent ion. This is particularly marked in aromatic molecules, in which most of the delocalization energy is retained in the positive ion, and the loss of bond energy is divided among all the ring bonds. For the same reasons, although doubly charged ions are rare in the mass spectra of aliphatic compounds (cyclohexane, cyclooctane 0.06, 0.03 %) they are common in those of aromatics, especially those containing condensed rings. Examples are given in Table X. In all cases P^+ is the most

TABLE X

Mass Spectral Data for Aromatic Hydrocarbons[a]

Compound	Benzene[b]	Naphthalene[c]	Pyrene[d]	Cyclooctatetraene[e]
$P^+(\%)$	42.6	44.1	45.2	18.6
$P^+/(P—H)^+$	6.9	10.1	8.8	1.6
Doubly charged ions (%)	0.8	6.3	6.3	0.9

[a] Data from "Mass Spectral Data," Res. Proj. No. 44, Spectrum Nos. as follows: [b] 175; [c] 410; [d] 599; [e] 690. Am. Petrol. Inst., Pittsburgh.

abundant ion, and the high stability is evidently increased in the condensed ring hydrocarbons, though lowered in cyclooctatetraene, in which delocalization is limited both for geometrical reasons and because the π system does not conform to the Hückel $4n + 2$ rule. Again, fluorine substitution has a destabilizing effect, $C_6F_6^+$ constituting 33% of the total intensity in the mass spectrum of C_6F_6.[116] In borazines, containing a delocalized system based on the B_3N_3 ring, the parent molecular ion is less abundant ($B_3N_3H_6^+$, 10.2%),[117] but it is still a major component of the spectrum. High yields of doubly charged ions are also found.[118]

[114] "Mass Spectral Data," Res. Proj. No. 44, Spectrum Nos. 201 and 735. Am. Petrol. Inst., Pittsburgh; J. R. Majer, *Advan. Fluorine Chem.* **2**, 55 (1961).

[115] F. W. McLafferty, *in* "Mass Spectrometry of Organic Ions" (F. W. McLafferty, ed.), p. 313. Academic Press, New York, 1963.

[116] V. H. Dibeler, R. M. Reese, and F. L. Mohler, *J. Chem. Phys.* **26**, 304 (1957).

[117] A. Cornu and R. Massot, "Compilation of Mass Spectral Data." Heyden, London, 1966.

[118] W. Snedden, *Advan. Mass Spectry.* **2**, 456 (1963).

The mass spectra of a series of phosphonitrilic bromides,[119] chlorides,[120,121] and fluorides[122] have been determined, and are comparable to those of organic compounds, in that (1) the fragment species are of types familiar in normal phosphonitrilic chemistry, and (2) their detailed nature and distribution suggests a substantial degree of electronic delocalization. Phosphonitrilic chlorides and bromides are normally prepared by ammonolysis of the phosphorus pentahalides, and in both reactions the presence of linear ionic intermediates containing cations of the type $(N_{n-1}P_nX_{2n+2})^+$ ($X = Cl$, Br) has been recognized.[123,124] Similarly, the mass spectra show not only the cyclic fragments $N_nP_nX_{2n-x}^{m+}$ ($X = Br$, Cl, F) (of ring size up to and including that of the parent), but also the linear series $N_{n-1}P_nX_{2n+2-x}^{m+}$. Again the even-electron species (with some important exceptions discussed below) are more abundant than their odd-electron counterparts.

In other respects there are large differences between carbocyclic and phosphonitrilic compounds. Some numerical information is given in Table XI. An especially important point of difference is the great stability of the phosphonitrilic ring. In the chlorides and the bromides this can be seen in the high yields of the parent series. The bromides $N_3P_3Br_6$ and $N_4P_4Br_8$ give rise to the complete series $N_3P_3Br_x^+$ ($x = 0$–6) and $N_4P_4Br_x^+$ ($x = 0$–8) and to most of the corresponding doubly charged ions. These high yields in part reflect the low dissociation energy of the P—Br bond (~ 63 kcal)[125]; the parent ion itself is scarce, the most abundant ion in the spectra of $(NPBr_2)_{3-5}$ being $(Parent-Br)^+$. Stability is not confined to the cyclic derivatives of the parent series; $(NPBr_2)_6$ breaks down into trimeric ions, the most abundant ion in the spectrum being $N_3P_3Br_5^+$. All the bromides give a high yield of doubly charged ions.

The spectra of the fluorides provide more extensive information. Since $E(P—F)$ is high (117 kcal in PF_3)[126] and $E(P—N)$ is lower (~ 73 kcal),[48] breakage of the ring bonds, in the smaller phosphonitrilic fluorides at least, tends to occur more readily than in the bromides or chlorides. Nevertheless the ring stability is much greater than that of the cyclofluoroalkanes, even though there is here a smaller disparity in bond energies [$E(C—F) = 116$ kcal, $E(C—C) = 83$ kcal].[127] Also, as the ring size is increased, both the total yield of cyclic fragments and the yield of the parent series steadily increase, to an extent which is not matched by any other series of compounds. At the same time,

[119] G. E. Coxon, T. F. Palmer, and D. B. Sowerby, *J. Chem. Soc.*, A p. 1568 (1967).
[120] C. E. Brion and N. L. Paddock, *J. Chem. Soc.*, A p. 388 (1968).
[121] C. D. Schmulbach, A. G. Cook, and V. R. Miller, *Inorg. Chem.* 7, 3463 (1968).
[122] C. E. Brion and N. L. Paddock, *J. Chem. Soc.*, A p. 392 (1968).
[123] M. Becke-Goehring and W. Lehr, *Ber.* 94, 1591 (1961).
[124] G. E. Coxon, D. B. Sowerby, and G. C. Tranter, *J. Chem. Soc.* p. 5697 (1965).
[125] T. Charnley and H. A. Skinner, *J. Chem. Soc.* p. 450 (1953).
[126] E. Neale and L. T. D. Williams, *J. Chem. Soc.* p. 2485 (1955).
[127] T. L. Cottrell, "The Strengths of Chemical Bonds." Butterworth, London and Washington, D.C., 1958.

TABLE XI

MASS SPECTRAL DATA FOR CYCLIC PHOSPHONITRILIC DERIVATIVES $(NPX_2)_n$

$n =$	3	4	5	6^b	7	8
$X = Br^a$:						
Total cyclic fragments, %	78.2	80.2	87.7	90.3	—	—
Total cyclic fragments in parent series $N_nP_nBr_{2n-x}^{m+}$, %	78.2	56.6	50.4	30.5	—	—
Parent ion P^+, %	1.1	1.0	1.2	0.3	—	—
$P^+/(P—Br)^+$	0.024	0.027	0.040	—	—	—
Doubly charged ions, %	12.9	19.3	33.9	33.1	—	—
$X = Cl^c$:						
Total cyclic fragments, %	78.4	72.4	82.9	78.0	85.6	92.4
Total cyclic fragments in parent series $N_nP_nCl_{2n-x}^{m+}$, %	78.4	65.8	59.0	12.0	7.1	25.4
Parent ion P^+, %	16.5	13.2	10.3	3.0	5.3	22.9
$P^+/(P—Cl)^+$	0.30	0.29	0.24	0.87	6.7	56.8
Doubly charged ions, %	3.8	8.2	9.8	17.7	15.2	8.5
$X = F^d$:						
Total cyclic fragments, %	29.4	32.9	80.8	66.2	71.3	88.7
Total cyclic fragments in parent series $N_nP_nF_{2n-x}^{m+}$, %	29.4	18.5	72.1	23.5	33.3	64.8
Parent ion P^+, %	14.4	7.6	54.8	19.2	30.2	64.5
$P^+/(P—F)^+$	1.1	1.2	3.6	4.4	29.4	191
Doubly charged ions, %	2.4	4.5	2.1	0.2	2.4	1.2

$X = F^d: n =$	9	10	11	12	13	14
Total cyclic fragments, %	88.5	89.8	90.7	92.3	91.2	92.6
Total cyclic fragments in parent series $N_nP_nF_{2n-x}^{m+}$, %	67.7	71.9	76.1	68.0	72.0	75.6
Parent ion P^+, %	67.0	71.8	74.8	67.9	69.4	70.7^e
$P^+/(P—F)^+$	3500	2300	1375	1100	875	850
Doubly charged ions, %	0.9	0.5	2.0	0.5	2.6	4.9

[a] G. E. Coxon, T. F. Palmer, and D. B. Sowerby, *J. Chem. Soc.*, A, p. 1568 (1967).
[b] Not directly comparable with results for $(NPBr_2)_{3-5}$; see footnote a.
[c] C. E. Brion and N. L. Paddock, *J. Chem. Soc.*, A, p. 388 (1968).
[d] C. E. Brion and N. L. Paddock, *J. Chem. Soc.*, A, p. 392 (1968).
[e] Difference from 75.6% due to doubly charged parent ion.

the dominance of the parent ion itself increases, in sharp contrast to the behavior of the aliphatic series, whether linear or cyclic, but similar to that of benzene and its homologs. The stability of the larger ring phosphonitrilic fluorides is manifestly greater than that of even the condensed-ring aromatic hydrocarbons.

The stability of phosphonitrilic ions, and the points of similarity in fragmentation behavior to benzenoid aromatics, give a compelling indication of delocalization. The effects, however, are not directly comparable to those in benzene, particularly on account of the occurrence of the π_s system in addition to the π_a, both in providing an additional mechanism for spreading the charge, and making the π overlap, and consequently the charge distribution, independent of conformation. Similarly, the high intensity of $C_4H_2O_2{}^{2+}$ in the spectrum of maleic anhydride is probably attributable[128] to the structure

$$^+O\equiv C-CH=CH-C\equiv O^+$$

and the otherwise surprising occurrence of the abundant doubly charged ions both here and in the spectra of linear methylsiloxanes[129] may be due to π-bonding arising from lone-pair delocalization.

Finally, we have argued throughout from the properties of a cation to those of the corresponding neutral species, by analogy with carbon compounds. Since the bonding properties of d-orbitals depend much more on ligand electronegativity than do those of the underlying s- and p-orbitals, a larger change in bonding on ionization is expected when the bonds involve d-orbitals. In other words, the delocalization in the ground state of a phosphonitrilic molecule would be less, relative to a carbocyclic aromatic molecule, than might appear from a comparison of their mass spectra. Nevertheless, the unprecedentedly high yields of cyclic ions, especially the parent ions, in the phosphonitrilic fluoride spectra, the high proportions of multiply charged ions, and the general nature of the fragmentation process, all suggest that cyclic delocalization plays an important part in determining the fragmentation pattern, and has an important influence on ground state properties.

2. Bond Length Inequalities

In homogeneously substituted phosphonitriles, the ring bond lengths in a particular molecule are all closely equal (Section IV,A). Inequalities in bond length are found either (a) when the substituents on one phosphorus atom are different from those on the others, or (b) when one nitrogen atom in the ring is protonated. We can expect two general results from a change in ligand

[128] J. H. Beynon, "Mass Spectrometry and its Applications to Organic Chemistry," p. 283. Elsevier, Amsterdam, 1960.

[129] V. H. Dibeler, F. L. Mohler, and R. M. Reese, *J. Chem. Phys.* **21**, 180 (1953).

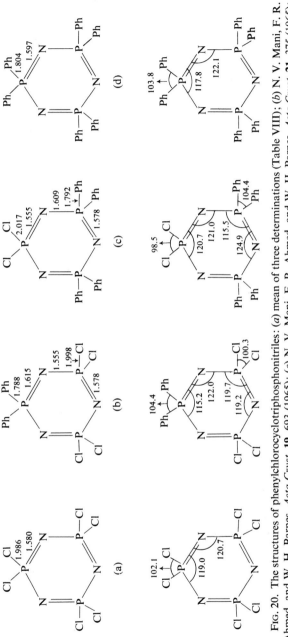

Fig. 20. The structures of phenylchlorocyclotriphosphonitriles: (a) mean of three determinations (Table VIII); (b) N. V. Mani, F. R. Ahmed, and W. H. Barnes, *Acta Cryst.* **19**, 693 (1965); (c) N. V. Mani, F. R. Ahmed, and W. H. Barnes, *Acta Cryst.* **21**, 375 (1965); (d) F. R. Ahmed, P. Singh, and W. H. Barnes, *Acta Cryst.* **B25**, 316 (1969). [Averaged dimensions (b–d).]

electronegativity at one center. As a consequence of the change in orbital size, and the resulting changes in π-bond (and to a smaller extent, σ-bond) strengths, the mean ring bond length will change, and we can expect the greatest change to occur in the two bonds which meet at the perturbed atom. Additionally, bond length inequalities can result from partial localization.

The structures of a series of phenylchlorotriphosphonitriles[130] and of 1,1-diphenylphosphonitrilic fluoride trimer[131] give evidence that the mean bond lengths in the series $N_3P_3Cl_6$, $N_3P_3Cl_4Ph_2$, $N_3P_3Cl_2Ph_4$ tend (Fig. 20) to increase with decreasing ligand electronegativity. The values are P—N, $(a-d)$ 1.580 (mean of three determinations), 1.584, 1.582, 1.597 Å, and 1.572 Å in

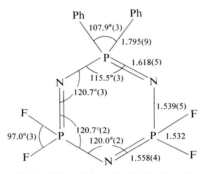

FIG. 21. The structure of 1,1-diphenylphosphonitrilic fluoride trimer. C. W. Allen, J. B. Faught, T. Moeller, and I. C. Paul, *Inorg. Chem.* **8**, 1719 (1969). (Averaged dimensions.)

$N_3P_3F_4Ph_2$; P—Cl, $(a-c)$, 1.986, 1.998, 2.017 Å; P—C $(b-d)$, 1.788, 1.792, 1.804 Å; 1.795 Å in $N_3P_3F_4Ph_2$.

Because all the bonds meeting at a given atom vary together in the same sense, changes in σ-hybridization are not now of major importance. The bond length inequalities shown in Fig. 20 and in Fig. 21 are therefore attributable to changes in the σ- and π-bonds. The importance of the latter is shown by the alternation of the bond lengths; for a pure σ-inductive effect a steady change away from the perturbed atom would be expected.

All three inhomogeneously substituted molecules show the same type of variation, in that only the four ring bonds nearest to the perturbed atom are affected, the remote pair being close to the average length. The bonds to the more electronegative phosphorus atom are shortened and the adjacent pair is lengthened, the difference (0.06 Å in $N_3P_3Cl_4Ph_2$) being significantly greater than that expected from a comparison of the structures of $N_3P_3Cl_6$ and $N_3P_3Ph_6$ ($\sim$0.02 Å).

[130] N. V. Mani, F. R. Ahmed, and W. H. Barnes, *Acta Cryst.* **19**, 693 (1965); **21**, 375 (1966); F. R. Ahmed, P. Singh, and W. H. Barnes, *ibid.* **B25**, 316 (1969).

[131] C. W. Allen, J. B. Faught, T. Moeller, and I. C. Paul, *Inorg. Chem.* **8**, 1719 (1969).

Similar variations in $N_3P_3F_4Ph_2$ have been interpreted in terms of greater donation of lone-pair electrons to the bond to the more electronegative phosphorus atom,[131] and also (but including both π_a and π_s contributions) for the unequal bond lengths in $NPCl_2(NSOCl)_2$.[132] In Table XII, allowance is made for the electronegativity effect by comparing individual bonds lengths to

TABLE XII

COMPARISON OF BOND LENGTH INEQUALITIES IN $N_3P_3X_4Ph_2$ MOLECULES WITH MODEL CALCULATIONS[a]

$N_3P_3Cl_4Ph_2$: Successive bonds			
Lengths from Ph_2P (Å)[b]	1.615	1.555	1.578
Difference from lengths in parent compound ($N_3P_3Cl_6$,[c] $N_3P_3Ph_6$[d])	+0.020	−0.025	−0.002
$N_3P_3F_4Ph_2$: Successive bonds			
Lengths from Ph_2P (Å)[e]	1.618	1.539	1.558
Difference from lengths in parent compound ($N_3P_3F_6$,[f] $N_3P_3Ph_6$[d])	+0.023	−0.021	−0.002
Homomorphic delocalization			
Deviation of bond order from mean	−0.036	+0.025	+0.011
Heteromorphic delocalization			
Deviation of bond order from mean	−0.038	+0.026	+0.012

[a] K. A. R. Mitchell, unpublished data (1969). Hückel calculations, $\alpha_N = \alpha_P + 2\beta$, phenyl substitution simulated by perturbation $\alpha'_P = \alpha_P - 0.5\beta$. Bond orders in unperturbed molecules 0.507 (homomorphic), 0.480 (heteromorphic); mean bond orders in perturbed molecules 0.491 (homomorphic), 0.469 (heteromorphic).

[b] N. V. Mani, F. R. Ahmed, and W. H. Barnes, *Acta Cryst.* **19**, 693 (1965).

[c] P—N = 1.580 Å (mean of three values in Table VIII).

[d] P—N = 1.597 Å, F. R. Ahmed, P. Singh, and W. H. Barnes, *Acta Cryst.* **B25**, 316 (1969).

[e] C. W. Allen, J. B. Faught, T. Moeller, and I. C. Paul, *Inorg. Chem.* **8**, 1719 (1969).

[f] P—N = 1.560 Å, M. W. Dougill, *J. Chem. Soc.* p. 3211 (1963).

the values they would have in a homogeneously substituted derivative carrying the same substituents. The range of variation is somewhat reduced by this adjustment, but is almost identical in the two compounds. The type of variation is also paralleled by the calculations, the first bond being most affected, the second nearly as much, but in the opposite direction. Further, the difference between the calculated bond orders of the first and second bonds (homomorphic, 0.061, heteromorphic 0.064) is much greater than the reduction in mean bond order as a result of the perturbation (0.016, 0.011); this, too, is a

[132] J. C. van de Grampel and A. Vos, *Acta Cryst.* **B25**, 651 (1969).

feature of the real molecules. The bond length inequalities in $N_3P_3Cl_2Ph_4$, and the corresponding calculations, are similarly related, and we can conclude that the patterns of bond lengths found are a result of π-electron interactions, though they do not allow a distinction to be made between homomorphic and heteromorphic types, or the restricted "island"-type partial delocalization.

Perturbation at nitrogen by protonation produces structural effects comparable to those resulting from a change in ligand electronegativity at phosphorus, as illustrated by the structure[133] of $N_3P_3(NHPr^i)_4Cl_2H^+\cdot Cl^-$ (Fig. 22). The tendency for electrons to accumulate near nitrogen, already present in the

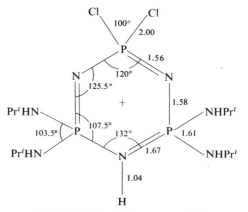

FIG. 22. The structure of $N_3P_3(NHPr^i)_4Cl_2H^+$; averaged dimensions [after N. V. Mani and A. J. Wagner, *Chem. Commun.* p. 658 (1968), by permission].

neutral molecule, is accentuated by protonation, and since the bonds meeting at the protonated nitrogen atom (N^+) have a planar configuration, only the p_z orbital is then used in π-bonding. The bonds to neighboring phosphorus atoms are consequently long, and are succeeded by shorter bonds to the next nitrogens, which are unprotonated and can use both p_z and p_y in bonding. The structure is important for directly showing the importance of the π_s system, and, as is discussed in Section IV,C, in relation to the conjugation of the ring and exocyclic bonds. The bond lengths suggest that the contributions made by π_a- and π_s-bonding in the neutral molecule are comparable.

The structures of eight-membered, homogeneously substituted protonated rings are more informative.[134] Two of them, which occur in the same crystal associated with the $CoCl_4^{2-}$ ion are shown in Fig. 23. In both rings there are four distinct pairs of NP bonds, long and short bonds alternating with increasing distance from N^+, a type of variation characteristic of a cyclic π-system

[133] N. V. Mani and A. J. Wagner, *Chem. Commun.* p. 658 (1968).
[134] J. Trotter, S. H. Whitlow, and N. L. Paddock, *Chem. Commun.* p. 695 (1969).

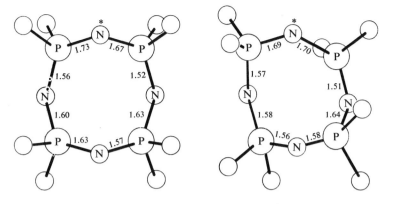

Fɪɢ. 23. The structures of the cations in $(N_4P_4Me_8H^+)_2CoCl_4^{2-}$ [J. Trotter, S. H. Whitlow, and N. L. Paddock, *Chem. Commun.* p. 695 (1969)]. The protonated nitrogen atoms are marked with asterisks.

perturbed at one center. The conformations of the two cations deviate in opposite senses from that of the parent octamethylcyclotetraphosphonitrile, the distinction between "tub" and "saddle" forms becoming much more marked, although, in view of the bond length variations, the molecules have no symmetry. The mean displacements from the mean plane of the sets of phosphorus and nitrogen atoms are shown in the following tabulation:

	$N_4P_4Me_8H^+$ (tub)	$N_4P_4Me_8$	$N_4P_4Me_8H^+$ (saddle)
P (Å)	0.40	0.21	0.02
N (Å)	0.59	0.54	0.61

In all cases, as would be expected from a consideration of steric interactions alone, the phosphorus atoms are the nearer to the plane, but the conformation difference is large, and the coexistence of cations with such different shapes shows that there can be no great energy difference between them. Apart from the conformations, the detailed geometry of the two cations is very similar. Not only is the mean length of corresponding pairs of bonds the same for both conformations, but, within either cation, the two bonds of a pair are also equal in length, though not required to be by symmetry. Protonation again ensures that only the p_z orbital at N^+ takes part in the bonding, and equal interactions between it and the orbitals on the neighboring phosphorus atoms (and hence equal bond lengths to them) are to be expected for the saddle conformation, in which the dihedral angles between the p_z axis and the normals to the neighboring NPN planes are nearly equal (39°, 43°). In the tub cation, the corresponding dihedral angles are much more unequal (15°, 60°), and the

p_z orbital overlaps principally with a π_a orbital on one side and π_s orbital on the other. Equality is no longer required by near-symmetry, so that, although the structure is not highly accurate $[\sigma(\text{P—N}) = 0.03 \text{ Å}]$ the experimental equality of the bonds to N^+ in the "tub" cation shows that the π_a and π_s contributions are of approximately equal importance. A similar conclusion has been reached from a study[135] of the structure of the neutral compound $N_6P_6(NMe_2)_{12}$, in which steric interactions play an important part in determining the conformation. The tendency toward ring planarity noted in Section IV,A evidently depends on a fine balance of steric and π-electron interaction.

The bond length variations are somewhat larger than in the isopropylamino derivative referred to earlier (Fig. 22). The average bond lengths of the two cations are given in Table XIII, where it can be seen that the behavior shown

TABLE XIII

COMPARISON OF BOND LENGTHS OF $N_4P_4Me_8H^+$ IONS WITH CALCULATED BOND ORDERS

Lengths of successive bonds from N^+ (Å)[a]	1.695	1.538	1.614	1.582
Estimated π-contraction (Å)[b]	0.02	0.17	0.08	0.12
Bond order[c]				
Saddle[d]	0.49	1.00	0.88	0.92
Tub[e]	0.51	1.02	0.89	0.95

[a] Each length is the average of the four bonds equidistant from N^+ (Fig. 23).

[b] See Section IV,A,3,b.

[c] K. A. R. Mitchell, unpublished (1969). Hückel calculations, $\hat{N} = 130°$, $\hat{P} = 120°$, equiexponent d_{xz}, $d_{x^2-y^2}$, and p_z, p_y, except p_y omitted at N^+.

[d] Phosphorus atoms coplanar.

[e] Phosphorus, nitrogen atoms equidistant from mean plane. The calculated orders of pairs of bonds equidistant from N^+ are in general unequal, by <0.04, and are averaged.

by the phenylchlorophosphonitriles, in which the bond lengths alternate with increasing distance from the perturbed center, is here continued to a further bond, and we can again attribute the bond length changes to variations in the π-interactions. In the six-membered rings, it was not possible to distinguish between the "island" model and those models involving more extensive delocalization, the expected length of the third bond from the perturbed phosphorus atom being about the same in all cases. In the simple "island" model, the influence of the protonation would be expected to extend no further than the nearest phosphorus atoms. As a second order effect, the increased electronegativity of these atoms might cause the second bond to be short,

135 A. J. Wagner and A. Vos, *Acta Cryst.* **B24**, 1423 (1968).

followed by a further steady increase in the lengths of the third and fourth bonds; in fact, the fourth bond is shorter than the third. The qualitative pattern of bond length variations is similar to that calculated for a π system which includes both π_a- and π_s-interactions, and does not depend greatly on the conformation of the ring (Table XIII).

Quantitatively, the bond length variations seem large whether the observed lengths or the estimated π-contractions are used as a basis for comparison (protonation of pyridine causes no significant change in the lengths of corresponding ring bonds).[136] The difference in behavior may be due in part to the large polarizability of d-orbitals, but, although the calculations are based on an oversimplified model, unrealistic assumptions about orbital exponents would need to be made to bring the calculated bond orders into line with the bond length variations. It seems that perturbation of the delocalized system at nitrogen tends to localize the bonds to an extent greater than simple calculation would suggest, and it is possible that a detailed explanation of the structural results will require consideration of the σ-compression energies also.[41, 42] In confirmation of the general conclusions of this section, there is a close parallelism between the observed inequalities in bond length found in the structure of $N_4P_4F_6Me_2$ and the calculated bond-atom polarizabilities of a delocalized π-system based on an eight-membered ring.[93] The polarizabilities simulate the π-inductive effects of an entering methyl group; the chemical effects of such interactions are considered in Section IV,B,4.

3. *Ionization Potentials*

Measurements of ionization potentials give important information supplementary to the indications of delocalized π-bonding from bond length inequalities (Section IV,B,2). They discriminate more clearly between the symmetry types and between the π_a and π_s systems. The first ionization potentials of a number of phosphonitrilic derivatives are given in Table XIV, together with those of the hydrides of the exocyclic groups, where available. In all cases except that of the phenoxy derivatives, the ionization potentials of the phosphonitriles are less than those of the parent hydrides, so that we may in most cases assume that ionization takes place from the ring atoms, rather than the exocyclic groups.

There is a formal similarity between the phosphonitrilic group $\geqslant P{=}N{-}$ and the carbonyl group $\geqslant C{=}O$, in that both are unsaturated, and both the nitrogen and oxygen atoms carry unshared pairs of electrons in the valence shell; in the former case they are expected to contribute to the π-bonding. The ionization of aliphatic ketones and amides involves the removal of a non-

---

[136] B. Bak, L. Hansen-Nygaard, and J. Rastrup-Andersen, *J. Mol. Spectry.* **2**, 361 (1958); C. Rérat, *Acta Cryst.* **15**, 427 (1962).

bonding π-electron,[137, 137a] and the ionization potentials show the expected substituent effects; comparison of the ionization potentials of formaldehyde (10.86 eV[138]) acetaldehyde (10.25 eV[139]), acetone (9.67 eV[140]), methyl acetate (10.27 eV[141]) and dimethylformamide (9.12 eV[141]), shows that electrons are released decreasingly to the carbonyl group in the order $Me_2N > CH_3 > OMe$. As expected from these results, and from general chemical experience, the ionization potentials of the phosphonitrilic derivatives[142] $(NPX_2)_{3,4}$ also

TABLE XIV

FIRST IONIZATION POTENTIALS OF PHOSPHONITRILIC DERIVATIVES (eV)[a]

X	$N(CH_3)_2$	CH_3	OC_6H_5	OCH_3	OCH_2CF_3
$(NPX_2)_3$[b]	7.85	8.35	8.83	9.29	10.43
$(NPX_2)_4$[b]	7.45	7.99	8.70	8.83	10.01
HX	8.36[d]	12.98[e]	8.50[f]	10.83[d]	—

n	3	4	5	6	7	8
$(NPCl_2)_n$[b,g]	10.26	9.80	9.83	9.81	9.80	—
$(NPF_2)_n$[c,g]	11.4	10.7	11.4	10.9	11.3	10.9

[a] G. R. Branton, C. E. Brion, D. C. Frost, K. A. R. Mitchell, and N. L. Paddock, J. Chem. Soc., A p. 151 (1970).

[b] By electron impact; reproducibility ±0.05 eV.

[c] By photoelectron spectroscopy; estimated uncertainty ±0.01 eV.

[d] M. I. Al-Joboury and D. W. Turner, J. Chem. Soc. p. 4434 (1964).

[e] C. R. Brundle and D. W. Turner, Chem. Commun. p. 314 (1967).

[f] K. Watanabe, T. Nakayama, and J. Mottl, J. Quant. Spect. & Radiative Transfer 2, 369 (1962).

[g] The first ionization potentials of HCl and HF are 12.75 and 16.06 eV, respectively [D. C. Frost, C. A. McDowell, and D. A. Vroom, J. Chem. Phys. 46, 4255 (1967)].

increase correspondingly in the order $X = N(CH_3)_2 < CH_3 < OC_6H_5 < OCH_3 < Cl < OCH_2CF_3 < F$. This order is not precisely what would be

[137] R. S. Mulliken, J. Chem. Phys. 3, 564 (1935).

[137a] H. D. Hunt and W. T. Simpson, J. Am. Chem. Soc. 75, 4540 (1953).

[138] C. R. Brundle and D. W. Turner, Chem. Commun. p. 314 (1967).

[139] H. Hurzeler, M. G. Inghram, and J. D. Morrison, J. Chem. Phys. 28, 76 (1958).

[140] M. I. Al-Joboury and D. W. Turner, J. Chem. Soc. p. 4434 (1964).

[141] K. Watanabe, T. Nakayama, and J. Mottl, J. Quant. Spectry. Radiative Transfer 2, 369 (1962).

[142] G. R. Branton, C. E. Brion, D. C. Frost, K. A. R. Mitchell, and N. L. Paddock, J. Chem. Soc., A p. 151 (1970).

expected for simple inductive interactions, (the electronegativities of nitrogen and chlorine are equal on Pauling's scale), but, in view of the comparability with carbonyl compounds, no specific d-orbital effects are necessarily involved. These become apparent on considering the effect of ring size on ionization potential. For every substituent, the ionization potential decreases from trimer to tetramer, though the difference is barely significant for the phenoxides. The subsequent change with increasing ring size is insignificant for the chlorides, but a pronounced alternation is found for the fluorides (Table XIV).

An interpretation of these results, and of the inner ionization potentials of $(NPF_2)_n$, is possible in terms of the simple theory described earlier, the treatment being expanded so as to include $p\pi$ orbitals on fluorine as well as s-p_y hybrids at nitrogen.[142] Even in carbonyl compounds, the nonbonding oxygen orbital and the bonding π-orbital are energetically close together,[137a] and the difficulty of distinguishing between the corresponding orbitals in phosphonitrilic derivatives is increased by the possibility of the delocalization of the formally unshared electrons on the nitrogen atom into vacant d-orbitals on phosphorus. If the electrons involved were strictly a nonbonding pair, no difference in the ionization potentials of similar trimeric and tetrameric molecules would be expected. Constancy of ionization potential would also be expected if the electrons were ionized from isolated π_a-type PNP "islands," or, to take the extreme case, if all five d-orbitals were involved equally. The primary interest of the ionization potentials of the series of phosphonitrilic fluorides is that their variability excludes ionization either from a localized π_s nonbonding pair or from a π_a "island," and shows also that the d-orbitals are used unequally; the detailed form of the variation defines the symmetry type of the uppermost orbitals.

The type of π system formed by a particular d-orbital depends, as explained in Section III,A,2, on its behavior in the local molecular site group, C_{2v}; the π_a system includes d_{xz}, d_{yz} (heteromorphic and homomorphic respectively), and the π_s systems d_{z^2}, $d_{x^2-y^2}$ (homomorphic), and d_{xy} (heteromorphic). For the simplest assumptions, the arrangement of π-electron levels for the two types of interaction is shown in Fig. 24. To a first approximation, the ionization potential (i.p.) of a heteromorphic π system should be independent of ring size, whereas that of a homomorphic π system should oscillate. A higher approximation would allow for the effect of ionic charge on d-orbital size, and through it on the delocalization energy of the ion. We expect it to stabilize the ions, and thus to diminish all ionization potentials, to a degree that goes down with increasing ring size without alternation.

The observed alternation is therefore good evidence that the uppermost π system is of the homomorphic type, and the conclusion is confirmed by the comparison of the calculated and observed inner levels of $(NPF_2)_n^+$. Simple overlap considerations, reinforced by the inclusion of exchange interactions,[36] suggest that the two orbitals chiefly involved are $d_{x^2-y^2}(\pi_s)$ and $d_{xz}(\pi_a)$, related, but not identical, to the d_y pair which provide the strongest π-bonding in regular tetrahedral molecules,[11] though it is not possible to tell, in advance, which π levels lie deeper.

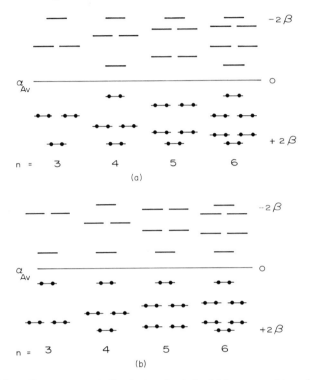

FIG. 24. Schematic arrangement of π-electron levels for (a) homomorphic and (b) heteromorphic interactions. HMO calculations, $\alpha_B = \alpha_A + \beta$.

The number of observed levels which can be attributed to π orbitals is greater than that possible for one π system, but too few for two π systems, so that the two systems evidently overlap. The calculations, which included the interactions of two $2p\pi$ orbitals at each fluorine atom and at each nitrogen atom, were calibrated by assuming an ionization potential of 15.8 eV to correspond to the nonbonding electrons of fluorine (the first i.p. of HF is 16.06 eV[143]), and that the first ionization potentials of both the heteromorphic and homomorphic π systems were 11.4 eV. Since levels closer than 0.5 eV were not resolved, calculated levels closer than this were averaged, and are compared with the observed levels in Fig. 25.

[143] D. C. Frost, C. A. McDowell, and D. A. Vroom, *J. Chem. Phys.* **46**, 4255 (1967).

We can draw the following conclusions. (1) The energy levels fall into two well-defined groups, 10–13 eV and ≥15. 8eV. One intermediate level, which is not reproduced by the calculations, is attributed[142] to ionization from P—N σ-bonds. The energy difference between the σ and π levels seems sufficiently great to justify their separate consideration. (2) In agreement with the conclusions arrived at from the study of bond length inequalities (Section IV,B,2),

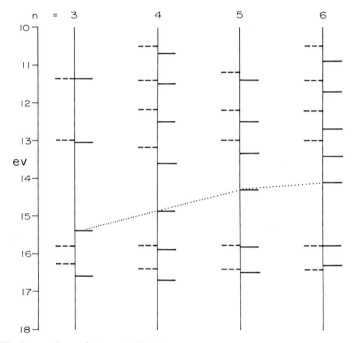

FIG. 25. Comparison of observed (full-lines) and calculated (dashed-line) energy levels of $(NPF_2)_n^+$. For details, see text and G. R. Branton, C. E. Brion, D. C. Frost, K. A. R. Mitchell, and N. L. Paddock, *J. Chem. Soc., A* p. 151 (1970). The dotted line connects levels attributed to ionization from P—N σ-bonds.

the two π systems have similar energies. The higher one is identified as homomorphic, from the behavior of the first i.p. with increasing ring size. The conclusion that the lower system is heteromorphic is consistent with the inner ionization potentials, but because the levels overlap, its type is less firmly established. For a particular choice (in the calculations) of relative orbital electronegativities of fluorine and nitrogen, the results are fitted by $\alpha_P = 7.0$, $\beta = 2.2$ eV for the homomorphic system, and for the heteromorphic system by $\alpha_P = 5.2$, $\beta = 2.6$ eV. Since the model used is oversimplified, in ignoring, for example, the possible inadequacy of the single-configuration approximation in both the molecules and the ions produced from them, the numerical values are not very significant, and no attempt has been made to optimize agreement with experiment. In particular, much lower values of β are likely to be appropriate to the neutral molecules.

Correspondence between calculated and experimental behavior is likely to be closest for the fluorides, because the planar model applies accurately to $(NPF_2)_{3,4}$ and nearly to $(NPF_2)_{5,6}$. For other derivatives, applicability of the simple theory will be limited by nonplanarity, but since the symmetry of the structures of the tetrameric derivatives frequently approaches D_{2d}, and for other reasons,[142] the restriction is not serious. The ionization potentials of tetramers are lower than trimers, and a difference in this sense is still interpretable in terms of a homomorphic π system. The differences are reduced, as compared with that between the fluorides, because the ligands are less electronegative, and β consequently smaller.

4. *Reactivity*

We expect that in phosphonitrilic chemistry reactivity will depend on π-electron distribution, and that its study will provide evidence about electronic delocalization. In the following discussion reactivity is interpreted in terms of permanent effects rather than polarizability, specifically the charge density distributions characteristic of homomorphic and heteromorphic systems, and the perturbing effects on them of a substituent group.

Phosphonitrilic halides react with many nucleophiles, such as primary and secondary amines, alcohols, phenols, and halide ions, usually without change in ring size. The reactions are similar to those of phosphoryl compounds, the formal bonding schemes of the two series being also similar, and to some extent the phosphoryl reactions can be used as models for those of the phosphonitriles. Many displacements at a phosphoryl center are bimolecular, and optical inversion has been established for the transesterification of a phosphinate ester PhMeP(O)OMe.[144] Reaction is retarded by electron-releasing groups, such as the amino group,[145] through conjugative $p\pi–d\pi$ interactions involving the unshared electron pair on nitrogen.[146]

The kinetics of the reactions of phosphonitrilic derivatives with nucleophiles have not been examined in the same detail, and, in particular, there has yet been no demonstration of inversion on substitution. Qualitatively, the same considerations apply as to the reactions of phosphoryl compounds; phosphonitrilic derivatives are in general less reactive, but electron-releasing groups such as Me_2N again strongly retard reaction with hydroxide ion. Quantitatively, the reaction of $N_3P_3Cl_6$ with secondary amines such as piperidine

[144] M. Green and R. F. Hudson, *Proc. Chem. Soc.* p. 307 (1962).

[145] E. W. Crunden and R. F. Hudson, *J. Chem. Soc.* p. 3591 (1962).

[146] R. F. Hudson, *Advan. Inorg. Chem. Radiochem.* **5**, 347 (1963).

follows mixed second- and third-order kinetics, the simple bimolecular reaction being catalyzed by base.[147] The rate of a second substitution in the same molecule is much less, and the rate of substitution is also very much greater (by a factor of about 500) in the eight-membered than in the six-membered ring. The retardation of the reaction on successive substitution (and the nongeminal orientation pattern) is explicable in terms of either the electrostatic or the conjugative properties of the substituent, both giving a partial negative charge to the substituted phosphorus atom. The kinetics of the aminolysis of the bromides $(NPBr_2)_{3,4}$ and the fluorides $(NPF_2)_{3,4}$, although investigated in less detail,[148] show that reactivity increases with decreasing ligand electronegativity, and is greater for the eight- than the six-membered ring. The greater rate of reaction of the larger ring may be due in part to its greater flexibility, since its deformation involves only torsional motion. Changes in bond angles, and therefore greater energies, are required in the six-membered ring to open the phosphorus center to attack.

Although these factors are important they are inadequate, because π-electron effects are not directly considered. Experimentally, the inadequacy is shown by the results of Sowerby[149] on the rate of exchange of chloride ion with the series of phosphonitrilic chlorides $(NPCl_2)_{3-6}$. The activation energy for this reaction is an oscillating function of ring size, being high for the six-membered and low for the eight-membered ring. In Fig. 26 the results are shown, in comparison with a model calculation of π-electron densities, the π system being assumed to be of the homomorphic π_s-type involving the $d_{x^2-y^2}$ orbital at phosphorus, and an sp_y orbital at nitrogen. The general correspondence seems good evidence that the highest occupied orbitals, identified by the determination of ionization potentials, are, as might have been expected, those concerned in chemical reactions also.

The quantitative information on successive substitution is less good, and refers to a different reaction. The phosphonitrilic chlorides react with anionic fluorinating agents such as potassium fluorosulfite to give a series of geminally substituted phosphonitrilic chloride–fluorides.[150, 151] In contrast to dimethylamination, successive substitution takes place as close as possible to the center first attacked, and substitution accelerates reaction; determination of the relative yields of the successive chloride–fluorides shows that the PFCl group is more reactive than PCl_2. From a study of the heterogeneous reaction between $N_3P_3Cl_6$ and $N_4P_4Cl_8$ with potassium fluorosulfite in benzene, the

[147] B. Capon, K. Hills, and R. A. Shaw, J. Chem. Soc. p. 4059 (1965); see also C. D. Schmulbach and V. R. Miller, Inorg. Chem. 7, 2189 (1968).

[148] T. Moeller and S. G. Kokalis, J. Inorg. & Nucl. Chem. 25, 1397 (1963).

[149] D. B. Sowerby, J. Chem. Soc. p. 1396 (1965).

[150] A. C. Chapman, D. H. Paine, H. T. Searle, D. R. Smith, and R. F. M. White, J. Chem. Soc. p. 1768 (1961).

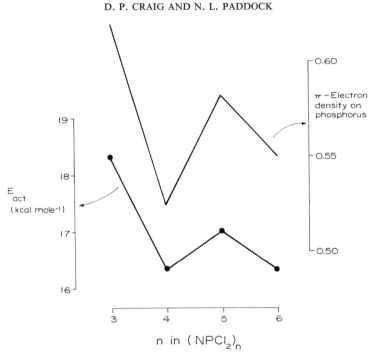

FIG. 26. Activation energies for the reaction $Cl^- + (NPCl_2)_n \rightarrow (NPCl_2)_n + Cl^-$ [D. B. Sowerby, *J. Chem. Soc.* 1396 (1965)]. The upper curve shows calculated (HMO) π-electron densities at phosphorus, assuming $\alpha_N = \alpha_P + \beta$.

relative rates of the first two steps

$$PCl_2 \xrightarrow{\ k_2\ } PClF \xrightarrow{\ k_2\ } PF_2$$

have been estimated, the approximate values of the ratios of the pseudo first-order rate constants being $k_2/k_1 = 8.0$, ~100, for the trimeric and tetrameric series, respectively, and although they cannot be accurate, they show a clear difference between the two ring sizes.[151]

The orientation pattern and the acceleration on substitution are explicable in electrostatic terms, since the more electronegative fluorine atom F in a PFCl group would induce a partial positive charge on phosphorus, and so facilitate nucleophilic attack. Although flexibility could again account for the greater reactivity of the eight-membered rings, the difference in the relative rates of successive substitution steps cannot be so explained; both flexibility and electrostatic effects would affect the two steps equally. The difference in

[151] J. Emsley and N. L. Paddock, *J. Chem. Soc., A* p. 2590 (1968).

behavior can, however, be understood in terms of changes in π-electron density arising from substitution, and involving the same π system as the exchange experiments referred to above.

Of the various possible types of π-interaction, only the inductive effect of a substituent on the π system has been examined, i.e., in model calculations using Hückel molecular orbitals, a perturbation $\delta\alpha_P$ is applied to one phosphorus atom, to simulate the effect of the increased electronegativity of the phosphorus orbitals, as a consequence of substitution of chlorine by fluorine. The general effect of the increase in electronegativity is to concentrate π-electron density on the perturbed atom. That such a concentration actually occurs is shown by the work of Heatley and Todd.[152] Two examples are shown in Fig. 27, in which the

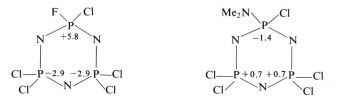

FIG. 27. Deviations from mean ^{31}P chemical shift (ppm) in monosubstituted chloro-cyclotriphosphonitriles [F. Heatley and S. M. Todd, *J. Chem. Soc.* 1152 (1966)].

opposite effects of fluorine and the dimethylamino group in concentrating or dispersing electron density are evident. Such effects are opposite in sign to the expected electrostatic effects of these groups, and, in the case of fluorine, for example, would offset the normal accelerative effect on substitution. The π-inductive effect, causing an accumulation of charge at phosphorus, is important because it differs, as explained below, according to whether the π system is homomorphic or heteromorphic.

The introduction of a single substituent into the six-membered phosphonitrilic ring leaves only a single vertical plane of symmetry, at most, so that the molecular orbitals are either symmetric or antisymmetric to this plane. These are illustrated in Fig. 28 for an in-plane homomorphic system, and, although d-orbitals are used, the formal similarity to benzene is clear; the lowest occupied orbital contains no nodes, the upper pair (which would be degenerate in the absence of the perturbation) one node each. One molecular orbital has a node passing through the perturbed phosphorus atom, and so is unchanged; the energies of the other two (A') are lowered. In the heteromorphic system, on the other hand, there are missing atomic orbitals rather than molecular nodes, and now only one molecular orbital is perturbed (A''). As a consequence, the effect of the perturbation is less on a heteromorphic than on a homomorphic system. For an eight-membered ring (Fig. 29) there is no

[152] F. Heatley and S. M. Todd, *J. Chem. Soc., A* p. 1152 (1966).

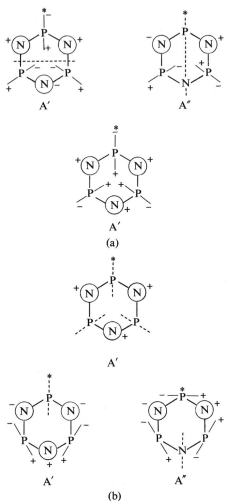

FIG. 28. Schematic formation of bonding molecular orbitals (π_a) in an N_3P_3 molecule: (a) homomorphic (p_z–d_{yz}); (b) heteromorphic (p_z–d_{xz}). The signs refer to the atomic orbital lobes above the molecular planes; molecular nodes (homomorphic) or missing orbitals (heteromorphic) are shown by dashed line. The perturbation referred to in the text is applied at the atom marked with an asterisk.

energetic distinction between a homomorphic and a heteromorphic π system, and in both, two molecular orbitals are perturbed by the substitution, the effect of which lies between the two possibilities for the six-membered ring. Although, as previously noted (Table XII) the observed bond length variations in inhomogeneously substituted N_3P_3 rings do not allow us to distinguish between

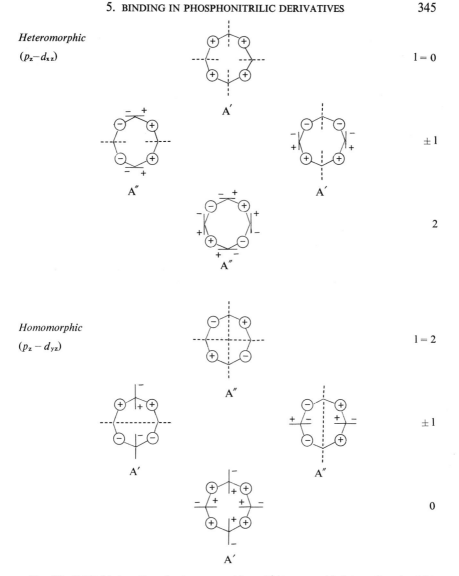

FIG. 29. Orbital interactions for homomorphic and heteromorphic interactions in eight-membered rings, orbitals and nodes, same as Fig. 28.

homomorphic and heteromorphic interactions, the effect of difference in symmetry type on charge density is greater. It is also more apparent, because in a comparison of reactivities we are concerned with differences in charge density between six- and eight-membered rings, rather than absolute values.

The charge densities at a phosphorus atom perturbed by an electronegative substituent (here simulated by taking $\delta\alpha_P = 0.5\beta$) are shown in Fig. 30 for 6-, 8-, and 10-membered rings, for both homomorphic and heteromorphic interactions. It can be seen that, insofar as reactivities are determined by π-electron density at phosphorus, we can expect the relative rates of the second fluorination step to be greater in the trimer than in the tetramer, if the higher levels are of the heteromorphic type, and less if they are homomorphic. The latter type of behavior is observed, in agreement with the more detailed kinetic

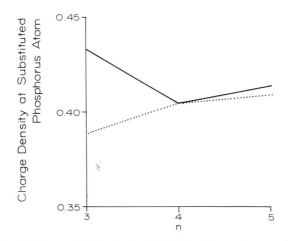

FIG. 30. Calculated charge density at substituted phosphorus atom as a function of ring size. HMO calculations, $\alpha_N = \alpha_P + 2\beta$, $\alpha_{P'} = \alpha_P + 0.5\beta$. The full line refers to homomorphic; the dotted line to heteromorphic interactions.

information on chloride ion exchange and with the measurements of ionization potentials. Figure 30 also suggests that, especially if the energy levels involved are of the homomorphic type, substitution rates in the N_5P_5 series should be less than for N_4P_4. Qualitatively, this is found to be so.[153]

In the fluorination reaction, electrostatic effects are dominant, and determine the orientation pattern; π-electron effects control only the relative rates of formation of the successive chloride–fluorides. Other points emerge from the calculations of π-inductive effects embodied in Fig. 13a. The charge withdrawal is especially small at the unsubstituted phosphorus atom in the six-membered ring, so that a third substitution, necessarily at that atom, is hardly more favored than in $N_3P_3Cl_6$ itself. In the eight-membered ring electron withdrawal is especially high at the first unsubstituted atom, and the π-inductive and electrostatic effects reinforce each other, in agreement with experience.

[153] J. Serreqi, unpublished results (1969).

For other reactions, only the qualitative knowledge of orientation is available, and even this is unsystematic. As the electrostatic effects become less extreme, π-inductive effects (acting in the opposite sense to electrostatic interactions) may influence orientation pattern. Several instances have been observed in which geminal and nongeminal substitution by amines takes place simultaneously,[154] and it is possible that the π-inductive effect (which would tend to disperse π-electron density from the atom first attacked) is sufficient to counteract the electrostatic effect, a second substituent then being directed to the same center as the first.

The reactions of the phosphonitrilic fluorides with carbanions are interesting, because, in view of the high electronegativity of fluorine, electrostatic considerations alone would require successive substitution to occur nongeminally, whatever the substituent. In fact, methylation of $N_3P_3F_6$, $N_4P_4F_8$, and $N_5P_5F_{10}$ by methyllithium proceeds geminally, and the third and fourth methyl groups enter the tetramer antipodally to the first two.[155] This orientation pattern has not been previously recognized, but, with the assumption that the methyl group is an electropositive substituent which would reduce the electrostatic effect of fluorine, this result is consistent with Fig. 13a, if "charge supply" is read for "charge withdrawal"; substitution at P(2) is less discouraged than at P(1). Arylation of $N_3P_3F_6$ by phenyllithium, on the other hand, proceeds nongeminally,[156] possibly because the phenyl group can transfer some negative charge to the phosphonitrilic ring by conjugation.

The calculation of π-charge densities according to the simple model adopted here can evidently be helpful in discussing reactivity, and we end this section by pointing out another feature of Fig. 13[(b) and (c)] which shows that the perturbing effect of a substituent falls off rapidly as the difference between the electronegativities of the ring atoms decreases. Electronic influences are therefore likely to be transmitted much more effectively round the ring in the phosphonitrilic fluorides than in, for example, the bromides, in which substituent effects will be more localized. The point is a general one; the evidence for cyclic delocalization is especially strong for the phosphonitrilic fluorides, but, because of the "insulating" effect of the nitrogen atoms, it is a less important feature in some other derivatives.

5. Magnetic Behavior

Aromaticity in organic systems has important magnetic effects arising from the ring currents induced by an external magnetic field. For homomorphic [4n + 2]-systems, the ring current is diamagnetic, and $\chi_\perp$ exceeds (numerically) the sum of the atomic contributions χ_A. For homomorphic [4n]-systems the

[154] S. K. Das, R. Keat, R. A. Shaw, and B. C. Smith, J. Chem. Soc., A p. 1677 (1966).
[155] N. L. Paddock, T. N. Ranganathan, and S. M. Todd, Canad. J. Chem. 49, 164 (1971).
[156] C. W. Allen and T. Moeller, Inorg. Chem. 7, 2177 (1968).

ring current is paramagnetic,[157] with opposite consequences for $\chi_\perp$. Measurements of magnetic anisotropy can therefore distinguish the contributions of ring currents to the magnetic susceptibility, and may indicate aromatic character. The influence of the ring current, in its opposite effects on proton chemical shifts of hydrogen atoms inside and outside both [4n]- and [4n + 2]-annulenes, has proved to be a useful tool in the discussion of the stereochemistry of these compounds.[158] Although the existence of ring currents of both types seems adequately demonstrated, the quantitative calculation of magnetic anisotropy is less satisfactory. Hoarau[159] has shown that, besides the ring current due to the delocalized π-electrons, there are contributions to the

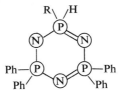

FIG. 31. Hydridocyclotriphosphonitriles [A. Schmidpeter and J. Ebeling, *Angew. Chem. Intern. Ed. Engl.* 7, 209 (1968)].

anisotropy from the σ-bonds and from the atomic π-orbitals, both tending to reduce the in-plane susceptibilities. Allowance for them reduces the ring-current anisotropy, $K_3 - \sum \chi_A$, to 5.2 in 1,3,5-trichloroborazine[160, 161] and to −3.4 in 2,4,6-trichloro-1,3,5-triazine[161, 162] (units, -10^{-6} cgs). These values are too small to be significant, and probably reflect the rapid attenuation of the effects of delocalization as the electronegativity difference between alternate atoms is increased.

Similar difficulties occur for phosphonitrilic derivatives. On any basis delocalization is less than in benzenoids, and on the view that there are both π_a heteromorphic and π_s homomorphic systems of electrons there will be some cancellation of magnetic contributions. Hydridophosphonitriles of the type shown in Fig. 31 have recently been synthesized,[163] and proton resonances for

[157] G. Berthier, B. Mayot, and B. Pullman, *J. Phys. Radium* 12, 717 (1951); J. A. Pople and K. G. Untch, *J. Am. Chem. Soc.* 88, 4811 (1966).

[158] F. Sondheimer, I. C. Calder, J. A. Elix, Y. Gaoni, P. J. Garratt, K. Grohmann, G. di Maio, J. Mayer, M. V. Sargent, and R. Wolovsky, *Chem. Soc. (London), Spec. Publ.* 21, 75 (1967).

[159] J. Hoarau, *Ann. Chim. (Paris)* [13] 1, 544 (1956).

[160] K. Lonsdale and E. W. Toor, *Acta Cryst.* 12, 1048 (1959).

[161] D. P. Craig, M. L. Heffernan, R. Mason, and N. L. Paddock, *J. Chem. Soc.* p. 1376 (1961).

[162] K. Lonsdale, *Z. Krist.* 95, 471 (1936).

[163] A. Schmidpeter and J. Ebeling, *Angew. Chem. Intern. Ed. Engl.* 7, 209 (1968).

the P—H group found at $\tau 2.64$, 2.35, 2.42 (R = OPh, Me, Et). It is hard to judge the significance of these results. The shifts lie at the bottom end of the long range ($\tau 2.5$–8) given by Mavel[164] for hydrogen attached to phosphorus, and are not much lower than the values for the related compounds $(EtO)_2P(O)H$ ($\tau 3.0$[165]), $(MeO)_2P(O)H$, ($\tau 3.23$), $CCl_3PH(O)(OH)$ ($\tau 3.15$[166]), and $(PhO)_2P(O)H$ ($\tau 2.62$), [though for $MeHP(O)(OH)$, the proton resonance occurs at $\tau 7.9$[167]]. There is therefore no shift which can be ascribed with certainty to ring currents.

The phosphonitrilic derivatives, $N_3P_3Cl_6$[161] and $N_4P_4(NMe_2)_8$[79] are magnetically anisotropic in the same sense as benzene, but to a much smaller extent. Correction for the (probable) σ anisotropy in $N_3P_3Cl_6$ leaves a small paramagnetic contribution,[161] consistent with the preponderance of a hetero-morphic π system, but the effect ($K_3 - \sum \chi_A = 11.7 \times 10^{-6}$ cgs units) is too small for confidence, and depends critically on the assumed value of χ_P. The total diamagnetic susceptibility of the series $(NPCl_2)_n$ ($n = 3$–7) varies linearly with n,[161] suggesting the near-compensation of para- and diamagnetic ring currents; any residual deviation from linearity would be minimized by the large electronegativity difference involved. While the magnetic evidence is not in disagreement with earlier conclusions, it provides no good test of them. A study of the fluoride series might be more informative, and especially of the tetrameric fluoride $(NPF_2)_4$, in which the paramagnetic contributions of both π_a and π_s systems might together be significant.

6. Ultraviolet Spectra

The ultraviolet spectra of benzene and its homologs have been the principal source of information about energy levels, and the interpretation of the spectra of heterocyclic aromatic compounds is also well advanced.[168] In the inorganic field, the existence of a π system in borazine derivatives directly analogous to that in benzene has been demonstrated [169] and confirmed by a study of the steric inhibition of conjugation with exocyclic phenyl groups.[170]

Phosphonitrilic derivatives show no bands in similar positions. A study[171] of the ultraviolet spectra of a series of compounds $Y_3P{=}NX$ has shown that, although the groups X and Y influence each other through the medium of the

[164] G. Mavel, in "Progress in Nuclear Magnetic Resonance Spectroscopy" (J. W. Emsley, J. Feeney, and L. H. Sutcliffe, eds.), Vol. 1, p. 251. Pergamon Press, Oxford, 1966.

[165] P. R. Hammond, J. Chem. Soc. p. 1365 (1962).

[166] F. Nixon, J. Chem. Soc. p. 2471 (1964).

[167] D. Fiat, M. Halmann, L. Kugel, and J. Reuben, J. Chem. Soc. p. 3837 (1962).

[168] S. F. Mason, Quart. Rev. (London) 15, 287 (1961).

[169] C. W. Rector, G. W. Schaeffer, and J. R. Platt, J. Chem. Phys. 17, 460 (1949).

[170] H. J. Becher and S. Frick, Z. Physik. Chem. (Frankfurt), [N.S.] 12, 241 (1957).

[171] A. E. Lutskii, Z. A. Sherchenko, L. I. Smarai, and A. M. Pinchuk, Zh. Obshch. Khim. 37, 2034 (1967).

$P{=}N$ bond, no transitions assignable to the $P{=}N$ electrons were found, so that conjugative interactions evidently occur to only a small extent. The spectra of phosphonitrilic derivatives all show an intense band, $\epsilon \sim 10^4$, at a wavelength dependent on the ligand, and many of the results are summarized in Table XV. Other experimental work has shown that the transitions in $N_3P_3Cl_6$, $N_3P_3Br_6$ are insensitive to protonation,[172, 173] and, at least up to $(NPCl_2)_8$, are almost independent of ring size.[172] These results rule out, for these derivatives,

TABLE XV

ULTRAVIOLET SPECTRA OF PHOSPHONITRILIC DERIVATIVES $(NPX_2)_n$

$n = 3; X =$	F^a	$Cl^{b, c}$	Br^b	$N_3{}^b$	NCS^b	OMe^b
λ_{max} (nm)	149.4	175	200	~ 197.0	196	< 192
$\log_{10} \epsilon_{max}$	> 4.0	4.0	4.4	4.0	5.09	> 2.7

$n = 4$ (F, Cl, Br); $n = 3$ (others), $X =$	F^a	$Cl^{b, c}$	Br^b	$NMe_2{}^b$	Ph^d	OPh^e
λ_{max} (nm)	147.5	173	199	193.5	260	272
$\log_{10} \epsilon_{max}$	> 4.0	4.0	4.5	3.95		3.18

[a] R. Foster, L. Mayor, P. Warsop, and A. D. Walsh, *Chem. & Ind.* (*London*) p. 1445 (1960).
[b] B. Lakatos, Á. Hesz, Zs. Vetéssy, and G. Horváth, *Acta Chim. Acad. Sci. Hung.* **60**, 309 (1969).
[c] Estimated values.
[d] R. A. Shaw and F. B. G. Wells, *Chem. & Ind.* (*London*) p. 1189 (1960).
[e] H. R. Allcock, *J. Am. Chem. Soc.* **86**, 2591 (1964). Other λ_{max} and ($\log_{10} \epsilon_{max}$) at: 266 (3.31), 259 (3.23).

excitation of an electron from a nitrogen lone-pair orbital (approximately a π_s-orbital), and suggest strongly that the transitions are localized on the exocyclic groups or in their bonds to phosphorus. Although the spectra are in most cases recognizably those of the ligands,[174] wavelengths and intensities are both affected (as are the spectra of the phosphinimines[171]), by interaction of the ligands, possibly inductively, with the phosphorus atom. In no case has absorption arising from the $P{=}N$ bond been identified. Since the ionization potentials of many phosphonitrilic derivatives are high (Section IV,B,3), the appearance of $\pi \rightarrow \pi^*$ transitions in the near-ultraviolet region is not necessarily to be expected. Moreover the lowest transition in the heteromorphic

[172] D. R. Smith, unpublished (1958), quoted in N. L. Paddock and H. T. Searle, *Advan. Inorg. Chem.* **1**, 347 (1959).
[173] B. Lakatos, Á. Hesz, G. Holly, and G. Horváth, *Naturwissenschaften* **49**, 493 (1962).
[174] B. Lakatos, Á. Hesz, Zs. Vetessy, and G. Horváth, *Acta Chim. Acad. Sci. Hung.* **60**, 309 (1969).

scheme can be shown with the help of the entries in Table II to be electronically forbidden and is unlikely to be induced by vibrations. In fact none of the present bonding theories implies strong absorption in the near-ultraviolet arising from delocalized electrons. Current investigations[174a] show that the methylphosphonitriles (which have low ionization potentials and which have no unshared electrons in the exocyclic groups) have an intense absorption band at $\lambda < 200$ nm, which is destroyed (reversibly) on acidification, and which must therefore belong to the π_s system. Its dependence on ring size should be interesting, since the present results give no certain evidence of a π system at all, whether partly or completely localized.

C. EXOCYCLIC CONJUGATION

Earlier references to conjugation between the phosphonitrilic ring and the exocyclic groups in connection with thermochemistry, base strength measurements, reactivity, and electronic spectra are now supplemented by a consideration of its mechanism. Conjugative interactions appear to be weak and to have effects often difficult to distinguish convincingly from inductive effects on account of the properties of the d-orbitals involved. An electronegative substituent will contract all d-orbitals at the substituted center,, and so strengthen all bonds to which the d-orbitals contribute. Conjugation, however need not be increased by such changes, and may even be decreased. For example, in phosphoryl compounds X_3PO, increasingly electronegative ligands X are associated, through changes in σ-hybridization, with decreasing angles XPX. The possibilities of conjugation are at the same time reduced on account of less favorable steric relations between π-overlapping orbitals, and perhaps also by contraction of the acceptor d-orbitals. Similarly, angular variations within a phosphonitrilic N_2PX_2 group can tend to separate endocyclic and exocyclic π-bonding, so that again bonds can be mutually strengthened without effective conjugation. Nevertheless conjugative interactions occur, and symmetry arguments are helpful in distinguishing them.

The two p-orbitals with π character with respect to the P—X bond can be conveniently chosen (Fig. 32) to be symmetric and antisymmetric with respect to the XPX plane. To avoid confusion with the π_s and π_a orbitals of the ring we use superscripts, designating the π-orbitals at X to be π^s and π^a. π^a overlaps maximally with the d_{xz} orbital at phosphorus and π^s with $d_{x^2-y^2}$ (and d_{z^2}). One orbital (d_{xz}) is common to the ring π_a and exocyclic π^a systems, and $d_{x^2-y^2}$ and d_{z^2} are common to the π_s and π^s systems. The d_{xy} and d_{yz} orbitals which are believed to interact less strongly would mix the ring and exocyclic π systems which are differently related to the defining planes, for example, π_s (exo) with π_a (endo). The d_{xz} and $d_{x^2-y^2}$, d_{z^2} orbitals provide a basis for

[174a] T. N. Ranganathan, unpublished results (1969).

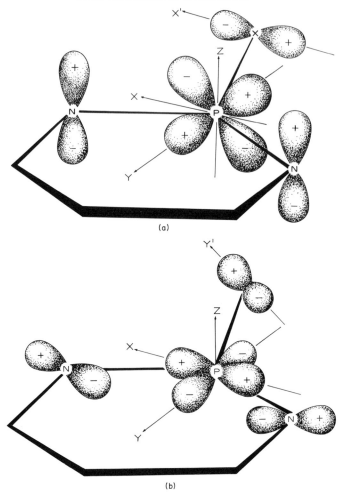

Fig. 32. Conjugation of (*a*) an exocyclic $p\pi^a$ orbital with a heteromorphic (d_{xz}) ring π_a system, and (*b*) an exocyclic $p\pi^s$ orbital with a homomorphic ring π_s system.

conjugation of the ring with the exocyclic groups, the different contributions being distinguishable in principle insofar as interaction symmetries differ.

In phosphine oxides R_3PO, conjugation is strongest when R is an electron-releasing group, particularly N-methyl pyrryl,[175] and similar behavior is to be expected for phosphonitriles. Consistently, the derivatives which show structural evidence of exocyclic π-bonding are those which release electrons to phosphorus by lone-pair delocalization, viz., alkoxy and alkylamino

[175] C. E. Griffin and R. A. Polsky, *J. Org. Chem.* **26**, 4772 (1961).

derivatives. In three methoxy derivatives $[NP(OMe)_2]_{4,6,8}$[81-83] all the POC angles are close to 120°, as in other phosphate esters[176] whereas the COC angle is about 112° in organic esters. The P—O bonds are also short (1.58–1.60 Å) compared with an expected 1.71 Å for a P—O single bond. The slight correlated increase in strengths of ring and exocyclic bonds with increasing ring size is possibly a conjugative effect, but a small one. A clearer indication of the orbitals involved is given by the structures of the dimethylamides $N_4P_4(NMe_2)_8$[79] and $N_6P_6(NMe_2)_{12}$.[80] Information about the geometry of the exocyclic groups is given in Table XVI. Electron release to phosphorus is

TABLE XVI

GEOMETRY OF THE EXOCYCLIC GROUPS IN
$N_4P_4(NMe_2)_8$[a] AND $N_6P_6(NMe_2)_{12}$[b]

P—N bond length (Å)	$N_4P_4(NMe_2)_8$[c]	$N_6P_6(NMe_2)_{12}$[c]
Bond (1)	1.671 (358.5°)	1.663 (357.5°)
Bond (2)	1.686 (345.5°)	1.675 (348.7°)
Average	1.678 (352.0°)	1.669 (353.1°)
NPN angle	103.8°	102.9°

[a] G. J. Bullen, *J. Chem. Soc.* p. 3193 (1962).

[b] A. J. Wagner and A. Vos, *Acta Cryst.* **B24**, p. 1423 (1968).

[c] The figures in parentheses are the sums of angles around the exocyclic nitrogen atoms.

indicated both by the large angles around the exocyclic nitrogen atoms and by the shortness of the P—N bonds; it is strikingly confirmed by their further contraction, to 1.61 Å, when the ring is protonated[133] (Fig. 22). In the neutral compounds, the two dimethylamino groups attached to the same phosphorus atoms differ in detail. For steric reasons, they are differently aligned, and therefore interact through different d-orbitals. In both molecules, that group which is the better oriented for π-interaction with the π_s orbitals $d_{x^2-y^2}$ and d_{z^2} is the more strongly bound, as judged by the sum of angles around the nitrogen atom; the smaller difference in orientation in the hexameric, as compared with the tetrameric dimethylamide, is correlated with a smaller difference in the nitrogen angles. The corresponding differences in P—N bond length although they are not in themselves significant follow the same pattern, and it is also consistent with the dominance of the $d_{x^2-y^2}$, d_{z^2} interactions that the smaller exocyclic angle is associated with the shorter mean P—N bond. These structures show that the π_a, π_s orbitals of phosphorus are used unequally in

[176] See, e.g., M. Calleri and J. C. Speakman, *Acta Cryst.* **17**, 1097 (1964).

exocyclic π-bonding, and suggest strongly that the more important are the π_s orbitals $d_{x^2-y^2}$ and d_{z^2}, although the involvement of d_{xz} need not be much less. This conclusion is in agreement with the structure of $N_3P_3Ph_2F_4$, in which the phenyl groups, unrestricted sterically by neighboring PF_2 groups, are both

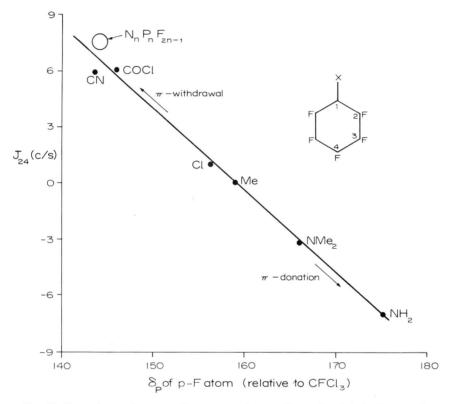

FIG. 33. Dependence of o–p-coupling constant in pentafluorophenyl derivatives on the p-^{19}F chemical shift. Based on M. G. Hogben, R. S. Gay, and W. A. G. Graham, *J. Am. Chem. Soc.* **88**, 3457 (1966).

optimally aligned for π overlap with the π_s orbitals. Since these orbitals, especially $d_{x^2-y^2}$, are used in the ring π-bonding, conjugation necessarily occurs, and the structural investigations sharpen the conclusions arrived at from the study of base strengths and substitution behavior.

Conjugation has also been studied through its effect on the ^{19}F nuclear magnetic resonance (NMR) spectra of fluorophenyl groups attached to the phosphonitrilic ring.[177] Two methods of interpretation give consistent results.

[177] T. Chivers and N. L. Paddock, *Chem. Commun.* p. 704 (1968).

Graham and his co-workers[178] have established a linear relationship between δ_P and J_{24} (the *ortho–para* fluorine coupling constant) for a large number of pentafluorophenyl compounds. By using it, a distinction can be made between π-donor and π-acceptor substituents. A typical plot is shown in Fig. 33, which also shows results for the series of compounds $N_nP_nF_{2n-1} \cdot C_6F_5$ ($n = 3$–8).[177] As a class, the fluorophosphonitriles are evidently strong π-acceptors, comparable to the cyano group, but their differences are not made clear in this way.

TABLE XVII

THE ^{19}F CHEMICAL SHIFTS OF THE
PENTAFLUOROPHENYLFLUOROPHOSPHONITRILES[a]

	δ_p	δ_m	$\delta_m - \delta_p$
$N_3P_3(C_6F_5)F_5$	143.8	159.1	15.3
$N_4P_4(C_6F_5)F_7$	144.6	159.3	14.7
$N_5P_5(C_6F_5)F_9$	143.3	159.4	16.1
$N_6P_6(C_6F_5)F_{11}$	144.4	159.4	15.0
$N_7P_7(C_6F_5)F_{13}$	144.4	159.4	15.0
$N_8P_8(C_6F_5)F_{15}$	144.4	159.4	15.0

[a] Shifts (of the C_6F_5 group) in ppm relative to CCl_3F. J_{24}' in the range 7.3–7.7 Hz.

An alternative way of assessing π-interactions depends on the difference between the *m*- and *p*-shifts; δ_m reflects inductive, $\delta_m - \delta_p$ resonance effects.[179] The ^{19}F chemical shifts of the pentafluorophenylfluorophosphonitriles are shown in Table XVII. The large deshielding of the *p*-fluorine atom, relative to the *m*-fluorine atom, is an indication of a conjugative effect, the difference $\delta_m - \delta_p$ being much greater than for CH_3 (5.0 ppm[180]) or Cl, Br, I (5.3, 6.0, 7.1 ppm[180]), which interact mainly inductively. Further, the variability of δ_p, while δ_m remains constant, shows that the conjugative interactions, except for the largest rings, vary with ring size in the way expected for interaction of the homomorphic π system of the benzene ring with a homomorphic phosphonitrilic π system in agreement with the structural work on the dimethylamides and on $N_3P_3Ph_2F_4$.[131] Conjugation need not always take place in the same way. Monofluorophenylfluorophosphonitriles also show a difference between δ_m

[178] M. G. Hogben, R. S. Gay, and W. A. G. Graham, *J. Am. Chem. Soc.* **88**, 3457 (1966).
[179] See, e.g., R. W. Taft, E. Price, I. R. Fox, I. C. Lewis, K. K. Andersen, and G. T. Davis, *J. Am. Chem. Soc.*, **85**, 709 (1963).
[180] I. J. Lawrenson, *J. Chem. Soc.* p. 1171 (1965).

and δ_p in the same sense as the pentafluorophenyl derivatives,[181] but δ_p now varies smoothly with ring size. It is possible that the difference is caused by differences in the orientations of the two rings, as found in structural investigations; the mutual orientation of the phenyl groups are different in the two molecules $N_3P_3Ph_2F_4$,[131] $N_3P_3Ph_2Cl_4$,[130] some overlap with π_a orbitals occurring in the latter molecule and in the tetraphenyl derivative. If this were dominant, conjugation would occur through a heteromorphic π system and would be hard to detect since its effects vary smoothly with ring size.

V. Conclusion

The first theoretical treatment of the delocalized bonding in phosphonitrilic molecules in 1959 was an attempt to rationalize the then most obvious chemical facts, namely, that the chlorides formed a series of cyclic molecules of high and roughly equal stability. If purely ionic interactions are discounted, $p\pi$–$d\pi$ bonding is formally required in the ring bonds, and was initially considered in terms of what would now be called the π_a system. The heats of polymerization to a common product showed a trend, in the variation with ring size, that could not be explained if the delocalization were benzenoid, but was compatible with the dominant use of d_{xz} at phosphorus in a heteromorphic system. In the past decade, as a more detailed picture of the properties of these systems has been built up by experimental work, a further primary feature of the electronic structure has had to be recognized. This is the existence of in-plane (π_s) bonding first indicated by crystal structure determinations. To elucidate the roles of the π_a and π_s systems and to interpret the effects of substitution on them have been the main purposes of much of the later experiment. The frequent occurrence of planar or other highly symmetrical conformations suggests that the π_a system is stronger, but conformational preferences arising from π-bonding are sometimes overridden by steric interactions.

At the next level of theoretical interpretation there is the need to examine the relative importance of the various d-orbitals in the π_a and π_s systems. The most important evidence (Section IV,B,3) is from ionization potentials where, especially in the homologous series $(NPF_2)_n$, it becomes clear that the least tightly bound electron is from a homomorphic system. Just as d_{xz} was selected, on the basis of its better overlap, as the main π_a contributor, so for the same reason $d_{x^2-y^2}$ is believed to be the principal in-plane d-orbital. We are thus led to conclude that the homomorphic π system is of the π_s-type. Supporting evidence comes from the nature and properties of the protonated species (Section IV,B,2).

[181] T. Chivers and N. L. Paddock, unpublished (1969).

It is an essential feature of the theoretical model that the d-orbitals of phosphorus are engaged selectively, and that electron delocalization occurs to some extent. The best direct experimental evidence of both the fact of delocalization and of its character is in the ionization potentials (Section IV,B,3), and the alternation of bond lengths in the $N_4P_4Me_8H^+$ cations and in $N_4P_4F_6Me_2$ (Section IV,B,2). Delocalization also provides the basis for the interpretation of substituent and ring-size effects in a way which is not possible for simpler models.

The choice of model for the description of the bonding in phosphonitrilic derivatives depends on the precision required. A simple ionic picture explains many features, but not the detailed geometry. The approximation in which delocalization is limited to 3-center bonds improves the geometrical description, especially of homogeneously substituted derivatives, but gives no basis for the interpretation of ring size effects. Full cyclic delocalization, as indicated by the ionization potentials, is required to deal with ring-size effects, and the transmission of substituent influences, and is compatible with other properties.

In advancing models and interpretative schemes we have tried to avoid becoming too specific and detailed, and to avoid following too closely lines of argument applicable to benzene and its homologs in which π-electron influences are bigger and more definitely identifiable in particular fields, like UV spectra. The effects in phosphonitrilic molecules are sometimes smaller, often different in character, and more open to alternative explanations. The delocalized model in its simple form cannot deal quantitatively with individual results; the case for it rests upon its ability to give a generally plausible basis for understanding the range of chemical experience of phosphonitrilic systems. For this reason, there has so far seemed to be little point in taking model calculations beyond the simple Hückel stage, though this is technically possible.

Further advances are to be looked for. For example, further studies on electronic spectra, on magnetic properties, and on the conjugation of the ring and exocyclic groups would be useful. The direct analogies with the chemistry of the annulenes have yet to be explored in detail, and investigation of the kinetics and mechanism of substitution reactions offer scope for the application of the general methods and concepts of physical organic chemistry in a field which at present is still undeveloped.

ACKNOWLEDGMENTS

We are grateful to Dr. G. J. Bullen, Dr. R. A. Jacobson, and Dr. A. J. Wagner for permission to use figures from their papers, to Dr. J. Serreqi, Mr. T. N. Ranganathan, and Dr. M. W. Dougill for unpublished results, and to Mr. J. Christie and Dr. K. A. R. Mitchell for the results of unpublished calculations.

6

Cyclobutadiene-Metal Complexes

P. M. Maitlis and K. W. Eberius

I. Introduction and Historical Background

Although this subject has been reviewed fairly recently by Maitlis[1] and by Cava and Mitchell,[2] the large volume of work on cyclobutadiene-metal complexes which has appeared in the last 3 years and the enormous interest in this field has prompted us to update these reviews.

The recognition of the remarkable properties (aromaticity) of benzene very soon led workers to investigate similar compounds with a closed, cyclic system of conjugated double bonds.

Interest first centered on cyclooctatetraene, C_8H_8, and on cyclobutadiene, C_4H_4. The former was first synthesized by a classical route by Willstätter et al.[3] and, later, by the nickel cyanide-catalyzed tetramerization of acetylene by Reppe et al.[4] Willstätter in 1905 already attempted to synthesize cyclobutadiene,[5] but neither he nor any of the numerous later workers were successful in obtaining it or even a simple derivative. Considerable evidence for the intermediacy of cyclobutadienes in reactions has however accumulated. This has been reviewed by Baker and McOmie[6, 7] and by Cava and Mitchell.[2]

The instability of cyclobutadiene has been the subject of considerable discussion. It does not have $4n + 2$ π-electrons and therefore does not satisfy the simple Hückel criterion for aromaticity. It should behave as a conjugated diene with considerably enhanced reactivity owing to ring strain. However some molecular orbital treatments[8-11] suggest this is not the only reason. Two possibilities exist, one that cyclobutadiene is square planar, and two, that it is rectangular. The first implies considerable conjugation (interaction) of the double bonds, the latter does not. Most molecular orbital treatments have assumed a square planar geometry. In this case, combination of the four carbon

[1] P. M. Maitlis, *Advan. Organometal. Chem.* **4**, 95 (1966).

[2] M. P. Cava and M. J. Mitchell, "Cyclobutadiene and Related Compounds." Academic Press, New York, 1967.

[3] R. Willstätter and E. Waser, *Ber.* **44**, 3423 (1911); R. Willstätter and M. Heidelberger, *ibid.* **46**, 517 (1913).

[4] W. Reppe, O. Schlichting, K. Klager, and T. Toepel, *Ann.* **560**, 1 (1948).

[5] R. Willstätter and W. von Schmaedel, *Ber.* **38**, 1992 (1905).

[6] W. Baker and J. F. W. McOmie, *Chem. Soc. (London), Spec. Publ.* **12**, 49–67 (1958).

[7] W. Baker and J. F. W. McOmie, *in* "Non-Benzenoid Aromatic Compounds" (D. Ginsburg, ed.), pp. 43–106. Wiley (Interscience), New York, 1959.

[8] J. D. Roberts, A. Streitwieser, and C. M. Regan, *J. Am. Chem. Soc.* **74**, 4579 (1952); S. L. Manatt and J. D. Roberts, *J. Org. Chem.* **24**, 1336 (1959).

[9] C. A. Coulson, *Chem. Soc. (London), Spec. Publ.* **12**, 97 (1958).

[10] A. Streitwieser, "Molecular Orbital Theory for Organic Chemists," p. 261. Wiley, New York, 1961.

[11] H. E. Simmons and A. G. Anastassiou, in Cava and Mitchell.[2]

$2p_z$ orbitals gives four molecular orbitals, one bonding (ψ_1), two nonbonding (ψ_2, ψ_3), and one antibonding (ψ_4). Although there are a number of ways in which the four electrons can fill these orbitals, the one of lowest energy is probably **1** where the degenerate nonbonding orbitals are each singly filled, and in which these electrons have parallel spins. This lowest energy state is therefore a triplet.

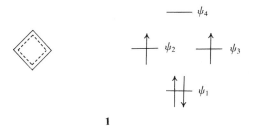

1

Although some evidence for a triplet square planar cyclobutadiene has been presented, especially for the tetrasubstituted molecule,[12-14] this evidence may be ambiguous.

SCF–molecular orbital calculations by Dewar and Gleicher[15] however suggested that the singlet, rectangular, unconjugated state was the lowest energy state for cyclobutadiene itself. This was supported by the work of Pettit and his co-workers, who generated cyclobutadiene from its iron tricarbonyl complex, and showed that it added stereospecifically to dienophiles.[16,17]

The idea that cyclobutadiene might be square in a triplet state did have a very useful consequence. Longuet-Higgins and Orgel in 1956 pointed out that the presence of the two unpaired electrons in the nonbonding orbitals (ψ_2, ψ_3), which were of correct symmetry to overlap with two metal orbitals, would lead to strong binding of the ring to the metal.[18] This implied that a cyclobutadiene should be stabilized by complexation with a metal. Longuet-Higgins and Orgel further suggested[18] that since the cyclopentadienyl radical has only one unpaired electron, as compared with two for cyclobutadiene, the C_4H_4-metal bond might well be even more stable than the C_5H_5-metal bond. Although the specific types of complexes which they predicted [based on d^8 and d^{10} metal ions, e.g., $C_4H_4AuCl_2^+$, $C_4H_4Ni(CO)_2$] have not, to date, been synthesized, the basic predictions of the theory were rapidly fulfilled by

[12] P. S. Skell and R. J. Petersen, *J. Am. Chem. Soc.* **86**, 2530 (1964).
[13] H. H. Freedman, *J. Am. Chem. Soc.* **83**, 2194 and 2195 (1961).
[14] W. D. Hobey and A. D. McLachlan, *J. Chem. Phys.* **33**, 1695 (1963).
[15] M. J. S. Dewar and G. J. Gleicher, *J. Am. Chem. Soc.* **87**, 3255 (1965).
[16] L. Watts, J. D. Fitzpatrick, and R. Pettit, *J. Am. Chem. Soc.* **88**, 623 (1966).
[17] L. Watts, J. D. Fitzpatrick, and R. Pettit, *J. Am. Chem. Soc.* **87**, 3253 (1965).
[18] H. C. Longuet-Higgins and L. E. Orgel, *J. Chem. Soc.* p. 1969 (1956).

preparation in 1959 of tetramethylcyclobutadienenickel chloride (2) by Criegee and Schröder[19] and tetraphenylcyclobutadieneiron tricarbonyl (3) by Hübel and his co-workers.[20,21] Recent concepts of the bonding in these complexes are more fully discussed in Section VI.

II. Preparation of Cyclobutadiene-Metal Complexes

The routes by which the first complexes 2 and 3 were prepared (from a metal carbonyl and 3,4-dihalocyclobutene, and an acetylene, respectively) still represent the most direct synthetic routes to cyclobutadiene complexes. However the reactions of acetylenes with metals still lead to rather unpredictable results and these reactions are only sometimes [e.g., with Pd(II) or Pt(II)] of real synthetic value.

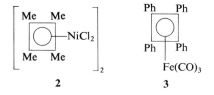

2   3

Synthesis of substituted cyclobutadiene complexes can often be accomplished by ligand-transfer reactions. These, however, appear in general to work better for cyclobutadienes with bulky substituents (phenyl).

In addition to the synthetic methods listed below, some reactions of the complexes in which the metal-cyclobutadiene bond is not broken are discussed in Sections IV,E–G.

A. FROM 3,4-DIHALOCYCLOBUTENES

Potentially the most straightforward synthetic route is dehalogenation of a 3,4-dihalocyclobutene with a metal in a low oxidation state. This route is limited only by the availability of the appropriate dihalocyclobutene, many of which are difficult to make. In some cases too, these reactions are not successful and dimers of cyclobutadiene, rather than complexes, result.

Criegee and Schröder[19] prepared tetramethylcyclobutadienenickel chloride (2) from *trans*-3,4-dichloro-1,2,3,4-tetramethylcyclobutene[22] (4) and nickel tetracarbonyl. The dibromo analog of 2 was prepared similarly,[23] and the

[19] R. Criegee and G. Schröder, *Ann.* **623**, 1 (1959); *Angew. Chem.* **71**, 70 (1959).

[20] W. Hübel and E. H. Braye, *J. Inorg.-Nucl. Chem.* **10**, 250 (1959).

[21] W. Hübel, E. H. Braye, A. Clauss, E. Weiss, U. Krüerke, D. A. Brown, G. S. D. King, and C. Hoogzand, *J. Inorg.-Nucl. Chem.* **9**, 204 (1959).

[22] R. Criegee and A. Moschel, *Ber.* **92**, 2181 (1959).

[23] J. P. Pfrommer, Ph.D. Dissertation, Karlsruhe, 1961.

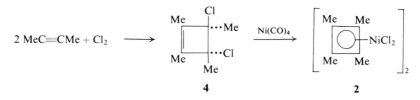

4   **2**

diiodotetramethylcyclobutene reacted with Raney nickel to give tetramethyl-cyclobutadienenickel iodide.[24]

Similar complexes **6** were obtained as shown from the 1,5-hexadiynes **5**.[25, 26]

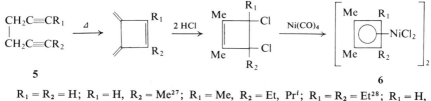

5

6

$R_1 = R_2 = H$; $R_1 = H$, $R_2 = Me$[27]; $R_1 = Me$, $R_2 = Et$, Pr^i; $R_1 = R_2 = Et$[28]; $R_1 = H$, $R_2 = Pr^i$[29].

An interesting modification of the reaction which avoids the use of nickel carbonyl was developed by Henrici-Olivé and Olivé.[30] They reacted **4** with a mixture of nickel bromide and lithium naphthalenide in THF at low temperatures and obtained a nearly quantitative yield of **2**.

Pettit and his co-workers adapted the reaction of Criegee and Schröder for the synthesis of the unsubstituted cyclobutadieneiron tricarbonyl (**8**).[31] This was obtained in good yield from cis-3,4-dichlorocyclobutene (**7**)[32] and diiron enneacarbonyl in pentane in a heterogeneous reaction.

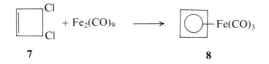

7   **8**

[24] R. Criegee, *Angew. Chem.* **74**, 703 (1962); *Angew. Chem. Intern. Ed. Engl.* **1**, 519 (1962).

[25] W. D. Huntsman and H. J. Wristers, *J. Am. Chem. Soc.* **85**, 3308 (1963); W. D. Huntsman, J. A. DeBoer, and M. H. Woosley, *ibid.* **88**, 5846 (1966); W. D. Huntsman, *ibid.* **82**, 6389 (1960).

[26] R. Criegee, W. Eberius, and H. A. Brune, *Ber.* **101**, 94 (1968).

[27] W. Eberius, Ph.D. Dissertation, Karlsruhe, 1967.

[28] H. A. Brune, personal communication (1969).

[29] H. A. Brune, H. P. Wolff, and H. Hüther, *Tetrahedron* **25**, 1089 (1969).

[30] G. Henrici-Olivé and S. Olivé, *Angew. Chem.* **79**, 897 (1967).

[31] G. F. Emerson, L. Watts, and R. Pettit, *J. Am. Chem. Soc.* **87**, 131 (1965).

[32] M. Avram, J. Dinulescu, M. Elian, M. Farcasiu, E. Marica, G. Mateescu, and C. D. Nenitzescu, *Ber.* **97**, 372 (1964).

The complexes **9, 10**, and **11** were prepared similarly[28, 31, 33-36]:

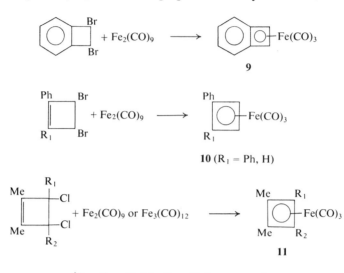

9

10 (R_1 = Ph, H)

11

($R_1 = R_2 = $ H, Me; $R_1 = $ H, $R_2 = $ Me, Et)

The properties of these complexes[37] and of the ligands liberated from them are discussed in Section IV,D and I.

Amiet *et al.*[38] have prepared cyclobutadiene complexes of other metals by reaction of the 3,4-dichlorocyclobutenes **4** and **7** with the appropriate anions. These were usually obtained from the neutral carbonyl by reaction with sodium amalgam. Two series, one of unsubstituted cyclobutadiene-metal carbonyls

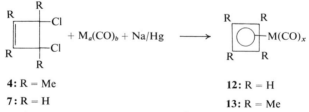

4: R = Me   **12:** R = H

7: R = H   **13:** R = Me

[33] L. Watts and R. Pettit, *Advan. Chem. Ser.* **62**, 549 (1967).

[34] H. A. Brune, W. Eberius, and H. P. Wolff, *J. Organometal. Chem. (Amsterdam)* **12**, 485 (1968).

[35] M. R. Churchill, J. Wormald, W. P. Giering, and G. F. Emerson, *Chem. Commun.* p. 1217 (1968); M. F. Churchill and J. Wormald, *Inorg. Chem.* **8**, 1936 (1969).

[35a] R. E. Davis, *Chem. Commun.* p. 1218 (1968).

[36] H. A. Brune and H. P. Wolff, *Tetrahedron* **24**, 4861 (1968).

[37] J. D. Fitzpatrick, L. Watts, G. F. Emerson, and R. Pettit, *J. Am. Chem. Soc.* **87**, 3254 (1965).

[38] R. G. Amiet, P. Reeves, and R. Pettit, *Chem. Commun.* p. 1208 (1967).

(12, M = Fe, Ru, $x = 3$; M = Mo, W, $x = 4$) and the other of tetramethyl-cyclobutadiene-metal carbonyls **(13**, M = Fe, Cr, $x = 3$; M = Mo, W, $x = 4$), were prepared, though in low yields, except for **(12**, M = Mo, $x = 4$).

Dichlorocyclobutene **(7)** also reacted with sodium tetracarbonylcobaltate **(14)** to give cyclobutadienedicobalthexacarbonyl formulated as **15**[39] by analogy with the tetramethyl derivative reported earlier by Maitlis and co-workers.[40] This was converted into the iodide **16** and then, by treatment with cyclopenta-diene in the presence of base, into cyclobutadiene(cyclopentadienyl)cobalt **(17)**, a complex isoelectronic with ferrocene.

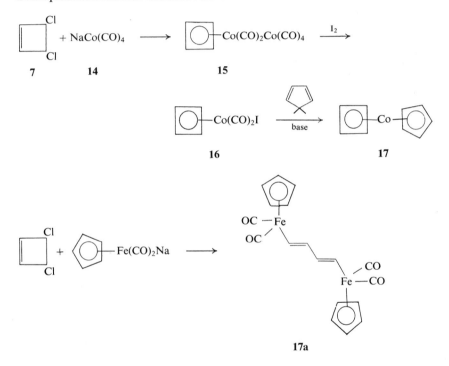

An interesting reaction which did not give a cyclobutadiene complex was described by Emerson and Churchill and their co-workers[35] and by Davis.[35a] The reaction of *cis*-3,4-dichlorocyclobutene with sodium cyclopentadienyliron dicarbonyl gave a complex, which was found by an X-ray crystal structure determination to be **17a**. However, the nuclear magnetic resonance (NMR) spectrum only showed two types of protons, one arising from the cyclopenta-dienyls; but see Anet and Abrams.[40a]

[39] R. G. Amiet and R. Pettit, *J. Am. Chem. Soc.* **90**, 1059 (1968).
[40] R. Bruce and P. M. Maitlis, *Can. J. Chem.* **45**, 2017 (1967).
[40a] F. A. L. Anet and O. J. Abrams, *Chem. Commun.* p. 1611 (1970).

The mechanism by which these cyclobutadiene complexes are formed from the dihalocyclobutenes remains obscure; however it is doubtful whether a free cyclobutadiene is ever produced as an intermediate.

B. FROM ACETYLENES AND TRANSITION-METAL CARBONYLS

The reactions of metal carbonyls with acetylenes have been extensively studied, and a wide range of products isolated. They include organic products such as benzenes and cyclopentadienones, as well as metal complexes containing di-, tri-, and tetramers of the acetylene, both with and without incorporation of carbon monoxide.

In 1959 Hübel et al.[21] showed that one of the products from the high-temperature reaction of diphenylacetylene was tetraphenylcyclobutadieneiron tricarbonyl (3). This was obtained in 16% yield, together with tetraphenyl-cyclopentadienoneiron carbonyl (18) as major product. At low temperatures no 3 was obtained, though it was formed in trace amounts from reaction of diphenylacetylene with $Fe_3(CO)_{12}$ at 90°C.[20] The reactions between diphenyl-

acetylene and molybdenum hexacarbonyl (19) or diglyme molybdenum tricarbonyl (24) gave a variety of products, including some, 21, 22, 23, and 25, claimed to be tetraphenylcyclobutadienemolybdenum complexes.[41]

Whitlock and Sandvick have reported the preparation of 25a from o-di(phenylethynyl)benzene and iron carbonyls. The crystal structure has been determined.[41a]

C. FROM ACETYLENES AND OTHER TRANSITION-METAL COMPLEXES

Other zero-valent iron complexes have been used in reactions with acetylenes to give the tetraphenylcyclobutadieneiron tricarbonyl complex, 3; for example,

[41] W. Hübel and R. Merényi, J. Organometal. Chem. (Amsterdam) 2, 213 (1964).

[41a] H. W. Whitlock and P. E. Sandvick, J. Am. Chem. Soc. 88, 4526 (1966); E. F. Epstein and L. F. Dahl, J. Am. Chem. Soc. 92, 493 (1970).

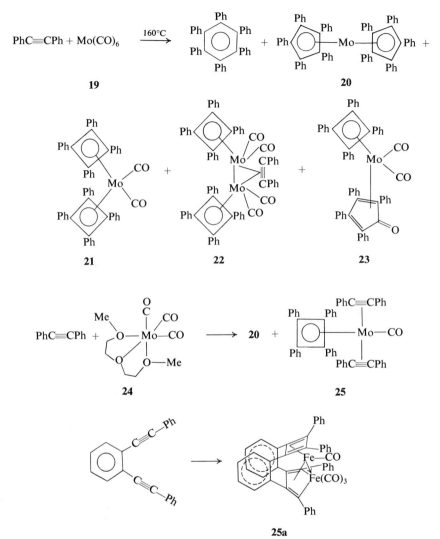

reaction of diphenylacetylene with cyclooctatetraeneiron tricarbonyl (**26**) gave a 4% yield of **3**.[42]

$$PhC{\equiv}CPh + C_8H_8Fe(CO)_3 \xrightarrow{196°C} Ph_4C_4Fe(CO)_3$$

$$\textbf{26} \qquad\qquad\qquad\qquad \textbf{3}$$

[42] A. Nakamura and N. Hagihara, *Nippon Kagaku Zasshi* **84**, 339 (1963).

Similarly Nakamura and Hagihara[43, 44] and Rausch and Genetti[45] obtained cyclopentadienyl(tetraphenylcyclobutadiene)cobalt (**30**) from a variety of cyclopentadienylcobalt complexes, including **27**, **28**, and **29**. An extension of

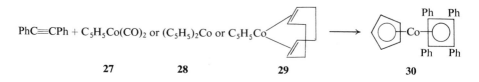

PhC≡CPh + C₅H₅Co(CO)₂ or (C₅H₅)₂Co or C₅H₅Co

    27   **28**   **29**   **30**

this reaction by Helling et al.[46] allowed the preparation of the tetrasubstituted cyclobutadiene(cyclopentadienyl)cobalt complexes **31** and **32** from **28** or **29** and the appropriate phenylacetylene. The complexes, **31** or **32**, with R = SiMe₃

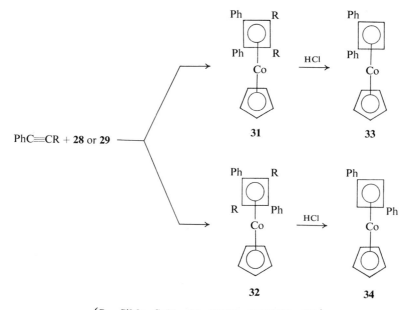

PhC≡CR + **28** or **29**

    31   **33**

    32   **34**

(R = SiMe₃, SnPh₃, Me, COMe, CH(OEt)₂, CF₃)

or SnPh₃, reacted with HCl to give the 1,2- (or 1,3-) diphenylcyclobutadiene-(cyclopentadienyl)cobalt complexes (**33**) or (**34**).

[43] A. Nakamura, *Mem. Inst. Sci. Ind. Res., Osaka Univ.* **19**, 81 (1962).

[44] A. Nakamura and N. Hagihara, *Bull. Chem. Soc. Japan* **34**, 452 (1961).

[45] M. D. Rausch and R. A. Genetti, *J. Am. Chem. Soc.* **89**, 5502 (1967).

[46] J. F. Helling, S. C. Rennison, and A. Merijan, *J. Am. Chem. Soc.* **89**, 7140 (1967).

Yamazaki and Hagihara[47] in a remarkable series of reactions demonstrated one way in which tetraphenylcyclobutadiene complexes and hexaphenylbenzene could be built up from diphenylacetylenes. The starting material here was cyclopentadienyl(diphenylacetylene)triphenylphosphinecobalt (**36**) prepared from **35** and diphenylacetylene in the presence of isopropylmagnesium bromide (as a reducing agent). The acetylene complex **36** reacted readily with

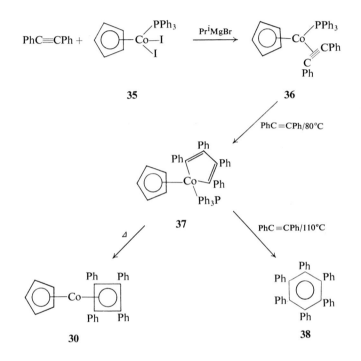

one additional mole of diphenylacetylene on heating to 80°C to give the cobaltacyclopentadiene complex **37**. On strong heating in the absence of further acetylene the complex **37** underwent isomerization with loss of triphenylphosphine (and effectively a decrease of two in the formal oxidation state of the metal) to give the complex **30**. Complex **37** also reacted with an excess of the acetylene at higher temperature to give hexaphenylbenzene (**38**). However, the full paper by these authors[47a] does not mention this last reaction.

Palladium chloride reacts readily with diphenylacetylenes to give a number of products, the nature of which depend on the exact conditions (solvent, rate

[47] H. Yamazaki and N. Hagihara, *J. Organometal. Chem.* (*Amsterdam*) **7**, 22 (1967).
[47a] H. Yamazaki and N. Hagihara, *J. Organometal. Chem.* (*Amsterdam*) **21**, 431 (1970).

of addition, etc.) of the reaction. This is further discussed in Sections II,E and IV,H. In aprotic solvents, especially benzene or chloroform, tetraphenyl-cyclobutadiene complexes **39** and hexaphenylbenzenes were formed, the former being favored when the acetylene was added slowly to the bis(benzo-nitrile)palladium chloride complex.[48-50] The complexes **39** were converted

$RC{\equiv}CR + (PhCN)_2PdCl_2 \longrightarrow$

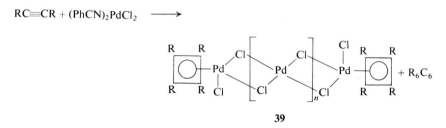

39

$R = Ph, p\text{-}ClC_6H_4, p\text{-}MeOC_6H_4, p\text{-}MeC_6H_4$

into the normal cyclobutadienepalladium halide complexes **40** by treatment with DMF followed by HCl or HBr.[49, 50]

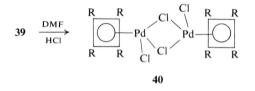

40

Chini *et al.* have prepared the platinum analog of **40** (R = Ph) by heating $Pt(CO)_2Cl_2$ with diphenylacetylene in an appropriate solvent.[50a] Müller *et al.* have reported the preparation of what was claimed to be a biscyclobutadiene complex of palladium from *o*-di(phenylethynyl)benzene.[50b]

Hosokawa and Moritani[50c] have reported the formation of 1,2-di-*t*-butyl-3,4-diphenylcyclobutadienepalladium chloride (dimer) from reaction of *t*-butylphenylacetylene and bis(benzonitrile)palladium chloride in benzene.

[48] P. M. Maitlis, D. F. Pollock, M. L. Games, and W. J. Pryde, *Can. J. Chem.* **43**, 470 (1965).
[49] P. M. Maitlis and D. F. Pollock, *J. Organometal. Chem. (Amsterdam)*, in press (1971).
[50] R. Hüttel and H. J. Neugebauer, *Tetrahedron Letters* p. 3541 (1964).
[50a] P. Chini, F. Canziani, and A. Quarta, *Proc. Inorg. Chim. Acta Symp.*, Venice, 1968 p. D8.
[50b] E. Müller, K. Munk, P. Ziemek, and M. Sauerbier, *Ann.* **713**, 40 (1968).
[50c] T. Hosokawa and I. Moritani, *Tetrahedron Letters* p. 3021 (1969).

However, di-*t*-butylacetylene did not give a cyclobutadiene complex under these conditions.

Nesmeyanov and his collaborators[51-53] obtained a novel tetraphenylcyclobutadieneniobium complex, **43**, by the route indicated from cyclopentadienylniobium tetracarbonyl **(41)** and cyclopentadienylbis(diphenylacetylene)niobium monocarbonyl **(42)**.

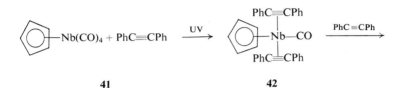

41   **42**

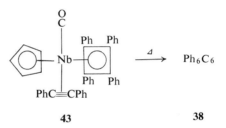

43   **38**

The crystal structure of **43** shows the presence of the separate, but complexed, acetylene and cyclobutadiene. However, the plane of the cyclobutadiene carbons is parallel to the line of the acetylenic carbons; this seems to indicate an easy pathway for reaction between the acetylene and the cyclobutadiene, presumably first to a Dewar benzene and finally to the benzene. On heating **43**, in fact, hexaphenylbenzene was formed.

Little is known for certain about the mechanism by which acetylenes form cyclobutadiene-metal complexes, and a number of possible reaction paths can be suggested. One such, a stepwise path, is discussed briefly in Section II,E. Another possibility for which there is some evidence (for example, from the

[51] A. N. Nesmeyanov, K. N. Anisimov, N. E. Kolobova, and A. A. Pasynskii, *Ivz. Akad. Nauk SSSR, Ser. Khim.* p. 100 (1968).

[52] A. N. Nesmeyanov, A. I. Gusev, A. A. Pasynskii, K. N. Anisimov, N. E. Kolobova, and Y. T. Struchkov, *Chem. Commun.* p. 277 (1969).

[53] A. N. Nesmeyanov, A. I. Gusev, A. A. Pasynskii, K. N. Anisimov, N. E. Kolobova, and Y. T. Struchkov, *Chem. Commun.* p. 739 (1969).

work of Yamazaki and Hagihara,[47] Nesmeyanov *et al.*,[51–53] and Hübel and his collaborators,[54]) is:

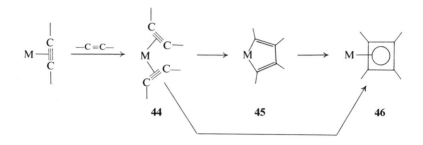

44   **45**   **46**

In this, a bisacetylene complex **44** is formed, which then undergoes a rearrangement, either directly to the cyclobutadiene-metal complex **46** or via a metalacyclopentadiene complex **45**.

Molecular orbital calculations by Mango and Schachtschneider[55] on the direct reaction, **44** → **46**, suggest that this cannot be a concerted process unless two (or more) metal atoms participate. In fact there is considerable evidence (*a*) for the participation of two or more metal atoms and also (*b*) for a stepwise process (e.g., via **45**) in this transformation.

D. From 1-Heterocyclopentadienes and from α-Pyrone

One of the earliest "convenient" routes to a cyclobutadiene complex (**48**) was that devised by Freedman, who used as starting material the stannole (**47**).[13]

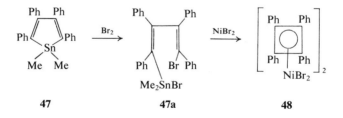

47   **47a**   **48**

[54] W. Hübel, *in* "Organic Syntheses via Metal Carbonyls" (I. Wender and P. Pino, eds.), pp. 273 and 288. Wiley (Interscience), New York, 1968.
[55] F. D. Mango and J. H. Schachtschneider, *J. Am. Chem. Soc.* **91**, 1030 (1969).

Somewhat analogous reactions have been utilized by Hübel and his collaborators,[56] e.g.,

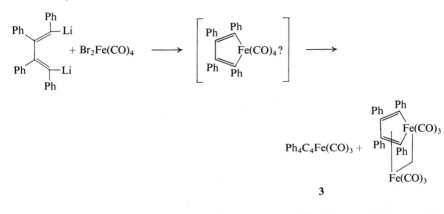

3

Another example of this type of reaction is that of Hagihara and Yamazaki referred to in the previous section.

Rosenblum and his co-workers[57,58] have described a very novel and potentially useful route to the unsubstituted cyclobutadiene-metal complexes. This involves irradiating α-pyrone (**49**) or its photoisomer (**50**)[58a] in the presence of a metal carbonyl, e.g.,

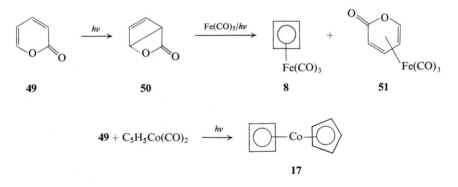

When the irradiation was carried out in the presence of iron pentacarbonyl, the cyclobutadiene complex **8** and the α-pyrone complex **51** were obtained. This gave only rather poor yields of these complexes since both decomposed

[56] W. Hübel, *in* "Organic Syntheses via Metal Carbonyls" (I. Wender and P. Pino, eds.), p. 294. Wiley (Interscience) New York, 1968.

[57] M. Rosenblum and C. Gatsonis, *J. Am. Chem. Soc.* **89**, 5074 (1967).

[58] R. Rosenblum and B. North, *J. Am. Chem. Soc.* **90**, 1060 (1968).

[58a] This can be thought of as a Diels–Alder adduct of cyclobutadiene and carbon dioxide.

under the conditions (irradiation) used for their formation. The cyclobutadiene-(cyclopentadienyl)cobalt **17** was synthesized analogously; no yields were reported.

E. FROM π-CYCLOBUTENYL-METAL COMPLEXES

Malatesta et al.[59] first reported the reaction between diphenylacetylene and palladium chloride in ethanol to give $(Ph_4C_4OEtPdCl)_2$. Subsequent investigations by a number of authors[60-63] and a crystal structure determination showed the complex to have the structure **52** with the ethoxy group endo to the metal.[64]

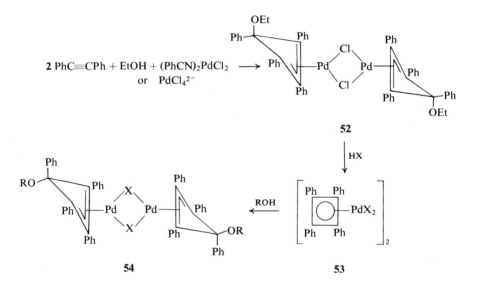

A number of complexes similar to **52** with other bridging ligands (Br, I), alkoxy groups (MeO, HO), and also with p-chlorophenyl and p-tolyl in place of phenyl have been prepared.[48,49,61,63] All these complexes reacted with hydrogen halides to give the cyclobutadiene complexes **53** (X = Cl, Br, I;

[59] L. Malatesta, G. Santarella, L. M. Vallarino, and F. Zingales, *Atti Accad. Nazl. Lincei, Rend., Classe Sci. Fis., Mat. Nat.* [8] **27**, 230 (1959); *Angew. Chem.* **72**, 34 (1960).

[60] A. T. Blomquist and P. M. Maitlis, *J. Am. Chem. Soc.* **84**, 2329 (1962).

[61] L. M. Vallarino and G. Santarella, *Gazz. Chim. Ital.* **94**, 252 (1964).

[62] R. C. Cookson and D. W. Jones, *J. Chem. Soc.* p. 1881 (1965).

[63] P. M. Maitlis and M. L. Games, *Can. J. Chem.* **42**, 183 (1964).

or with p-ClC_6H_4 in place of Ph). The latter complexes reacted with alkoxide ion under very mild conditions to give the alkoxytetraphenylcyclobutenyl complexes **54**. These are isomeric with **52** but the alkoxy group is now exo to the metal.[60, 64] The reaction **53** → **54** is reversed on treatment of the complexes **54** with acid. A similar reversible reaction occurred with cyclopentadienyl-(tetraphenylcyclobutadiene)nickel and -palladium cations[65, 66]:

$$[Ph_4C_4MC_5H_5]^+ \underset{H^+}{\overset{OR^-}{\rightleftarrows}} exo\text{-}ROC_4Ph_4MC_5H_5$$

M = Ni, Pd
R = alkyl, H

The formation of the *endo*-alkoxycyclobutenyl complexes **52** or the cyclo-butadiene complexes **40** from L_2PdCl_2 (L = PhCN, Cl⁻) and a diphenyl-acetylene, depending on the conditions, is an intriguing reaction. Only acetylenes with two extremely bulky substituents (e.g., phenyl) appear to undergo this reaction; others with only one bulky substituent give trimers,[67] while acetylenes such as dimethylacetylene give trimers and higher oligomers.[68] A simple explanation[68] is that reaction occurs by (*i*) formation of a π-complex; (*ii*) cis "insertion" of the acetylene into Pd—X to give a σ-vinyl complex, (this step is slow for X = Cl, fast for X = OR); (*iii*) formation of a new π-complex; and (*iv*) cis "insertion" of the acetylene into the Pd—vinyl bond. This step is very fast, and the overall process of π-complex formation and cis insertion is repeatable indefinitely and limited largely by steric considera-tions. With large R groups, reaction does not proceed beyond the insertion of three acetylenes and a significant amount of reaction stops after insertion of two. Rearrangements of these products then leads to the benzene (*v*) or the cyclobutenyl-metal complex (*vi*). For electronic or possibly steric reasons, when X = alkoxy and the acetylenic substituents are very large (phenyl) reaction effectively ceases after two acetylenes have been inserted (**55**); little or no trimer is ever detected under these conditions.

[64] L. F. Dahl and W. E. Oberhansli, *Inorg. Chem.* **4**, 629 (1965).

[65] P. M. Maitlis, A. Efraty, and M. L. Games, *J. Organometal. Chem.* (*Amsterdam*) **2**, 284 (1964).

[66] P. M. Maitlis, A. Efraty, and M. L. Games, *J. Am. Chem. Soc.* **87**, 719 (1965).

[67] H. Dietl and P. M. Maitlis, *Chem. Commun.* p. 481 (1968).

[68] H. Reinheimer, H. Dietl, J. Moffat, D. Wolff, and P. M. Maitlis, *J. Am. Chem. Soc.* **90**, 5321 (1968); H. Dietl, H. Reinheimer, J. Moffat, and P. M. Maitlis, *J. Am. Chem. Soc.* **92**, 2276 (1970); H. Reinheimer, J. Moffat, and P. M. Maitlis, *J. Am. Chem. Soc.* **92**, 2285 (1970).

$$L_2PdCl_2 + ROH \rightarrow L_2PdCl(OR) + HCl$$

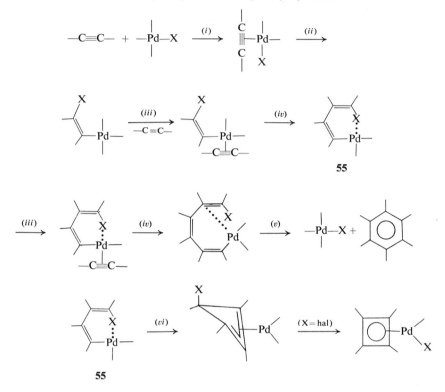

55

F. From Other Cyclobutadiene-Metal Complexes by Ligand-Transfer Reactions

The preparation of cyclobutadiene complexes of different metals from some of the easily accessible cyclobutadiene complexes of palladium and nickel has become possible through the ligand-exchange and transfer reactions developed by Maitlis and co-workers.

As described below (Section IV,E), Maitlis and Stone observed that the tetraphenylcyclobutadiene ligand could be detached from the complex **53** by reaction with tertiary phosphines.[69] The product was the cyclobutadiene dimer, octaphenylcyclooctatetraene, and the bis(*t*-phosphine)palladium halide.

It has long been recognized that the phosphine ligands in $(R_3P)_2NiX_2$ are labile; on reaction of these complexes with tetraphenylcyclobutadiene-palladium halide complexes, virtually complete exchange of the ligands

[69] P. M. Maitlis and F. G. A. Stone, *Proc. Chem. Soc.* p. 330 (1962).

between the two metals occurs.[70] The reaction can be used synthetically to prepare the tetraphenylcyclobutadienenickel complexes **56**, if a phosphine which gives a soluble palladium complex (such as tri-*n*-butyl phosphine) is employed.

$$[R_4C_4PdX_2]_2 + 2\ (Bu_3)P_2NiX_2 \rightarrow [R_4C_4NiX_2]_2 + 2\ (Bu_3P)_2PdX_2$$

56

R = Ph, *p*-ClC$_6$H$_4$, *p*-MeC$_6$H$_4$, and *p*-MeOC$_6$H$_4$

It appears doubtful that a free tetraphenylcyclobutadiene is an intermediate in this reaction; more probably a complex in which the cyclobutadiene is bonded to both the nickel and the palladium is the true intermediate. This mechanism has been discussed elsewhere.[1,70]

Unfortunately this reaction is very limited in scope; attempts to transfer the cyclobutadiene onto other metals (Fe, Co, Cu, etc.) have failed. The usual product then is the cyclooctatetraene.

A synthetic method of wider applicability is the ligand-transfer reaction which occurs when the tetraphenylcyclobutadienepalladium dihalides (the bromides were most frequently employed) react with metal carbonyls.[71-77] Most of the reactions were carried out on tetraphenylcyclobutadienepalladium bromide, but the tetrakis(*p*-tolyl)- and tetrakis(*p*-chlorophenyl)cyclobutadienepalladium bromides (and probably other *p*-substituted tetraphenylcyclobutadienepalladium complexes) react in the same way.[75] In one reaction at least, tetraphenylcyclobutadienenickel bromide reacted similarly.[75] One example of the reverse reaction, in which the cyclobutadiene was transferred from iron onto palladium, is also known.[75] These reactions are summarized in Scheme 1. The reactions were usually carried out in a solvent such as xylene, toluene, or benzene at reflux. Yields varied from near quantitative for the iron complex **3**[73] to a few percent for the molybdenum and tungsten complexes **61** and **74**. The molybdenum complex **58** when originally prepared was formulated as the tricarbonyl dimer [Ph$_4$C$_4$Mo(CO)$_3$Br]$_2$[74] on the basis of a complete elemental analysis. A recent X-ray determination has indicated it to be as shown.[78] However it is not yet quite clear that this material is identical to that originally synthesized. In the reaction with nickel carbonyl, two products were

[70] P. M. Maitlis and D. F. Pollock, *Can. J. Chem.* **44**, 2673 (1966).

[71] P. M. Maitlis and A. J. Efraty, *J. Organometal. Chem.* (*Amsterdam*) **4**, 172 (1965).

[72] P. M. Maitlis, A. Efraty, and M. L. Games, *J. Organometal. Chem.* (*Amsterdam*) **2**, 284 (1964).

[73] P. M. Maitlis and M. L. Games, *J. Am. Chem. Soc.* **85**, 1887 (1963).

[74] P. M. Maitlis and M. L. Games, *Chem. & Ind.* (*London*) p. 1624 (1963).

[75] D. Pollock, Ph.D. Thesis, McMaster University, 1969.

[76] P. M. Maitlis and A. Efraty, *J. Am. Chem. Soc.* **89**, 3744 (1967).

[77] P. M. Maitlis, *Ann. N.Y. Acad. Sci.* **159**, 110 (1969).

[78] M. Mathew and G. J. Palenik, *Can. J. Chem.* **47**, 705 (1969).

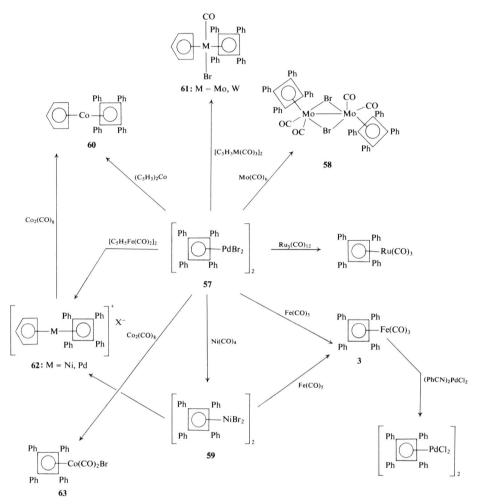

Scheme 1. Synthesis of tetraphenylcyclobutadiene-metal complexes by ligand-transfer reactions.

obtained; when the reaction was run in refluxing benzene the major product was tetraphenylcyclopentadienone. At 40°C hardly any of this was formed and the major product was the nickel complex **59**.

The cobalt complex **60** could be obtained both from reaction of the palladium complex **57** with cobaltocene,[73] or (in low yields) by reaction of the cationic complex **62** (X = Br) with dicobalt octacarbonyl.[66] The latter involved the transfer of both a cyclopentadienyl and a cyclobutadiene ligand, probably in a stepwise manner.

The cyclopentadienylmolybdenum and -tungsten tricarbonyl dimers reacted with **57** to give **61** in which a tetraphenylcyclobutadiene group had been transferred.[65,66] In contrast, cyclopentadienyliron dicarbonyl dimer (or the bromide) reacted with **57** to give **62** (X = FeBr$_4$) in which the *cyclopentadienyl* group had transferred. The nickel complex **59** underwent the same reaction.[76]

Metal (palladium) was always formed in these reactions; other by-products obtained included tetraphenylcyclopentadienone and, or, octaphenylcyclooctatetraene. The reaction apparently proceeded in two steps, the first involving decomposition of the palladium complex by the carbonyl (or other low valent species). In the second step, two (or more) carbonyl groups were displaced by the entering ligand. This could be either tetraphenylcyclobutadiene itself or a closely related species (for example, a halotetraphenylcyclobutenyl radical[109]). The second step is the critical one, since decomposition (detected by the formation of palladium metal) always occurred on reaction of a carbonyl with **57** and its analogs. However a cyclobutadiene complex was not obtained in all cases. Even when a complex was formed (as, for example, **58**), the major products were often the cyclooctatetraene and the cyclopentadienone. Mechanistic suggestions have been made elsewhere.[1]

The reaction was also very dependent on the substituents on the cyclobutadiene ring. With one notable exception, tetramethylcyclobutadienenickel halides reacted less readily and less cleanly than did the tetraphenyl complexes. Nevertheless, a number of reactions of this type were carried out by Bruce and Maitlis,[40,79] as shown in Scheme 2.

The reactions of **64** with iron carbonyls proceeded readily enough but several different complexes, in addition to organic products, were usually obtained. The cyclobutadiene complex **65** is still, however, obtained most conveniently by this route. The interesting bimetallic cyclobutadiene complex **66** is related to **25a** and has had its structure determined by Epstein and Dahl.[80] This is a red, crystalline, and rather inert substance formed to the extent of 5% in the reaction of **64** with triiron dodecacarbonyl.[79]

The reaction of **64** with dicobalt octacarbonyl was more successful and gave rise to only one product, the dicobalt hexacarbonyl complex **67** (no bridging carbonyls). On reaction with iodine at low temperatures, the Co—Co bond in **67** was oxidized to give **69** quantitatively. This complex was also obtained quantitatively from the cyclobutadienenickel iodide complex **68** by ligand-transfer. Interestingly, no metal was precipitated in either of these reactions. The nickel was converted to nickel carbonyl, and the overall reaction is better described as a ligand exchange, rather than as a ligand transfer, i.e.,

$$2 \, [Me_4C_4NiI_2]_2 + 3 \, Co_2(CO)_8 \rightarrow 4 \, Me_4C_4Co(CO)_2I + Ni(CO)_4 + 2 \, CoI_2$$

[79] R. Bruce, K. Moseley, and P. M. Maitlis, *Can. J. Chem.* **45**, 2011 (1967).
[80] E. F. Epstein and L. F. Dahl, *J. Am. Chem. Soc.* **92**, 502 (1970).

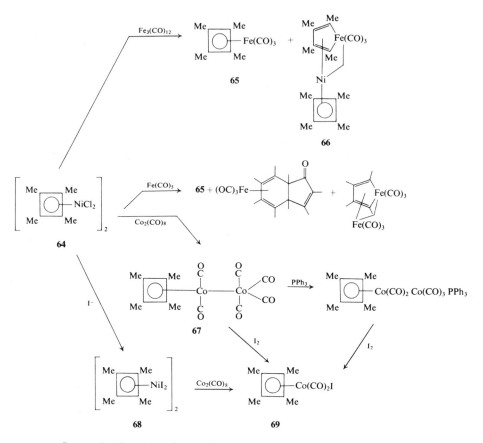

Scheme 2. Ligand-transfer reactions of tetramethylcyclobutadiene complexes.

G. Unsuccessful Approaches to Cyclobutadiene-Metal Complexes

A great many attempts to form cyclobutadiene-metal complexes have failed, even though, in some cases, they represented merely the extension of an established procedure.

For example, while diphenylacetylene and palladium chloride could be reacted to give a cyclobutadiene complex, this was not the case with most other acetylenes, where trimers and tetramers of the acetylene were the usual products.[68, 81] Only acetylenes with very bulky substituents appear to undergo this reaction (see footnote 50c).

Similarly, although 3,4-dichlorotetramethylcyclobutene (4) reacted with nickel carbonyl to give cyclobutadiene complex 2 in good yield, other 3,4-

[81] F. Zingales, *Ann. Chim.* (*Rome*) 52, 1174 (1962).

dihalocyclobutenes did not. 3,4-Dibromo-1,2-diphenylcyclobutene only gave polymers and nickel bromide.[82] Attempts to prepare cyclobutadienenickel chloride from 3,4-dichlorocyclobutene in a variety of solvents all failed.[83] Similarly 3,4-difluorocyclobutene,[27] hexafluorocyclobutene[83] did not react with diiron enneacarbonyl to give iron tricarbonyl complexes. In fact, even the tetramethylcyclobutadieneiron tricarbonyl complex (65) was obtained only in very poor yield from the dichlorocyclobutene (4) and diiron enneacarbonyl.[79]

Again, whereas the tetraphenylcyclobutadienenickel bromide complex 48 was readily accessible from nickel bromide and (4-bromo-1,2,3,4-tetraphenyl-cis,cis-1,3-butadienyl)dimethyltin bromide (47a), this reaction did not work when palladium bromide was substituted for NiBr₂, despite the proven stability of the tetraphenylcyclobutadienepalladium bromide complex 57.[13] Attempts to prepare iron(III) and copper(II) complexes by this method also failed.[13]

The reported preparation of cyclobutadienesilver nitrate[84-87] was later shown to be wrong.[88] The complex was in fact the tricyclooctadiene bis-silver nitrate adduct.

Attempts to prepare a cyclopentadienyl(tetramethylcyclobutadiene)nickel complex from 2 and sodium cyclopentadienide failed.[89,90] Two molecules of cyclopentadienide were added and a cyclopentadienyl(cyclobutenyl) complex was obtained.[91]

McOmie and Bullimore[92] attempted to prepare the nickel complex 48 by pyrolysis of tetraphenylthiophene-1,1-dioxide in the presence of nickel bromide.

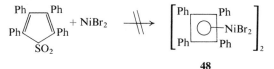

48

Only organic products (1,2,3-triphenylnaphthalene, tetraphenylthiophene, and 3,6-diphenyl-1,2:3,4-dibenzopentalene) were obtained.

[82] A. T. Blomquist and E. A. LaLancette, *J. Org. Chem.* **29**, 2331 (1964).

[83] J. D. Fitzpatrick, *Dissertation Abstr.* **27**, 405B (1966).

[84] M. Avram, W. P. Fritz, H. Keller, G. Mateescu, J. F. W. McOmie, N. Sheppard, and C. D. Nenitzescu, *Tetrahedron* **19**, 187 (1963).

[85] M. Avram, E. Marica, and C. D. Nenitzescu, *Tetrahedron Letters* p. 21 (1961).

[86] H. P. Fritz, J. F. W. McOmie, and N. Sheppard, *Tetrahedron Letters* p. 35 (1960).

[87] N. Martologu and D. Mumuianu, *Rev. Roumaine Chim.* **6**, 303 (1961).

[88] H. P. Fritz, *Z. Naturforsch.* **16b**, 415 (1961); *Advan. Organometal. Chem.* **1**, 260 (1964).

[89] R. Criegee, F. Förg, H. A. Brune, and D. Schönleber, *Ber.* **97**, 3461 (1964).

[90] R. Criegee and P. Ludwig, *Ber.* **94**, 2038 (1961).

[91] W. Oberhansli and L. F. Dahl, *Inorg. Chem.* **4**, 150 (1965).

[92] J. F. W. McOmie and B. K. Bullimore, *Chem. Commun.* p. 63 (1965).

III. Cyclobutadiene-Metal Complexes as Reaction Intermediates

Several reactions are known in which the intermediacy of a cyclobutadiene-metal complex has (or can be) reasonably postulated. However, concrete evidence for their participation has so far been lacking.

For example, on treatment of a tetraphenylcyclobutadienepalladium bromide (57) with nickel carbonyl in benzene at 80°C, good yields of tetraphenyl cyclopentadienone were obtained by Maitlis et al.[73, 75] Similarly a cyclobutadiene complex *could* be an intermediate in the formation of 3,4-bis-(trimethylsilyl)-2,5-diphenylcyclopentadienone (70) from diphenylacetylene-dicobalt hexacarbonyl and bis(trimethylsilyl)acetylene.[93]

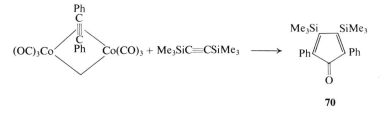

70

However, there is no real evidence for the intermediacy of cyclobutadiene-metal complexes in the metal carbonyl-catalyzed formation of cyclopentadienones from acetylenes and carbon monoxide.[20, 21, 93, 94]

Perhaps the most intriguing suggestion has been that cyclobutadiene complexes are intermediates in the cyclotrimerization of acetylenes to benzenes[18, 93, 95–97] and in their cyclotetramerization to cyclooctatetraenes.[98, 98a]

Blomquist and Maitlis showed that although hexaphenylbenzene and tetra-phenylcyclobutadienepalladium chloride were formed in the reaction of palladium chloride with diphenylacetylene, the cyclobutadiene complex itself did not react with diphenylacetylene to give hexaphenylbenzene.[60] This appears true of most other cyclobutadiene complexes which are rather inert, especially to Diels–Alder reactions of this type.

[93] U. Krüerke and W. Hübel, *Ber.* **94**, 2829 (1961).

[94] G. N. Schrauzer, *Chem. & Ind.* (*London*) p. 1404 (1958).

[95] E. M. Arnett and J. M. Bollinger, *J. Am. Chem. Soc.* **86**, 4729 (1964).

[96] T. L. Cairns, V. A. Engelhardt, H. L. Jackson, G. H. Kalb, and J. C. Sauer, *J. Am. Chem. Soc.* **74**, 5636 (1952); J. C. Sauer and T. L. Cairns, *ibid.* **79**, 2659 (1957); T. J. Bieber, *Chem. & Ind.* (*London*) p. 1126 (1957).

[97] M. Tsutsui and H. Zeiss, *J. Am. Chem. Soc.* **82**, 6255 (1960).

[98] E. D. Bergmann, "Chemistry of Acetylenes and Related Compounds," p. 93. Wiley (Interscience), New York, 1948.

[98a] W. Reppe, "Neue Entwicklungen auf dem Gebiete der Chemie des Acetylenes und Kohlenoxyds," p. 68. Springer, Berlin, 1949.

An ingenious test for the intermediacy of cyclobutadiene-metal complexes in the cyclotrimerization of 2-butyne has been devised by Whitesides and Ehmann.[99] They trimerized 2-butyne-1,1,1-d_3 (71) in the presence of triphenyl-tris(tetrahydrofuran)chromium, dimesitylcobalt, dicobalt octacarbonyl, bis-(acrylonitrile)nickel, a Ziegler catalyst, bis(benzonitrile)palladium chloride, and aluminum chloride. If the reaction proceeded by path (i), involving either a concerted trimerization or a stepwise head-to-head or head-to-tail trimerization, *none* of the isomer 74 should be formed. On the other hand,

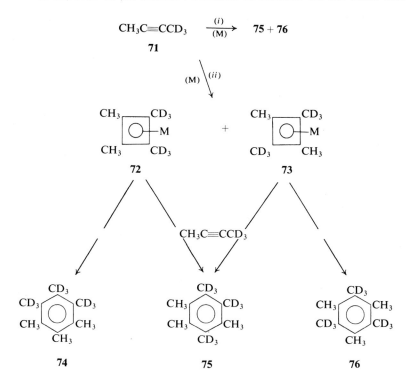

if a cyclobutadiene-metal complex (or a complex with the same symmetry) was an intermediate (ii), this could be formed in two ways, to give 72 and 73. Both of them could react further to give the benzenes, possibly via the Dewar benzenes or their complexes. Assuming deuterium and hydrogen exert the same steric and electronic effect, this path would yield a statistical distribution of the isomers: 12.5% 74, 62.5% 75, and 25% 76.

Whitesides and Ehmann showed that none of the above catalysts, except aluminum chloride and the palladium catalyst, gave any appreciable amount of

[99] G. M. Whitesides and W. J. Ehmann, *J. Am. Chem. Soc.* 90, 804 (1968); 91, 3800 (1969).

74. Aluminum chloride as catalyst gave exactly the 12.5 % of **74** predicted, whereas palladium chloride gave 9.5 %. However, for both these cases it appears probable that intermediates which are *not* cyclobutadienes are present.[68,99] This method appears to be very useful for establishing the absence of a cyclobutadiene intermediate, but a positive result does not exclude a cyclobutadiene *not* being an intermediate.

The intermediacy of cyclobutadienes in the formation of cyclooctatetraenes was first postulated by Bergmann[98] to explain the Reppe synthesis of cyclooctatetraene from acetylene and a nickel(II) catalyst.[98a] Longuet-Higgins and Orgel[18] suggested that a cyclobutadiene-metal complex was an intermediate in this reaction to account for the low yield of the thermodynamically more stable benzene.

Schrauzer *et al.*[100] in a study of the Reppe reaction found the rate of cyclooctatetraene formation was highest for kinetically labile nickel complexes and for weakly coordinating solvents. In the presence of a strong monodentate ligand (triphenylphosphine) no cyclooctatetraene, only benzene, was formed. Chelating ligands (dipyridyl, etc.) completely poisoned the catalyst. Schrauzer *et al.* suggested that only complexes with four coordination sites available around the (octahedral) nickel would function as cyclotetramerization catalysts. Blocking one such site by a phosphine resulted in the formation of benzene. They also suggested that the intermediate in the cyclooctatetraene synthesis was a complex with four acetylenes arranged in such a manner as to give the product by a "concerted" ring-closure. Cyclobutadiene was not an intermediate. Although this hypothesis has much to recommend it, it well may be too simple and naive, since recent studies have shown these types of reaction to be exceedingly complex.[68]

In contrast, a large number of presently known cyclobutadiene-metal complexes can react under suitable conditions to give tricyclooctadienes or cyclooctatetraenes; this also occurs in reactions where such complexes are reasonable intermediates, although not isolated.[12,24,31,69,101-106] A particularly interesting case where a cyclobutadiene-metal complex does appear to be an intermediate in the formation of hexaphenylbenzene is the niobium complex **43**[51-53] already discussed (Section II,C).

[100] G. N. Schrauzer, P. Glockner, and S. Eichler, *Angew. Chem. Intern. Ed. Engl.* **3**, 185 (1964).

[101] M. Avram, H. P. Fritz, H. J. Keller, C. G. Kreiter, G. Mateescu, J. F. W. McOmie, N. Sheppard, and C. D. Nenitzescu, *Tetrahedron Letters* p. 1611 (1963).

[102] M. Avram, E. Marica, and C. D. Nenitzescu, *Ber.* **92**, 1088 (1959).

[103] R. Criegee, J. Dekker, W. Engel, P. Ludwig, and K. Noll, *Ber.* **96**, 2362 (1963).

[104] H. H. Freedman and D. R. Petersen, *J. Am. Chem. Soc.* **94**, 2837 (1962).

[105] K. Nagarajan, M. C. Caserio, and J. D. Roberts, *J. Am. Chem. Soc.* **86**, 449 (1964).

[106] E. H. White and H. C. Dunathan, *J. Am. Chem. Soc.* **86**, 453 (1964).

IV. Reactions of Cyclobutadiene-Metal Complexes

Many of the reactions to be described were carried out with the aim of confirming the structure of the particular complex. However, structural evidence from chemical degradation is not really conclusive in cases where the reaction path is not known.

A. THERMOLYSIS

The thermal decomposition of a number of cyclobutadiene-metal complexes has been studied, in particular in relation to the cyclooctatetraene synthesis discussed above. Many cyclobutadiene complexes are decomposed to cyclo-octatetraenes thermally and in other ways, but the exact conditions are usually quite different from those of the direct cyclooctatetraene synthesis. For example, Criegee et al.[19, 24, 103] found that decomposition of the nickel complex 2 needed 185°C under high vacuum. Even then, the cyclooctatetraene 77 was only formed in 7–12% yield. The main products were the methylene-bicyclo[4.2.0]octadienes, 79 and 80, which probably arose from the tricyclo-octadienes, 78.[24, 107] The triphenylphosphine complex of 2 also gave 77, 79,

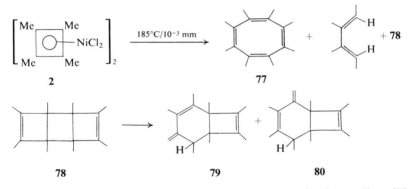

and 80 but the o-phenanthroline complex gave the anti-tricyclooctadiene (78a) and 3-methylene-1,2,4-trimethylcyclobutene, as well as 79 and 80.

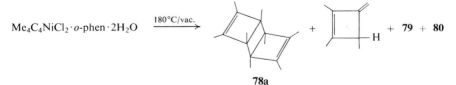

[107] R. Criegee and G. Louis, Ber. 90, 417 (1957).

The *syn*-tricyclooctadiene (**78b**) was obtained by hydrolysis of **2** at 100°C.[24] It is not clear whether tetramethylcyclobutadiene or a closely related substance is the true intermediate in these reactions.

Tetraphenylcyclobutadienenickel bromide (**48**) on pyrolysis gave largely octaphenylcyclooctatetraene (**81**) and nickel bromide.[13, 104, 108]

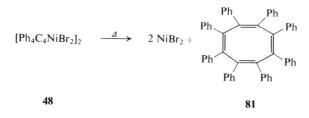

$[Ph_4C_4NiBr_2]_2 \xrightarrow{\Delta} 2\ NiBr_2 +$

48   **81**

In contrast, the tetraphenylcyclobutadienepalladium chloride complex gave two isomeric 1,4-dichloro-1,2,3,4-tetraphenylbutadienes and metal, as well as some diphenyldibenzopentalene.[60, 62] The dichlorobutadienes may

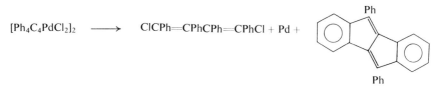

$[Ph_4C_4PdCl_2]_2 \longrightarrow ClCPh{=}CPhCPh{=}CPhCl + Pd +$

well arise from a chlorotetraphenylcyclobutenyl radical,[109] discussed below, and are formed partly because of the low thermal stability of palladium–halogen bonds.

The formation of tetraphenylcyclobutadiene (as $Ph_4C_4^+$) in the mass spectrum of the pyrolyzate of $(Ph_4C_4PdCl_2)_2$ has been claimed by Beynon *et al.*[110] However, Sandel and Freedman have questioned this.[109] The latter authors suggested that the peaks observed in the mass spectrum were due to the dichlorotetraphenylcyclobutene and the chlorotetraphenylcyclobutenyl radical, and the peak at 356 m/e ascribed to $Ph_4C_4^+$ arose from these species.

Tyerman *et al.* observed cyclobutadiene directly by kinetic mass spectrometry; it was generated by flash photolysis of the iron tricarbonyl complex **8**.[111] Li and McGee have reported that pyrolysis of **8**, followed by quenching at very low temperatures allowed the preparation of cyclobutadiene. Its mass

[108] G. S. Pawley, W. N. Lipscomb, and H. H. Freedman, *J. Am. Chem. Soc.* **86**, 4725 (1964); P. J. Wheatley, *J. Chem. Soc.* p. 3136 (1965).

[109] V. R. Sandel and H. H. Freedman, *J. Am. Chem. Soc.* **90**, 2059 (1968).

[110] J. H. Beynon, R. C. Cookson, R. R. Hill, D. W. Jones, R. A. Saunders, and A. E. Williams, *J. Chem. Soc.* p. 7052 (1965).

[111] W. J. R. Tyerman, M. Kato, P. Kebarle, S. Masamune, O. P. Strausz, and H. E. Gunning, *Chem. Commun.* p. 497 (1967).

spectrum and ionization potential were determined at $-105°C$ and compared with those of other C_4H_4 isomers[112]; however, see Appendix.

B. SOLVOLYSIS

Tetramethylcyclobutadienenickel chloride **2** is soluble in water (to give $[Me_4C_4Ni(H_2O)_2]^{2+}$?).[19] On heating to 100°C this solution gives *anti*-tricyclooctadiene (**78a**). The di- and the trimethylcyclobutadienenickel chloride complexes are more soluble in water than **2** but decompose rapidly even at 20°C.[27] Their solutions in hydrochloric or hydrobromic acid are more stable and the dimethyl complex could be recovered from the latter solvent.[28] This suggests that nucleophilic attack is easy on cyclobutadiene-nickel halide complexes which lack one or more substituents on the ring. The tetraphenylcyclobutadienenickel halide complexes are slowly decomposed by weakly coordinating solvents (ethers, alcohols, nitriles) and water but can be crystallized from methylene chloride. Complexing solvents give deep blue or green colors. However, treatment of a solution of the complex **48** in DMF with aqueous sodium acetate or sodium hydroxide apparently resulted only in exchange of the anionic ligands.[13]

C. REDUCTION

Catalytic hydrogenation or zinc–hydrochloric acid reduction of tetra-methylcyclobutadienenickel chloride (**2**) gave a 90% yield of all-*cis*-tetra-methylcyclobutane.[24]

Tetraphenylcyclobutadieneiron tricarbonyl (**3**) gave *cis,cis*-1,2,3,4-tetra-phenylbutadiene on reduction with lithium aluminum hydride[20,113] and tetraphenylbutene with sodium in liquid ammonia.[20] This latter reagent also reduced cyclopentadienyl(tetraphenylcyclobutadiene)cobalt (**30**) to tetra-phenylbutene.[44]

Lithium aluminum hydride reduced the nickel complex **48** to *cis*-tetraphenyl-cyclobutene, which isomerized at 50°C to *cis,trans*-tetraphenylbutadiene.[104, 113] Drastic conditions were needed to hydrogenate **48** catalytically (75°C/60 atm) to a hydrocarbon, originally identified as a tetraphenylcyclobutene,[13] but which appears to be a tetraphenylbutene.[62] This same hydrocarbon was also obtained by catalytic hydrogenation of tetraphenylcyclobutadienepalladium chloride. Under milder conditions,[49] or when using sodium borohydride[50, 61] of lithium aluminum hydride, [61, 62] *cis,cis*-tetraphenylbutadiene was obtained from the palladium complex.

[112] P. H. Li and H. A. McGee, *Chem. Commun.* p. 592 (1969).

[113] H. H. Freedman, G. A. Doorakian, and V. R. Sandel, *J. Am. Chem. Soc.* **87**, 3019 (1965).

D. OXIDATION

Mild oxidation of tetramethylcyclobutadienenickel chloride (2) with aqueous sodium nitrite gave cis-3,4-dihydroxy-1,2,3,4-tetramethylcyclobutene (82).[19] A similar reaction carried out on the tetraphenyl complex gave tetraphenylfuran (83).[13]

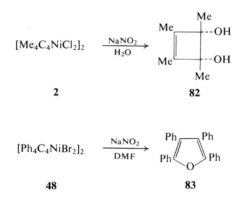

$[Me_4C_4NiCl_2]_2$ $\xrightarrow[H_2O]{NaNO_2}$

2   82

$[Ph_4C_4NiBr_2]_2$ $\xrightarrow[DMF]{NaNO_2}$

48   83

The furan 83 was also formed when the tetraphenylcyclobutadienepalladium halides were treated with a phosphine or phosphite in the presence of air.[62, 69] Oxidation of this palladium complex with nitric acid gave cis-dibenzoylstilbene (84).[50, 61]

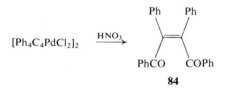

$[Ph_4C_4PdCl_2]_2$ $\xrightarrow{HNO_3}$

84

The diketone 84 was also obtained by air oxidation of a heated solution of 47a, and Sandel and Freedman[109] have proposed a mechanism to account for its formation there.

Pettit and his co-workers have opened a new branch of organic chemistry with reactions involving the oxidation of cyclobutadieneiron tricarbonyl 8.[31] Using iron(III) or cerium(IV) as oxidizers, cyclobutadiene (probably) is liberated, which then reacts further in a number of ways. In the presence of excess chloride ion the product is trans-3,4-dichlorocyclobutene (85),[31] whereas acetylenes give a variety of Dewar benzenes 86 which rearrange to the normal benzenes **87** on heating.[17,114] This represents a useful route to the simpler Dewar benzenes.

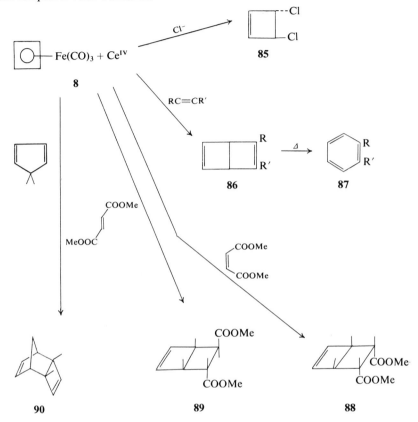

With maleate and fumarate it gives the *endo,cis*-5,6-dicarbomethoxybicyclo-[2.2.0]hexene (**88**) and *trans*-5,6-dicarbomethoxybicyclo[2.2.0]hexene (**89**), respectively.[16] These stereospecific reactions, and that with cyclopentadiene to give **90**, suggest that cyclobutadiene reacts as a singlet rather than a triplet both as a diene and a dienophile. However the dimerization of cyclobutadiene (in the absence of other reagents) is not entirely stereospecific and gives the *syn*- and the *anti*-tricyclooctadienes, (**91**) and (**92**), in a 5:1 ratio.

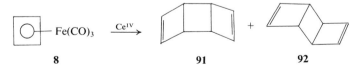

[114] R. Pettit and G. D. Burt, *Chem. Commun.* p. 517 (1965).

Pettit *et al.* have also developed a synthesis of cubane (**93**) starting from **8** and 2,5-dibromoquinone[115]:

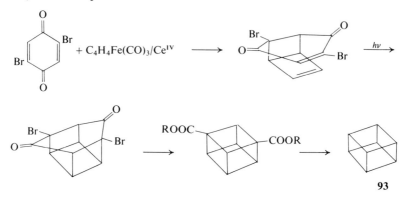

Among other recently reported syntheses based on cyclobutadiene was that of **94** from thiapyronedioxide by Paquette and Wise.[116]

In contrast to these reactions, the benzocyclobutadieneiron tricarbonyl (**9**) gave polymer on oxidation with cerium(IV) or iron(III). However, silver nitrate oxidation gave hydrocarbon **95** arising from dimerization of benzo-cyclobutadiene.[31]

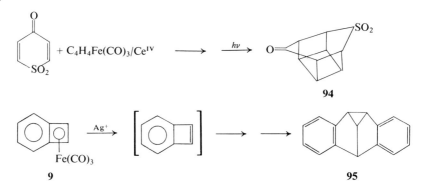

E. REACTIONS WITH DONOR LIGANDS

The cyclobutadiene-metal dihalide complexes of nickel and palladium form structures with halogen bridges linking two or more units.[19, 70, 117] This makes the metal atoms in these complexes formally 5-coordinate and allows them to

[115] R. Pettit, J. C. Barborak, and L. Watts, *J. Am. Chem. Soc.* **88**, 1328 (1966).

[116] L. A. Paquette and L. D. Wise, *J. Am. Chem. Soc.* **89**, 6659 (1967).

[117] J. D. Dunitz, H. C. Mez, O. S. Mills, and H. M. M. Shearer, *Helv. Chim. Acta* **45**, 647 (1962).

obey the Effective Atomic Number formalism. With donor ligands these bridges are broken with the formation of $R_4C_4MX_2L$.[24,103] These are sometimes very labile. Complexes such as $R_4C_4Fe(CO)_3$ are much more inert and will only react with triphenylphosphine; for example, at 140°C by substitution of one carbonyl.[31,118]

$$Ph_4C_4Fe(CO)_3 + PPh_3 \rightarrow Ph_4C_4Fe(CO)_2PPh_3$$

The synthesis and crystal structure of tricarbonyltriphenylphosphinecyclo-butadienemolybdenum has been reported.[38] The cyclobutadiene dicarbonyl halide complexes **96** reacted with triphenylphosphine to give **97**.[40,76]

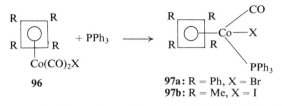

97a: R = Ph, X = Br
97b: R = Me, X = I

The enhanced solubility of complexes $(R_4C_4MX_2)_2$ (M = Ni, Pd; X = halogen) in donor solvents or in the presence of halide ion must be due to the formation of monomeric complexes $R_4C_4MX_2L$ (L = ligand, Cl^-, etc.). For example, tetraphenylcyclobutadienepalladium chloride is much more soluble in methylene chloride or chloroform saturated with dry hydrogen chloride than in the absence of HCl. This is presumably due to the formation of $Ph_4C_4PdCl_3^-$.[60]

The tetraphenylcyclobutadienepalladium halides (which are deep redbrown, occasionally nearly purple) react with tertiary phosphines and phosphites to give a deep green paramagnetic solution.[62,69,119] The adducts, $Ph_4C_4PdX_2PR_3$, are not known. Although this paramagnetic species has not been quite definitely identified, Sandel and Freedman have given evidence which suggests it to be the chloro- (or bromo-)tetraphenylcyclobutenyl radical **98**.[109]

$$[Ph_4C_4PdX_2]_2 + PR_3 \longrightarrow$$

Ph   Ph
Ph   Ph
X

$$+ (R_3P)_nPd^IX$$

98

$$2\ Ph_4C_4X \longrightarrow Ph_8C_8X_2$$

$$Ph_8C_8X_2 + 2\ (R_3P)_nPd^IX \longrightarrow Ph_8C_8 + 2\ (R_3P)_2PdX_2$$

81

[118] F. M. Chaudhari and P. L. Pauson, *J. Organometal. Chem. (Amsterdam)* **5**, 73 (1966).
[119] R. C. Cookson and D. W. Jones, *Proc. Chem. Soc.* p. 115 (1963).

In that case, the other product would have to be a palladium(I) species, a not unreasonable possibility in view of the existence of nickel(I) complexes of the type $(Ph_3P)_3NiX$.[120, 121] The green color persists for a long time at 20°C in benzene, but on heating to 80°C it fades rapidly with the formation of octaphenylcyclooctatetraene (**81**) and the bis(*tert*-phosphine)palladium(II) halide.[69] This reaction need not necessarily proceed via a free cyclobutadiene at all and a possible path is shown above. The stability of the radical **98** suggests that it is very selective in its reactivity.

Other cyclobutadienepalladium and -nickel complexes also slowly decomposed on heating with various ligands but the products have not, in general, been identified.

F. INTRODUCTION OF A SECOND ORGANIC LIGAND

Criegee and co-workers[89, 90] first attempted to obtain a cyclopentadienyl-cyclobutadiene complex from **2** by reaction with sodium cyclopentadienide. In fact, this reagent acted as a nucleophile toward carbon as well as toward nickel, and led to the isolation of **99** in which two C_5H_5 moieties had been incorporated.

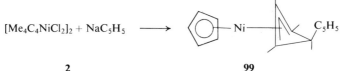

$$[Me_4C_4NiCl_2]_2 + NaC_5H_5 \longrightarrow$$

$$\qquad\qquad\qquad\text{2}\qquad\qquad\qquad\qquad\qquad\qquad\text{99}$$

An X-ray structural determination of **99** has been carried out by Oberhansli and Dahl.[91] The PMR spectrum of the complex suggests the presence of double bond isomers in the σ-bonded C_5H_5 ring.[89, 122]

Such reactions were possible using cyclopentadienyliron dicarbonyl dimer, $[C_5H_5Fe(CO)_2]_2$, or the bromide, $C_5H_5Fe(CO)_2Br$, as cyclopentadienylating agent, since this acted much more selectively toward the metal. Examples of reactions of this type, introduced by Maitlis and his collaborators,[40, 65, 66] are:

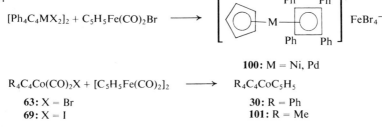

$$[Ph_4C_4MX_2]_2 + C_5H_5Fe(CO)_2Br \longrightarrow$$

100: M = Ni, Pd

$$R_4C_4Co(CO)_2X + [C_5H_5Fe(CO)_2]_2 \longrightarrow R_4C_4CoC_5H_5$$

63: X = Br	**30:** R = Ph
69: X = I	**101:** R = Me

[120] P. Heimbach, *Angew. Chem. Intern. Ed. Engl.* **3**, 648 (1964).
[121] L. Porri, M. C. Gallazzi, and G. Vitulli, *Chem. Commun.* p. 228 (1967).
[122] R. B. King, *Inorg. Chem.* **2**, 530 (1963).

The complex **101** was also prepared successfully from **69** and sodium cyclopentadienide.[40, 122a] Evidently the tetramethylcyclobutadiene here is not as sensitive to nucleophilic attack as it is in the nickel complex **2**. At least part of this may be due to the increased charge on the metal in **2**.

The paramagnetic tetrabromoferrates **100** were converted to the diamagnetic bromides **102** with aqueous potassium ferrocyanide. The cyclopentadienyl ring was cleaved from the palladium complex **102** to regenerate tetraphenyl-cyclobutadienepalladium chloride.[65, 66] Other reactions of **102** are discussed in Section IV,H.

$$[Ph_4C_4MC_5H_5]^+ FeBr_4^- + K_4Fe(CN)_6 \xrightarrow{H_2O} [Ph_4C_4MC_5H_5]^+ Br^- +$$

100   **102**

$$KFeFe(CN)_6 + 3\,KBr$$

$$\textbf{102}\,(M = Pd) + HBr \longrightarrow [Ph_4C_4PdBr_2]_2$$

The cyclobutadiene-cobalt complex **63** also underwent other reactions which resulted in the introduction of a second organic ligand, as observed by Efraty and Maitlis.[76] This could be σ-bonded, as in the reaction with pentafluoro-phenyllithium to give **103**, or π-bonded. The complex **63**, and its tetrakis-(p-tolyl) homolog, reacted with aromatic hydrocarbons under Friedel–Crafts conditions to give π-cyclobutadiene–π-arenecobalt complexes **104**.[76] Cyclo-heptatriene reacted analogously, but no catalyst was needed.

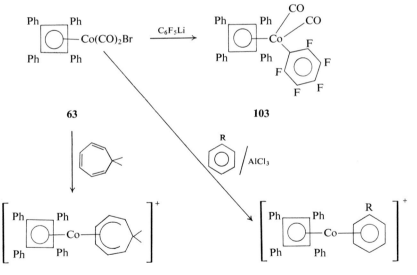

104: R = H, Me, Bu, NH₂, etc.

[122a] An analogous reaction was used to prepare **17** (Section II,A).

The complexes **104** were cationic and soluble in polar media. They were less reactive toward nucleophiles than were the complexes **100** (M = Pd, Ni), but under forcing conditions (eg. with butyllithium or sodium borohydride) reaction did occur to give **105** and **106** (entering substituent *exo* to the metal).[76]

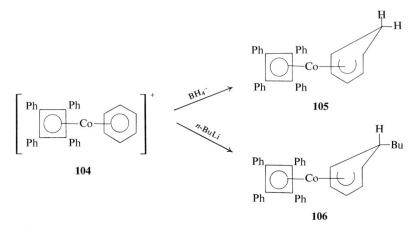

Efraty and Maitlis showed that *N*-bromosuccinimide in methanol acted as a hydride abstractor toward **105** and **106**, giving the cations **104** (*R* = H, *n*-Bu) once again.[76]

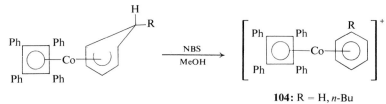

Wegner and Hawthorne[123] have prepared the dicarbollide complexes $Ph_4C_4PdB_9C_2H_{11}$ and $Ph_4C_4Pd(B_9C_2H_9Me_2)$ from $(Ph_4C_4PdCl_2)_2$ and $B_9C_2H_{11}^{2-}$ (or $B_9C_2H_9Me_2^{2-}$). A crystal structure determination of the latter complex has been carried out to confirm the structure.

G. ANION EXCHANGE REACTIONS

The nickel and palladium complexes $(R_4C_4MCl_2)_2$ readily exchange their chlorine atoms, both for other (heavier) halogens and other groups.[61] Criegee and his collaborators[23] showed that $(Me_4C_4NiCl_2)_2$ reacted directly with

[123] P. A. Wegner and M. F. Hawthorne, *Chem. Commun.* p. 861 (1966); *J. Am. Chem. Soc.* **90**, 889 (1968).

aqueous solutions of iodide, azide, tetraphenylborate, etc., to give the appropriate products [$Me_4C_4NiI_2$, $Me_4C_4Ni(N_3)_2$ and $Me_4C_4NiCl \cdot BPh_4$]. Other anions (nitrate, sulfate, oxalate, acetate, etc.) could be substituted by treatment of the complex with the appropriate silver salt.

H. NUCLEOPHILIC ADDITION TO THE CYCLOBUTADIENE RING

The first reaction of this type, discovered by Criegee et al., has already been discussed (Section IV,F). Shortly after this, Blomquist and Maitlis[60] showed that tetraphenylcyclobutadienepalladium chloride reacted with alcohol to give a complex, later identified by a crystal structure by Dahl and Oberhansli[91] to be the exo-ethoxytetraphenylcyclobutenylpalladium chloride dimer **107**

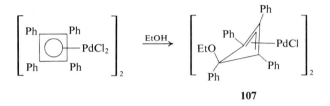

107

(Section II,E). Complexes of this type with other alcohols[61] and with p-chlorophenyl and p-tolyl in place of phenyl[48, 49, 75] have also been prepared by similar routes.

Maitlis and co-workers[66] also showed that the cationic cyclopentadienyl complexes **102** reacted similarly to give **108**.

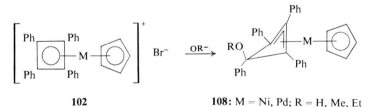

102   **108**: M = Ni, Pd; R = H, Me, Et

In all these cases, attack had occurred exo to the metal. In other words, it does not seem likely that metal participation occurred in the nucleophilic attack, but rather that the nucleophile attacked the molecule from the least hindered side. The cyclobutadiene complexes could be regenerated from **107** and **108** with hydrogen halides.

$$(ROPh_4C_4PdX)_2 + 2\,HX \rightarrow (Ph_4C_4PdX_2)_2 + 2\,ROH$$
$$ROPh_4C_4NiC_5H_5 + HX \rightarrow (Ph_4C_4NiC_5H_5)^+ + X^- + ROH$$
$$ROPh_4C_4PdC_5H_5 + 2\,HX \rightarrow \tfrac{1}{2}(Ph_4C_4PdX_2)_2 + ROH + C_5H_6$$

In the last case, the cyclopentadienyl ring was cleaved off as well.

These reactions are very reminiscent of those undergone by dienepalladium and -platinum halides,[124] e.g., **109**. Both complex **109** and tetraphenylcyclobutadienepalladium chloride reacted with malonate or acetylacetonate to give

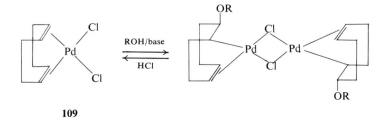

109

complexes with a new C—C bond. Tsuji and Takahashi formulated their product incorrectly[125]; it is most probably **110**.

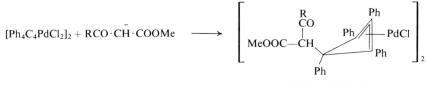

$$[Ph_4C_4PdCl_2]_2 + RCO \cdot CH \cdot COOMe \longrightarrow$$

110: R = MeO, Me

No reactions in which a cyclobutadiene ring is attacked by a nucleophile have been reported for complexes of the type $R_4C_4Fe(CO)_3$.

I. REACTIONS WITH ELECTROPHILIC REAGENTS

In contrast, the iron tricarbonyl complexes, particularly **8**, react very readily with electrophilic reagents to give substitution products.[37] In this sense cyclobutadieneiron tricarbonyl and cyclobutadiene(cyclopentadienyl)cobalt can be classed together with ferrocene[126] and other diene-iron tricarbonyl complexes[127, 128] and can be said to exhibit "aromatic" character.

[124] J. Chatt, L. M. Vallarino, and L. Venanzi, *J. Chem. Soc.* pp. 2496 and 3413 (1957).

[125] H. Takahashi and J. Tsuji, *J. Am. Chem. Soc.* **90**, 2387 (1968).

[126] M. Rosenblum, "Chemistry of the Iron Group Metallocenes." Wiley, New York, 1965.

[127] B. F. G. Johnson, J. Lewis, A. W. Parkins, and G. L. P. Randall, *Chem. Commun.* p. 595 (1969).

[128] G. B. Gill, N. Gourlay, A. W. Johnson, and M. Mahendran, *Chem. Commun.* p. 631 (1969).

Pettit and his co-workers have given a number of examples of such reactions, which are collected in Scheme 3.[33, 37] Typical reactions include Friedel–

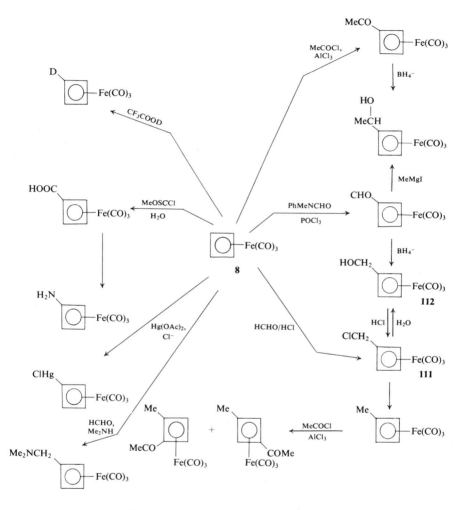

SCHEME 3. Electrophilic substitution and some related reactions of cyclobutadieneiron tricarbonyl.

Crafts acylation, mercuration, deuteration (D[+]), chloromethylation, and formylation. The substituents also undergo many normal organic reactions, without disruption of the complex, under nonoxidizing conditions.

Pettit and his co-workers[37] have also discussed the mechanism of these electrophilic substitution reactions, and have drawn a parallel with the electro-

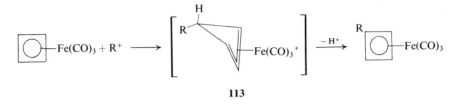

113

philic substitution of aromatic organic molecules. The reaction here is facilitated by stabilization of the probable intermediate π-cyclobutenyl complex **113**.

$$ClCH_2C_4H_3Fe(CO)_3 \underset{HCl}{\overset{H_2O}{\rightleftharpoons}} HOCH_2C_4H_3Fe(CO)_3$$

111   **112**

The ready reversibility of the above reaction parallels reactions in ferrocene chemistry, and is ascribed to the stability of the cation **114**. This could actually be isolated on treatment of the chloromethyl complex **111** with $SbCl_5$. An identical PMR spectrum was displayed by a solution of the alcohol **112** in concentrated sulfuric acid.[129] The actual electronic structure of **114** is still the subject of some dispute, but some form of metal-assisted stabilization undoubtedly occurs.

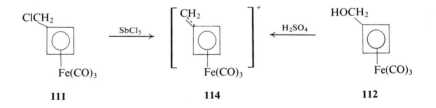

111   **114**   **112**

Amiet and Pettit[39] have also examined the reactivity of cyclobutadiene-(cyclopentadienyl)cobalt (**17**) toward electrophilic reagents. Acylation gave a low yield of the acetyl complex **115** and mercuration gave a high yield of the acetoxymercury complex **116**. In both cases substitution only occurred on the cyclobutadiene ring.

[129] J. D. Fitzpatrick, L. Watts, and R. Pettit, *Tetrahedron Letters* p. 1299 (1966).

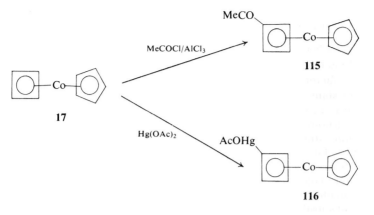

17 **115** **116**

On the other hand, cyclopentadienyl(tetraphenylcyclobutadiene)cobalt underwent electrophilic substitution in the cyclopentadienyl ring.[45]

J. HALOGENATION

Freedman *et al.*[130, 131] observed extensive decomposition on treatment of the nickel complex **48** with bromine. The use of pyridinium hydrobromide perbromide however gave a smooth reaction affording the *trans*-3,4-dibromo-tetraphenylcyclobutene (**117**) in high yield.

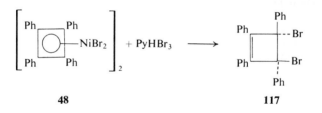

48 **117**

Cyclobutadieneiron tricarbonyl (**8**) reacted with bromine to give a mixture of three tetrabromocyclobutanes identified as the cis,cis,trans, the cis,trans,cis,trans, and the cis,cis,cis,trans isomers.[132]

Hübel and Merényi[41] have reported the bromination of some tetraphenyl-cyclobutadienemolybdenum-diphenylacetylene complexes. However the bromination products mostly arose from the acetylene, while the cyclobuta-diene apparently gave rise to oxidation products (*cis*-dibenzoylstilbene and tetraphenylfuran).

[130] H. H. Freedman and G. A. Doorakian, *Tetrahedron* **20**, 2181 (1964).
[131] H. H. Freedman and A. M. Frantz, *J. Am. Chem. Soc.* **84**, 4165 (1962).
[132] D. J. Severn and E. M. Kosower, *Tetrahedron Letters* p. 2119 (1968).

K. Summary

Cyclobutadiene-metal complexes undergo attack by both electrophilic and nucleophilic reagents at the cyclobutadiene. The extent of this is determined largely by factors such as the metal and its oxidation state. A higher formal oxidation state, Ni(II), Pd(II), favors nucleophilic attack, whereas a lower one, Fe(0), favors electrophilic attack.[132a] Presumably the other ligands (halide, carbonyl) also play a role in determining the type of reaction. For nucleophilic attack on sandwich cationic complexes of d^8 metals Maitlis and co-workers have proposed the order of decreasing reactivity: cycloheptatriene > benzene > tetraphenylcyclobutadiene > cyclopentadienyl.[71, 76]

Only one comparison of the reactivity toward electrophilic substitution has been made. Pettit and Amiet found that the cyclobutadiene in $C_4H_4CoC_5H_5$ underwent substitution in preference to the cyclopentadienyl.[39]

Much work still remains to be done in this area, but it appears that cyclobutadiene complexes are quite reactive to many types of reagent, given the right conditions.

V. Physical Properties of Cyclobutadiene-Metal Complexes

A. X-Ray and Electron-Diffraction Studies

The most detailed X-ray structure analyses have been carried out on tetramethylcyclobutadienenickel chloride (2) and tetraphenylcyclobutadiene-iron tricarbonyl (3).

The structure of 2 as the benzene solvate has been determined by Dunitz et al.[117] In this case the molecule exists as a chlorine-bridged dimer, with each nickel being formally 5-coordinate (assuming the cyclobutadiene to act as a bidentate ligand). A very similar geometry about the metal atom is present in 3 as shown by the studies of Dodge and Schomaker.[133] In both cases the cyclobutadiene ring is accurately square planar, with C—C bond lengths of 1.43 Å (2) and 1.46 Å (3). The metal is placed on the fourfold axis of the ring and all the metal–carbon bond lengths are equal, 2.02 Å (for 2), and 2.06 Å (for 3). In both structures the ring substituents are folded back from the plane of the ring.

[132a] However this is not the only criterion, the molybdenum(0) complex, $C_4H_4Mo(CO)_3PPh_3$, did not undergo electrophilic substitution.[38] It appears that most diolefin-iron tricarbonyl complexes undergo "aromatic"-type substitution reactions.

[133] R. P. Dodge and V. Schomaker, *Acta Cryst.* **18**, 614 (1965); *Nature* **186**, 798 (1960).

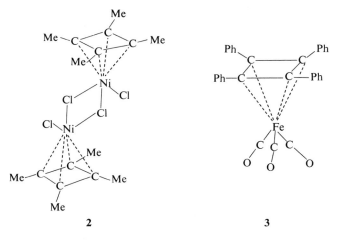

2   3

The other three ligands on the metal are equivalently disposed about the four-fold axis. In **2** the solvate benzenes are not within bonding distance, but they probably are important in determining the state of aggregation of the complex. Criegee and Schröder reported **2** to have a molecular weight corresponding to $(Me_4C_4NiCl_2)_{10}$ in bromoform **(19)**.

The C—C and C—metal bond lengths in both **2** and **3** are close to those found in ferrocene.[134] It is also interesting that the complexes $R_4C_4MX_2$, where M = Ni, Pd, and Pt, are all associated, making the metals effectively 5-coordinate. Since palladium and particularly platinum are rather reluctant to go to 5-coordination normally, this argues that the cyclobutadiene has a special feature to it. In fact, it is probably more correct to regard it as a very small bidentate ligand.[1]

Preliminary reports of the crystal structures of a number of other cyclo-butadiene complexes have appeared. These include $Me_4C_4Ni(\overline{Me_4C_4Fe(CO)_3})$, **(66)**[80]; $(C_6H_4(C_2Ph)_2)Fe_2(CO)_4$**(25a)**[41a]; $C_5H_5Nb(CO)(PhC_2Ph)(Ph_4C_4)$, **(43)**[53]; $Me_4C_4Mo(CO)_3PPh_3$[38]; $Ph_4C_4PdB_9C_2Me_2H_9$[123]; $[Ph_4C_4Mo(CO)_2Br]_2$ **(58)**.[78] The cyclobutadiene ring is square planar in all of these complexes.

In addition, an electron-diffraction study of cyclobutadieneiron tricarbonyl **(8)** has shown it to have a square-planar cyclobutadiene ring bonded to the iron in the same manner as in **3**. The carbon–carbon (1.44 Å) and carbon–ring (2.05 Å) bond-lengths are very close to those in **3**. The hydrogen atoms are found to be bent *toward* metal.[135] An electron-diffraction study of ferrocene showed the same effect.[136]

[134] J. D. Dunitz, L. E. Orgel, and A. Rich, *Acta Cryst.* **9**, 373 (1956).
[135] H. Oberhammer and H. A. Brune, *Z. Naturforsch.* **24a**, 607 (1969).
[136] R. K. Bohn and A. Haaland, *J. Organometal. Chem. (Amsterdam)* **5**, 470 (1966).

B. Nuclear Magnetic Resonance Spectra

Symmetrically tetrasubstituted cyclobutadiene-metal complexes exhibit one resonance in solution, indicating the magnetic equivalence of the four groups. This is probably due to rapid rotation about the metal-cyclobutadiene axis. Some typical values are given in Table I.

TABLE I

NMR Resonances of Some Cyclobutadiene-Metal Complexes

Complex	τ		Assignment	Reference
$C_4H_4Fe(CO)_3$	6.09		Ring protons	[a]
$MeCOC_4H_3Fe(CO)_3$	4.55		Ring protons	[a]
	5.60		Ring protons	[a]
$(Me_4C_4NiCl_2)_2$	8.63	(CDCl₃)	Methyl	[b]
$(Me_2C_4H_2NiCl_2)_2$	8.89	(HCl, aq.)	Methyl	[b]
	3.35		Ring protons	[b]
$Me_4C_4Fe(CO)_3$	8.27	(CDCl₃)	Methyl	[c]
$Me_4C_4Co_2(CO)_6$	8.33	(CDCl₃)	Methyl	[d]
$Me_4C_4Co(CO)_2I$	8.19	(CDCl₃)	Methyl	[d]

[a] J. D. Fitzpatrick, L. Watts, G. F. Emerson, and R. Pettit, *J. Am. Chem. Soc.* **87**, 3254 (1965).
[b] W. Eberius, Ph.D. Dissertation, Karlsruhe, 1967.
[c] R. Bruce, K. Moseley, and P. M. Maitlis, *Can. J. Chem.* **45**, 2011 (1967).
[d] R. Bruce and P. M. Maitlis, *Can. J. Chem.* **45**, 2017 (1967).

In general the tetraphenylcyclobutadiene-metal complexes do not give much structural information in their PMR spectra. However, Maitlis and co-workers[49, 75, 76] have successfully used the characteristic AB pattern of the aromatic protons in *p*-tolyl-substituted cyclobutadiene complexes as a test for the presence of a symmetrical cyclobutadiene ring.

The NMR spectrum of cyclobutadieneiron tricarbonyl in a nematic liquid crystal solvent has been determined by Yannoni *et al.*[137] who found that the lengths of the C—C bonds of the ring were unequal, in the ratio of 0.9977 ± 0.0045:1. This implies only a *very* small distortion from square symmetry.

Detailed NMR spectra, including ^{13}C—H coupling constant measurements, have been reported by Brune *et al.*[138, 139] for some simpler cyclobutadiene-complexes. 1-Monosubstituted complexes showed that the protons on carbons

[137] C. S. Yannoni, G. P. Ceasar, and B. P. Dailey, *J. Am. Chem. Soc.* **89**, 2833 (1967).
[138] H. A. Brune, H. P. Wolff, and H. Hüther, *Ber.* **101**, 1485 (1968).
[139] H. A. Brune, H. P. Wolff, and H. Hüther, *Z. Naturforsch.* **23b**, 1184 (1968).

2 and 4 were equivalent. Similarly, methyls in the 2- and 4-positions were equivalent. However a splitting observed in the ^{13}C—H couplings of cyclo-butadieneiron tricarbonyl (8) must be due to long-range coupling across the ring, according to Preston and Davis.[140] This was confirmed by Brune et al.[138] who found the splitting to be absent in 1,2-dimethylcyclobutadieneiron tricarbonyl. The coupling constant $J(^{13}C—H)$ decreased from 191.1 Hz for **8** to 186.7 Hz for the trimethylcyclobutadieneiron tricarbonyl. This has been interpreted in terms of an increase of the iron–carbon bond lengths.[138]

Maitlis and co-workers have found long-range P—H coupling in a tetra-methylcyclobutadiene complex.[141]

C. INFRARED AND ULTRAVIOLET SPECTRA

Except for a few special cases, infrared and ultraviolet spectra are of very limited utility in structure determinations. Blomquist and Maitlis[60] used the former to identify tetraphenylcyclobutadienepalladium complexes, since their spectra were almost identical, in the range in which the organic ligand absorbed, with those of the tetraphenylcyclobutadienenickel complexes prepared by Freedman.[13]

Maitlis and co-workers[79] also used the carbonyl stretching frequencies in some cyclobutadiene complexes as diagnostic aids. For example, methylation of the cyclobutadiene ring in the iron tricarbonyl complex led to a *decrease* in ν_{CO}. This is due to a decrease in back-bonding from iron to the permethylated ring and, hence, presumably an increase in the ring *to* metal binding by comparison with the unmethylated one. An analogous argument[40] was used to suggest that the binding of the tetramethylcyclobutadiene group to the metal was very strong in tetramethylcyclobutadienecobalt complexes, in particular **69**. Ring-metal bonding also seemed to be stronger than that in the analogous, and isoelectronic, cyclopentadienyliron dicarbonyl iodide.

Fritz has reported the ultraviolet spectra of some tetraphenylcyclobutadiene complexes and commented on some similarities.[88]

VI. Bonding in Cyclobutadiene-Metal Complexes

More detailed molecular orbital descriptions of these complexes than those originally given by Longuet-Higgins and Orgel[18] have been given by Coates[142] and Cotton,[143] among others.

---

[140] H. G. Preston, Jr. and J. C. Davis, Jr., *J. Am. Chem. Soc.* **88**, 1585 (1966).

[141] E. O. Greaves, R. Bruce, and P. M. Maitlis, *Chem. Commun.* p. 860 (1967).

[142] G. E. Coates, "Organometallic Compounds," 2nd ed., p. 330. Methuen, London, 1960.

[143] F. A. Cotton, "Chemical Application of Group Theory," pp. 132 and 180. Wiley, New York, 1963.

A schematic MO diagram according to Cotton's treatment for tetramethyl-cyclobutadienenickel chloride dimer (**2**) is given here (Fig. 1). Since the extent of the total interactions is not known, the nickel–chlorine and nickel–ring bonds are treated separately. The molecular orbitals are considered from the point of view of two local symmetries, one fourfold (Me₄C₄ ring), and the other threefold (Cl₃ group) with a collinear (z) axis. The nickel orbitals are shown in the center of the figure and their symmetries in C_3 and C_4 are on the right- and left-hand sides, respectively. All 18 available electrons can be accommodated in 9 bonding molecular orbitals, thus explaining the stability and diamagnetism of the complex.

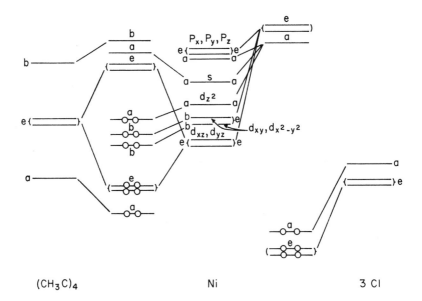

FIG. 1. Schematic MO energy diagram for the bonding to one nickel atom in tetramethylcyclobutadienenickel chloride dimer **2**. (From F. A. Cotton, "Chemical Application of Group Theory," pp. 132 and 180. Wiley, New York, 1963; reproduced with permission.)

The main metal-ring bonding is through interaction of the metal e orbitals (d_{xz}, d_{yz}) with the cyclobutadiene e orbitals (ψ_2, ψ_3 in **1**). According to this treatment ψ_1 and ψ_4 (in **1**) are largely nonbinding with respect to the metal-ligand interactions. This does not agree with the treatment by Coates,[142] but since numerical values are not available to check the relative energy levels and the overlap integrals, this is a rather philosophical point.

VII. Aromatic Character of Cyclobutadiene-Metal Complexes

Most cyclobutadiene-metal complexes are rather stable entities, and the unsubstituted cyclobutadieneiron tricarbonyl certainly undergoes "aromatic-type" electrophilic substitution reactions (Section III,J). However, other cyclic polyene-iron tricarbonyl complexes also undergo these reactions[127, 128] and it is not clear to what extent this character is unique for cyclobutadiene-metal complexes.

If aromaticity is defined as the ability to undergo substitution reactions with retention of the cyclic conjugated system, then these complexes may be said to be "aromatic." However, the definition of aromatic which is usually implied for organic compounds has no exact parallel for metal π-complexes. Considerable work in this area is necessary before this term can be accurately and meaningfully defined here.

Appendix

A number of papers which appeared too late for inclusion in the main body of the review are briefly mentioned here.

A new synthesis of 1,2-disubstituted-cyclobutadieneiron tricarbonyl complexes has been described by Roberts et al.[144] The key step is the photochemical synthesis of a 1,2-dicarbomethoxy-1,2,3,4-cyclobutane **118** from trans-1,2-dichloroethylene and dichloromaleic anhydride. The cyclobutane **118** on reaction with zinc and diiron enneacarbonyl gives the 1,2-dicarbomethoxycyclobutadieneiron tricarbonyl **119**, from which a number of other 1,2-disubstituted-cyclobutadieneiron tricarbonyl complexes are obtained by conventional methods.

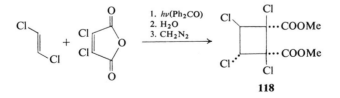

118

[144] B. W. Roberts, A. Wissner, and R. A. Rimerman, *J. Am. Chem. Soc.* **91**, 6208 (1969).

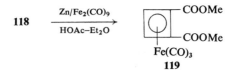

If the reported yields can be improved, this promises to be an attractive alternative to the older syntheses of cyclobutadieneiron tricarbonyl complexes (Section II,A).

Avram et al.[145] have also reported the synthesis of 1,2-di-t-butyl-3,4-diphenylcyclobutadienepalladium chloride (120) (see footnote 50c), while Hosokawa and Moritani have described the pyrolysis of this complex to 2-t-butyl-1,5-diphenyl-3,3,4-trimethyltricyclo[2.1.0.0^{2,5}]pentane (121).[146]

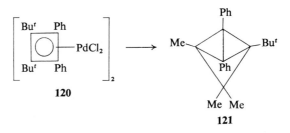

Interest continues in the decomposition of cyclobutadieneiron tricarbonyl and in the properties and structure of free cyclobutadiene. Reeves et al.[147] have prepared adducts, $C_4H_4Fe(CO)_2L$, where L = dimethyl maleate or fumarate, by the photochemical reaction of the unsaturated ester with the tricarbonyl 8. From their mode of decomposition and the products, the authors conclude that reaction does not occur in the adduct complex. The formation of the observed Diels–Alder adduct occurs between free cyclobutadiene and free dienophile. Interestingly, tetracyanoethylene, probably for steric reasons, adds 2,4- to 1,3-di-t-butylcyclobutadiene, formed by in situ oxidation of its iron tricarbonyl complex, to give 122.

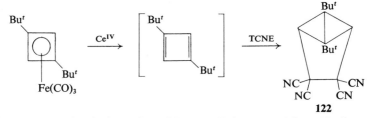

[145] M. Avram, I. G. Dinulescu, G. D. Mateescu, E. Avram, and C. D. Nenitzescu, Rev. Roumaine Chim. 14, 1181 (1969).

[146] T. Hosokawa and I. Moritani, Chem. Commun. p. 905 (1970).

[147] P. Reeves, J. Henery, and R. Pettit, J. Am. Chem. Soc. 91, 5888 (1969).

Other cyclobutadienes react normally (1,2-cycloaddition) with TCNE and Reeves *et al.*[148] have suggested that the reason why two isomeric adducts, **125** and **126**, arise from 1,2-diphenylcyclobutadiene, whereas benzoquinone only gives one isomer, **127**, is that 1,2-diphenylcyclobutadiene (**123**) is in equilibrium with its valence tautomer, 1,4-diphenylcyclobutadiene (**124**).

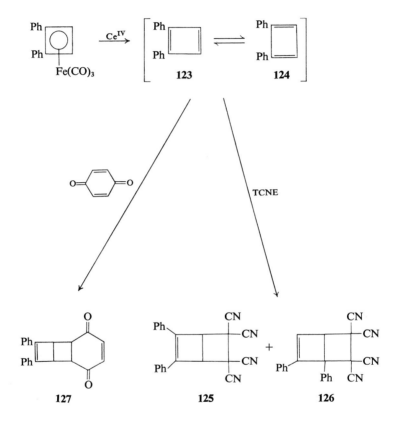

The more reactive dienophile (TCNE) is less discriminating and reacts with both isomers, while less reactive dienophiles react only with the less sterically hindered isomer **124**. These results are taken as evidence for the rectangular nature of cyclobutadiene and hence for its existence in a singlet state.

The vapor-phase flash photolysis of cyclobutadieneiron tricarbonyl gives benzene and acetylene; in the presence of other acetylenes *o*-disubstituted

[148] P. Reeves, T. Devon and R. Pettit, *J. Am. Chem. Soc.* **91**, 5890 (1969).

benzenes are formed.[149] These authors also presented evidence for the reversibility of the photochemical decomposition of **8**,

$$C_4H_4Fe(CO)_3 \underset{\text{hv}}{\rightleftharpoons} C_4H_4 + Fe(CO)_3$$

8

Evidence for the formation of cyclobutadiene by photolysis of the Diels–Alder adduct of cyclooctatetraene and dimethyl acetylenedicarboxylate, and by flash vacuum pyrolysis of photo-α-pyrone has been given.[150, 151] A useful source of substituted cyclobutadienes is by photolysis of the monoozonide of a Dewar benzene.[152, 153]

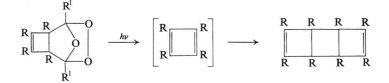

Amiet *et al.*[154] have shown that cyclobutadieneiron tricarbonyl (**8**) undergoes acetoxymercuration very easily to give an equilibrium mixture of all possible mono-, di-, tri-, and tetraacetoxymercury complexes. This equilibrium is also attained from the mono- and tetraacetoxymercury complexes. Reaction with KI_3 cleaves the mercury–carbon bond to give the appropriate iodo complex.

The X-ray structure of 4-carbomethoxybenzocyclobutadieneiron tricarbonyl has been briefly reported,[155] and shows some bond localization in the (uncomplexed) benzene ring. This conclusion is supported by NMR data.[156]

The photoelectron spectra of cyclobutadieneiron tricarbonyl and some of its monosubstituted derivatives have been reported. The first ionization is interpreted to involve the loss of an electron from a slightly perturbed atomic orbital of iron, while the second is said to arise from loss of an electron from a

[149] J. Font, S. C. Barton, and O. P. Strausz, *Chem. Commun.* p. 980 (1970).

[150] R. D. Miller and E. Hedaya, *J. Am. Chem. Soc.* **91**, 5401 (1969).

[151] E. Hedaya, R. D. Miller, D. W. McNeil, P. F. D'Angelo, and P. Schissel, *J. Am. Chem. Soc.* **91**, 1874 (1969).

[152] P. R. Story, W. H. Morrison, and J. M. Butler, *J. Am. Chem. Soc.* **91**, 2398 (1969).

[153] R. Criegee and R. Huber, *Chem. Ber.* **103**, 1855, 1862 (1970).

[154] R. G. Amiet, K. Nicholas, and R. Pettit, *Chem. Commun.* p. 161 (1970).

[155] R. E. Davis and R. Pettit, *J. Am. Chem. Soc.* **92**, 717 (1970).

[156] H. Gunther, R. Wenzl, and H. Klose, *Chem. Commun.* p. 605 (1970).

perturbed molecular orbital of the ligand.[157] The first and second ionization potentials of cyclobutadiene (8.51, 11.62 eV) have been estimated from these results. Good agreement with calculated values (SCF–MO) is reported.[157, 158]

[157] M. J. S. Dewar and S. D. Worley, *J. Chem. Phys.* **50**, 654 (1969).
[158] S. D. Worley, *Chem. Commun.* p. 980 (1970).

Author Index

Numbers in parentheses are reference numbers and indicate that an author's work is referred to although his name is not cited in the text.

A

Abraham, R. J., 205, 206
Abramowitch, R. A., 16
Adam, W., 210
Agahigian, H., 257
Ahmed, F. R., 315, 329, 330, 331, 356(130)
Aida, K., 305
Al-Joboury, M. I., 336
Ali, M. A., 93, 95(23)
Allen, C. W., 330, 331(131), 347, 355(131), 356(131)
Allendoerfer, R. D., 215, 216(26), 217(26)
Almenningen, A., 320
Altschuler, L., 24
Amako, Y., 77
Amiel, Y., 21, 140, 151(92), 236
Amiet, R. G., 364, 365, 391(38), 400(38, 39), 401, 408(38)
Anastassiou, A. G., 360
Andersen, K. K., 355
Anderson, A. G., Je, 20, 55
Anderson, M. E., 88
Anisimov, K. N., 371, 372(51, 52, 53), 384(51, 52, 53), 401(53)
Ansell, G. B., 309, 310, 315, 318, 353(81)
Anthoine, G., 235, 239(73)
Appel, R., 308
Arnett, E. M., 195, 382
Arnold, Z., 191
Asgar-Ali, M., 189
Atherton, N. M., 89, 137(15), 150, 151(104)
Atoji, M. R., 278
Audrieth, L. F., 276
Avram, E., 406
Avram, M., 363, 381, 384, 406
Ayscough, D. B., 83
Azatyan, V. D., 218, 219(47, 48), 222
Azumi, H., 111

B

Bach, H., 308
Backes, L., 256
Badger, G. M., 21, 152, 202, 205(84)
Baer, F., 188
Bailey, N. A., 149, 150(102)
Baird, J. C., 83
Bak, B., 335
Bak, D. A., 218
Baker, R., 260
Baker, W., 360
Balaban, A. T., 18, 21
Baldeschwieler, J. D., 177
Balogh, V., 255
Ban-yuań, U., 276
Bangert, K. F., 47
Banks, D., 251
Barber, M., 195
Barborak, J. C., 390
Barnes, W. H., 315, 329, 330, 331, 356(130)
Bart, J. C. J., 280
Bartell, L. S., 281
Barter, C., 177, 178
Barth, W., 256
Barton, S. C., 408
Bastiansen, O., 96, 320
Baudet, J., 75, 76
Bauer, S. H., 281
Bauer, W., 45
Baughan, E. C., 118
Bauld, N. L., 160, 161(119), 162(123), 247, 248, 251
Baumgarten, P., 54
Baumann, J. I., 189, 191(63)
Baxter, C. S., 251
Becher, H. J., 307, 349
Becke-Goehring, M., 276, 279, 326

411

Subject Index

A

Aceheptalene, exaltation, 188–189

Acenaphth[1,2-*a*]acenaphthylenes, 58–59, 67–68, 116–118
  electronic spectra, 58–59
  ESR, 67–68, 116–118
  radical ions, ESR, 116–118

Acenaphthylene, 53, 91, 110–113, 188–189
  electronic spectra, 53
  exaltation, 188–189
  irradiation, 112

Acenaphthylene radical anion, ESR, 91, 110–113

Acephthylene, molecular diagrams of ground and first excited states, 35

Acepleiadylene, 28, 67–68, 113–115, 118–120, 181–183, 187
  electrophilic nitration, 28
  ESR, 67–68, 113–115, 118–120
  exaltation, 181–183, 187

Acetylenes, reactions with metal complexes, 366–372, 374–384

Alkyl benzenes, magnetic susceptibilities, 182, 195–196

Alkylcyclooctatetraenes, ESR, 99–101

Alternate systems, 84–85, 90, 94–96, 128
  ESR, 94–95, 128
  HMO model, 84–85, 90, 95–96

Aminofluoranthenes, 24, 26–34, 78, *see also* 1-Aminofluoranthenes, 7-Aminofluoranthenes
  pk, 24, 26–34

1-Aminofluoranthenes, electronic spectra, 78

7-Aminofluoranthenes, electronic spectra, 78

Anisotrophy, 22–23, 168–169

Annulene(s), *see also* specific annulenes
  aromaticity, 21–23

[10]Annulene, *see also* 1,6-Bridged [10]-annulenes
  exaltation, 201

[12]Annulene, 271–272

[14]Annulene, 140–141, 271–272
  structure, 140–141

[16]Annulene, 181–183, 188, 204–205, 235–236
  dianion, 235–236
  exaltation, 181–183, 188, 204–205
  reduction, 235–236

[18]Annulene, 151–154, 205, 236
  ESR of 1,4 : 7,10 : 13,16-trisulfide radical anion, 151–154
  exaltations, 205

Antiaromaticity, 15, 18–23

Aromaticity, 1–80, 167–206, 208–272, 396–399, 404–405
  cata-condensed systems, 14–23
  chemical reactivity, 23–34
  complex "stability" index (P), 20–23
  cyclic ions, 207–272
  cyclobutadiene-metal complexes, 396–399, 404–405
  cyclooctatraene, 168
  definition, 13–23
  delocalization energy, 13–23
  diamagnetic susceptibility exaltation, 167–206
  effects of severe ring strain, 193–198
  electronic spectra, 36–60
  frontal orbital energies, 19–23
  hybridization, 13–23
  molecular orbital theory, 208–272
  nonalternate hydrocarbons, 1–80
  NMR, 21–23
  reactivity, 14–34
  stability, 1–2, 13–34, 47–48, 55–56, 60–66, 72
  symmetry, 16–23

ORGANIC CHEMISTRY

A SERIES OF MONOGRAPHS

EDITORS

ALFRED T. BLOMQUIST
Department of Chemistry
Cornell University
Ithaca, New York

HARRY WASSERMAN
Department of Chemistry
Yale University
New Haven, Connecticut

ORGANIC CHEMISTRY
A Series of Monographs